PEARSON EDEXCEL INTERNATIONAL A LEVEL

BIOLOGY
Student Book 2

Ann Fullick
with Frank Sochacki

Published by Pearson Education Limited, 80 Strand, London, WC2R 0RL.

www.pearsonglobalschools.com

Copies of official specifications for all Pearson Edexcel qualifications may be found on the website: https://qualifications.pearson.com

Text © Ann Fullick and Pearson Education Limited 2019
Edited by Deborah Webb and Penelope Lyons
Proofread by Penelope Lyons and Jess White
Indexed by Judith Reading
Designed by © Pearson Education Limited 2019
Typeset by © Tech-Set Ltd, Gateshead, UK
Original illustrations © Pearson Education Limited 2019
Illustrated by © Tech-Set Ltd, Gateshead, UK
Cover design by © Pearson Education Limited 2019
Cover images: Front: **Getty Images**: Michael Haegele
Inside front cover: **Shutterstock.com**: Dmitry Lobanov

The rights of Ann Fullick and Frank Sochacki to be identified as the authors of this work have been asserted by them in accordance with the Copyright, Designs and Patents Act 1988.

First published 2019

24
13

British Library Cataloguing in Publication Data
A catalogue record for this book is available from the British Library
ISBN 978 1 2922 4470 9

Endorsement statement
In order to ensure that this resource offers high-quality support for the associated Pearson qualification, it has been through a review process by the awarding body. This process confirmed that this resource fully covers the teaching and learning content of the specification at which it is aimed. It also confirms that it demonstrates an appropriate balance between the development of subject skills, knowledge and understanding, in addition to preparation for assessment.

Endorsement does not cover any guidance on assessment activities or processes (e.g. practice questions or advice on how to answer assessment questions) included in the resource, nor does it prescribe any particular approach to the teaching or delivery of a related course.

While the publishers have made every attempt to ensure that advice on the qualification and its assessment is accurate, the official specification and associated assessment guidance materials are the only authoritative source of information and should always be referred to for definitive guidance.

Pearson examiners have not contributed to any sections in this resource relevant to examination papers for which they have responsibility.

Examiners will not use endorsed resources as a source of material for any assessment set by Pearson. Endorsement of a resource does not mean that the resource is required to achieve this Pearson qualification, nor does it mean that it is the only suitable material available to support the qualification, and any resource lists produced by the awarding body shall include this and other appropriate resources.

Acknowledgements
We are grateful to the following for permission to reproduce copyright material:
(key: b-bottom; c-centre; l-left; r-right; t-top)

Figures
2-3 123RF: Twinsterphoto/123RF; 4, 15, 24(tr), 28, 29 (br), 30, 38, 44, 67, 71, 75 (l), 82, 144 (r), 162, 163 (bl), 227: Anthony Short; 8 Science Photo Library Ltd: DR.JEREMY BURGESS/Science Photo Library Ltd; 22-23 Alamy Stock Photo: Martin Harvey/Alamy Stock Photo; 24 Photodisc: (ml) PhotoDisc/Photolink; 29 Shutturstock: (bl) Andreas Juergensmeier/Shutterstock; 29 Shutterstock: (bc) Stefan Scharf/Shutterstock; 31 Science Photo Library Ltd: Alan Carey/Science Photo Library Ltd; 32 Alamy Stock Photo: Adrian Weston/Alamy Stock Photo; 34 123RF: (cr) MIKALAY VARABEY/123RF; 34 123RF: (br) Johan van Beilen/123RF; 35 123RF: (cl) Vaclav Volrab/123RF; 35 Shutterstock: (bl) Beneda Miroslav/Shutterstock; 35 Alamy Stock Photo: (br) Florida Images/Alamy Stock Photo; 36 Tui De Roy: Courtesy of Tui De Roy; 40 Edward Fullick; 46 Alamy Stock Photo: Francis Abbott/Nature Picture Library/Alamy Stock Photo; 48 Alamy Stock Photo: Martin Shields/Alamy Stock Photo; 50-51 123RF: Alberto Loyo/123RF;

61 Science Photo Library Ltd: Steve Gschmeissner/Science Photo Library Ltd; 63 Getty Images: Hindustan Times/Getty Images; 75 Science Photo Library Ltd: (r) PASCAL GOETGHELUCK/Science Photo Library Ltd; 77 123RF: Brian Kinney/123RF, Jan Lorenz/123RF; 78 Alamy Stock Photo: (tl) Thierry Vezon/Biosphoto/Alamy Stock Photo; 78 Alamy Stock Photo: (bl) Friso Gentsch/dpa/Alamy Stock Photo; 81 Shutterstock: (l) Hanmon/Shutterstock; 81 123RF: (r) Hagit berkovich/123RF; 86-87 Shutterstock: Matej Kastelic/Shutterstock; 89 Science Photo Library Ltd: (t) Norm Thomas/Science Photo Library Ltd; 89 Science Photo Library Ltd: (b) OMIKRON/Science Photo Library Ltd; 91 Science Photo Library Ltd: LEE D. SIMON/Science Photo Library Ltd; 93 Science Photo Library Ltd: DR TONY BRAIN/Science Photo Library Ltd; 96 Getty Images: (l) Rudigobbo/E+/Getty Images; 96 Science Photo Library Ltd: (r) Wolfgang Hoffmann/AgstockUSA/Science Photo Library Ltd; 98 Shutterstock: MyFavoriteTime/Shutterstock; 99 Getty Images: GerMan101/iStock/Getty Images; 100 Science Photo Library Ltd: MARTYN F. CHILLMAID/Science Photo Library Ltd; 102 Alamy Stock Photo: BSIP SA/Alamy Stock Photo; 103 123RF: Apatcha Muenaksorn/123RF; 104 Shutterstock: (r) Juan Gaertner/Shutterstock; 104 Science Photo Library Ltd: (l) A.B. DOWSETT/Science Photo Library Ltd; 107 Science Photo Library Ltd: (tl) A. DOWSETT, PUBLIC HEALTH ENGLAND/Science Photo Library Ltd; 107 Science Photo Library Ltd: (bl) STEVE GSCHMEISSNER/Science Photo Library Ltd; 108 Science Photo Library Ltd: GUSTOIMAGES/Science Photo Library Ltd; 110 Science Photo Library Ltd: NIBSC/Science Photo Library Ltd; 114 Science Photo Library Ltd: A. DOWSETT, HEALTH PROTECTION AGENCY/Science Photo Library Ltd; 116 Getty Images: GerMan101/iStock/Getty Images; 118,119 Alamy Stock Photo: Dimitar Todorov/Alamy Stock Photo; 121 123RF: (t) Zlikovec/123RF; 121 Shutterstock: (b) Freesoulproduction/Shutterstock; 125 Science Photo Library Ltd: PROF. S.H.E. KAUFMANN & DR J.R GOLECKI/Science Photo Library Ltd; 130 123RF: Richard Starkweather/123RF; 131 Science Photo Library Ltd: SPL/Science Photo Library Ltd; 134 Science Photo Library Ltd: Jim Varney/Science Photo Library Ltd; 138 Alamy Stock Photo: Owais Aslam Ali/Asianet-Pakistan/Alamy Stock Photo; 142-143 Shutterstock: Cristian Zamfir/Shutterstock; 144 Shutterstock: (l) Schankz/Shutterstock; 146 Alamy Stock Photo: Arthur Turner/Alamy Stock Photo; 148 Alamy Stock Photo: (tl) André Skonieczny/ImageBROKER/Alamy Stock Photo; 148 Science Photo Library Ltd: (tr) JOHN MITCHELL.Science Photo Library Ltd; 148 Alamy Stock Photo: (bl) Blickwinkel/Hecker/Alamy Stock Photo; 148 Science Photo Library Ltd: (br) STEVE GSCHMEISSNER/Science Photo Library Ltd; 149 Getty Images: (tl) Mlenny/E+/Getty Images; 149 Science Photo Library Ltd: (tr) SINCLAIR STAMMERS/Science Photo Library Ltd; 153 Science Photo Library Ltd: PASCAL GOETGHELUCK/Science Photo Library Ltd; 155 Biodiversity Institute of Ontario: Courtesy of Biodiversity Institute of Ontario.; 156 Shutterstock: Gopixa/Shutterstock; 160,161 Getty Images: BSIP/Universal Images Group/Getty Images; 163 Science Photo Library Ltd: (br) CNRI/Science Photo Library Ltd; 164 Science Photo Library Ltd: CNRI/Science Photo Library Ltd; 168 Science Photo Library Ltd: STEVE GSCHMEISSNER/Science Photo Library Ltd; 169 Alamy Stock Photo: Maurice Savage/Alamy Stock Photo; 171 Alamy Stock Photo: KEYSTONE Pictures USA/Alamy Stock Photo; 180,181 123RF: Puntasit Choksawatdikorn/123RF; 197 123RF: (c) Johan Swanepoel/123RF; 197 123RF: (b) Viktoria Makarova/123RF; 210,211 Shutterstock: Jose Luis Calvo/Shutterstock; 222 Alamy Stock Photo: Rick & Nora Bowers/Alamy Stock Photo; 225 Alamy Stock Photo: Arterra Picture Library/Alamy Stock Photo; 230 123RF: Luciano Mortula/123RF; 234,235 Shutterstock: Whitehoune/Shutterstock; 238 Science Photo Library Ltd: DANTE FENOLIO/Science photo Library Ltd; 245 Science Photo Library Ltd: DENNIS KUNKEL MICROSCOPY/Science Photo Library Ltd; 256 Alamy Stock Photo: HAL BERAL/VWPICS/Visual&Written SL/Alamy Stock Photo; 257 123RF: (l) Youkurna123/123RF; 257 Science Photo Library Ltd: (r) BSIP ASTIER/Science Photo Library Ltd; 260,261 Alamy Stock Photo: CAVALLINI JAMES/BSIP/Alamy Stock Photo; 265 Science Photo Library Ltd: (t) ADAM HART-DAVIS/Science Photo Library Ltd; 265 Science Photo Library Ltd: (b) MARTIN DOHRN/Science Photo Library Ltd; 270 Alamy Stock Photo: BSIP SA/Alamy Stock Photo; 271 Science Photo Library Ltd: (cl) ZEPHYR./Science Photo Library Ltd; 271 Science Photo Library Ltd: (tr) SOVEREIGN,ISM/Science Photo Library Ltd; 272 Alamy Stock Photo: Photo Researchers/Science History Images/Alamy Stock Photo; 276 Alamy Stock Photo: (l) Tim Gainey/Alamy Stock Photo; 276 Shutterstock: (r) Macro Uliana/Shutterstock; 277 Alamy Stock Photo: Nigel Cattlin/Alamy Stock Photo; 284 Alamy Stock Photo: Nigel Cattlin/Alamy Stock Photo; 290,291 Shutterstock: I Wei Huang/Shutterstock; 296 Science Photo Library Ltd: TEK IMAGE/Science Photo Library Ltd; 298 123RF: Praweena pratchayakupt/123RF; 304 Shutterstock: Roschetzky Photography/Shutterstock; 306 Shutterstock: Trabantos/Shutterstock; 309 123RF: Tim Hester/123RF

Text
18: Based on part of a poster produced by Science and Plants for Schools (SAPS) to introduce different types of photosynthesis to students; 68(t): Atmospheric Co2 at Mauna Loa Observatory, U.S department of commerce; 70(b): Global anthropogenic greenhouse gas emissions, United States Environmental Protection Agency; 72: Projections of Future Changes in Climate and Carbon Dioxide: Projected emissions and concentrations © IPCC. Used with Permission; 75: Vedder O, Bouwhuis S, Sheldon BC (2013) Quantitative Assessment of the Importance of Phenotypic Plasticity in Adaptation to Climate Change in Wild Bird Populations. PLoS Biol 11(7): e1001605. © 2013 Vedder et al.; 82: STC Programs: Research: Tortuguero Sea Turtle Program © 2018, Sea Turtle Conservancy. Used with Permission; 286: Quinn, Ben, "Paralysed man Darek Fidyka walks again after pioneering surgery", The Guardiam, 21 Oct 2014. © 2014, Guardian News & Media Limited.; 306: Food and Agriculture Organization of the United Nations, Madkour, Magdy, Status and Options for Regional GMOs Detection Platform: A Benchmark for the Region, [http://www.fao.org/3/al310e/al310e.pdf]. Reproduced with permission

CONTENTS

ABOUT THIS BOOK

This book is written for students following the Pearson Edexcel International Advanced Level (IAL) Biology specification. This book covers the second year of the International A Level (IAL) course.

The book contains full coverage of IAL units (or exam papers) 4 and 5. Each unit in the specification has two topic areas. The topics in this book, and their contents, fully match the specification. You can refer to the Assessment Overview on pages x–xi for further information. Students can prepare for the written Practical Skills Paper (unit 6) with the support of the IAL Biology Lab Book (see pages viii and ix of this book).

Each Topic is divided into chapters and sections to break the content down into manageable chunks. Each section features a mix of learning and activities supported by the features explained below.

Learning objectives
Each chapter starts with a list of key assessment objectives. **Cross references** to previous or following Student Book content help you navigate course content.

Specification reference
The exact specification references covered in the section are provided.

Exam hints
Tips on how to answer exam-style questions and guidance for exam preparation, including how to respond to **command words**.

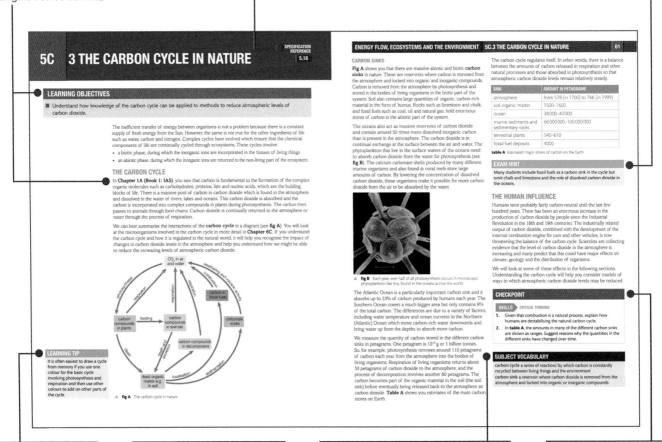

Learning Tips
These help you to focus your learning and avoid common errors.

Worked examples show you how to work through questions and set out calculations.
Did you know? boxes present interesting facts to help you remember key concepts.

Subject vocabulary
Key terms are highlighted in blue in the text. Clear definitions are provided at the end of each section for easy reference, and are also collated in the **glossary** at the back of the book.

Checkpoint
Questions at the end of each section check understanding of the key learning points. Certain questions allow you to develop **skills** which will be valuable for further study and in the workplace.

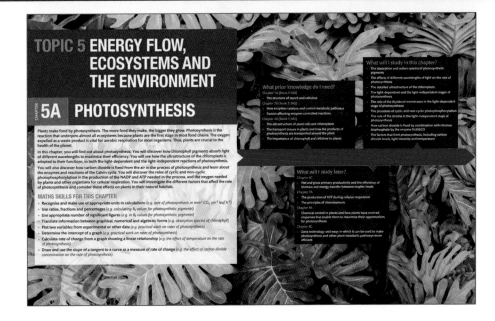

You should be able to put every stage of your learning in context, chapter by chapter.

- Links to other areas of Biology include previous knowledge that is built on in the chapter, and areas of knowledge and application that you will cover later in your course.
- Maths knowledge required is detailed in a handy checklist. If you need to practise the maths you need, you can use the **Maths Skills** reference at the back of the book as a starting point.

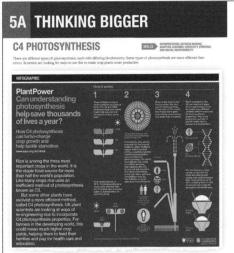

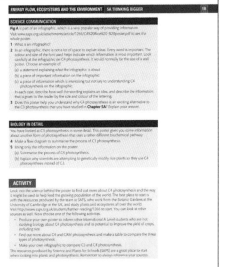

Thinking Bigger

At the end of most chapters there is an opportunity to read and work with real-life research and writing about science.

The activities help you to read authentic material that's relevant to your course, analyse how scientists write, think critically and consider how different aspects of your learning piece together.

These Thinking Bigger activities focus on **key transferable skills**, which are an important basis for key academic qualities.

Exam Practice

Exam-style questions at the end of each chapter are tailored to the Pearson Edexcel specification to allow for practice and development of exam-writing technique. They also allow for practice responding to the 'command words' used in the exams (see the **command words glossary** at the back of this book).

The **Preparing for your exams** section at the end of the book includes **sample answers** for different question types, with comments about the strengths and weaknesses of the answers.

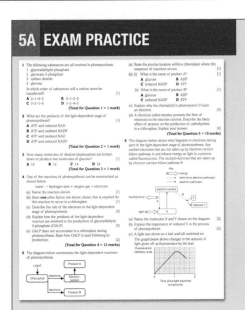

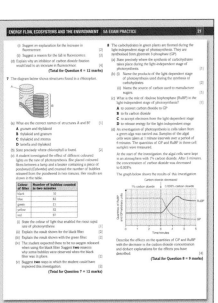

PRACTICAL SKILLS

Practical work is central to the study of biology. The second year of the Pearson Edexcel International Advanced Level (IAL) Biology course includes nine Core Practicals that link theoretical knowledge and understanding to practical scenarios.

Your knowledge and understanding of practical skills and activities will be assessed in all exam papers for the IAL Biology qualification.

- Papers 4 and 5 will include questions based on practical activities, including novel scenarios.
- Paper 6 will test your ability to plan practical work, including risk management and selection of apparatus.

In order to develop practical skills, you should carry out a range of practical experiments related to the topics covered in your course.

STUDENT BOOK TOPIC	IA2 CORE PRACTICALS	
TOPIC 5 ENERGY FLOW, ECOSYSTEMS AND THE ENVIRONMENT	CP10	Investigate the effects of light intensity, light wavelength, temperature and availability of carbon dioxide on the rate of photosynthesis using a suitable aquatic plant.
	CP11	Carry out a study of the ecology of a habitat, such as using quadrats and transects to determine the distribution and abundance of organisms, and measuring abiotic factors appropriate to the habitat.
	CP12	Investigate the effects of temperature on the development of organisms (such as seedling growth rate or brine shrimp hatch rates), taking into account the ethical use of organisms.
TOPIC 6 MICROBIOLOGY, IMMUNITY AND FORENSICS	CP13	Investigate the rate of growth of microorganisms in a liquid culture, taking into account the safe and ethical use of organisms.
	CP14	Investigate the effect of different antibiotics on bacteria.
TOPIC 7 RESPIRATION, MUSCLES AND THE INTERNAL ENVIRONMENT	CP15	Use an artificial hydrogen carrier (redox indicator) to investigate respiration in yeast.
	CP16	Use a simple respirometer to determine the rate of respiration and RQ of a suitable material.
	CP17	Investigate the effects of exercise on tidal volume, breathing rate, respiratory minute ventilation, and oxygen consumption using data from spirometer traces.
TOPIC 8 COORDINATION, RESPONSE AND GENE TECHNOLOGY	CP18	Investigate the production of amylase in germinating cereal grains.

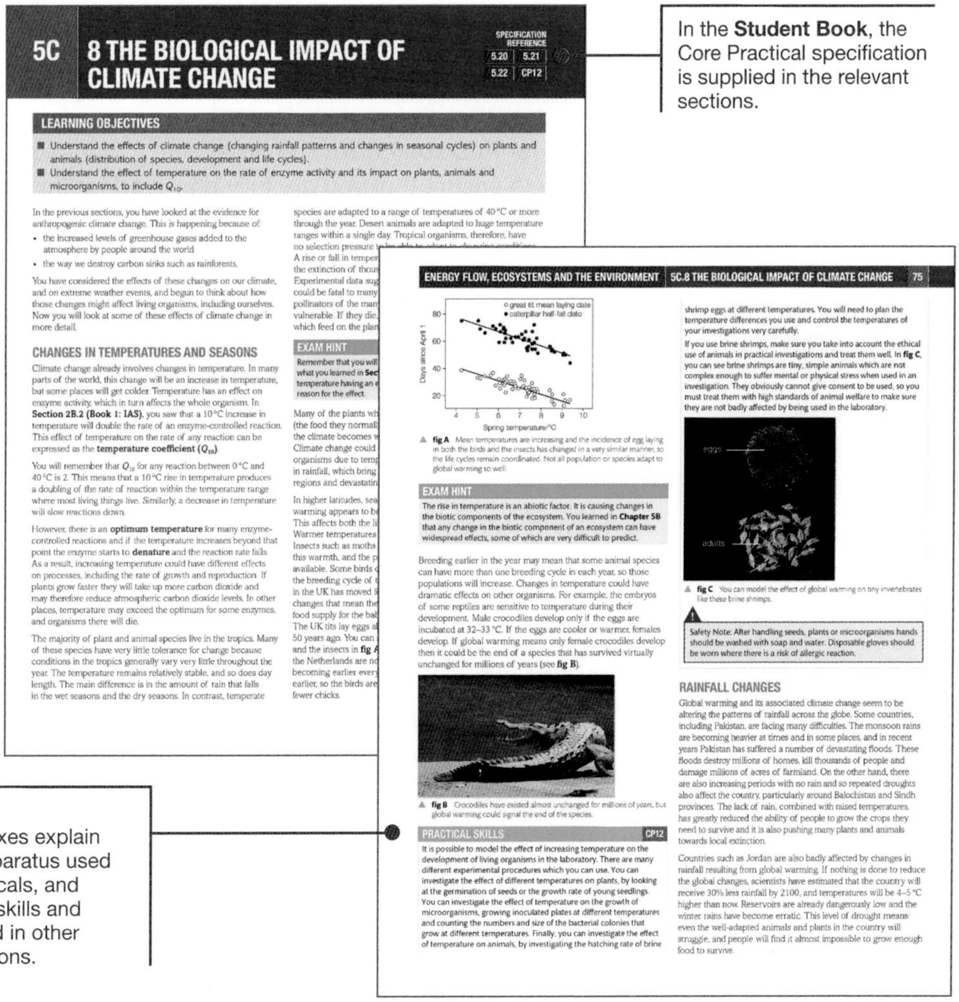

5C · 8 THE BIOLOGICAL IMPACT OF CLIMATE CHANGE

SPECIFICATION REFERENCE
5.20 · 5.21
5.22 · CP12

In the **Student Book**, the Core Practical specification is supplied in the relevant sections.

LEARNING OBJECTIVES

■ Understand the effects of climate change (changing rainfall patterns and changes in seasonal cycles) on plants and animals (distribution of species, development and life cycles).
■ Understand the effect of temperature on the rate of enzyme activity and its impact on plants, animals and microorganisms, to include Q_{10}.

In the previous sections, you have looked at the evidence for anthropogenic climate change. This *is* happening because of:
• the increased levels of greenhouse gases added to the atmosphere by people around the world
• the way we destroy carbon sinks such as rainforests.

You have considered the effects of these changes on our climate, and on extreme weather events, and begun to think about how those changes might affect living organisms, including ourselves. Now you will look at some of these effects of climate change in more detail.

CHANGES IN TEMPERATURES AND SEASONS

Climate change already involves changes in temperature. In many parts of the world, this change will be an increase in temperature, but some places will get colder. Temperature has an effect on enzyme activity, which in turn affects the whole organism. In **Section 2B.2 (Book 1: IAS)**, you saw that a 10 °C increase in temperature will double the rate of an enzyme-controlled reaction. This effect of temperature on the rate of any reaction can be expressed as the **temperature coefficient (Q_{10})**.

You will remember that Q_{10} for any reaction between 0 °C and 40 °C is 2. This means that a 10 °C rise in temperature produces a doubling of the rate of reaction within the temperature range where most living things live. Similarly, a decrease in temperature will slow reactions down.

However, there is an **optimum temperature** for many enzyme-controlled reactions and if the temperature increases beyond that point the enzyme starts to **denature** and the reaction rate falls. As a result, increasing temperature could have different effects on processes, including the rate of growth and reproduction. If plants grow faster they will take up more carbon dioxide and may therefore reduce atmospheric carbon dioxide levels. In other places, temperature may exceed the optimum for some enzymes, and organisms there will die.

The majority of plant and animal species live in the tropics. Many of these species have very little tolerance for change because conditions in the tropics generally vary very little throughout the year. The temperature remains relatively stable, and so does day length. The main difference is in the amount of rain that falls in the wet seasons and the dry seasons. In contrast, temperate

species are adapted to a range of temperatures of 40 °C or more through the year. Desert animals are adapted to huge temperature ranges within a single day. Tropical organisms, therefore, have no selection pressure to be able to adapt to changing conditions.

A rise or fall in temperature [...] the extinction of thous[...] Experimental data sugg[...] could be fatal to many [...] pollinators of the man[...] vulnerable. If they die, [...] which feed on the plan[...]

Many of the plants whi[...] (the food they normall[...] the climate becomes to[...] Climate change could [...] organisms due to temp[...] in rainfall, which bring[...] regions and devastating[...]

In higher latitudes, sea[...] warming appears to be [...] Warmer temperatures [...] Insects such as moths [...] this warmth, and the p[...] available. Some birds [...] the breeding cycle of [...] in the UK has moved [...] changes that mean the[...] food supply for the bab[...] The UK tits lay eggs ab[...] 50 years ago. You can [...] and the insects in **fig A**[...] the Netherlands are no[...] becoming earlier every [...] earlier, so the birds are[...] fewer chicks.

ENERGY FLOW, ECOSYSTEMS AND THE ENVIRONMENT · 5C.8 THE BIOLOGICAL IMPACT OF CLIMATE CHANGE · 75

▲ **fig A** Mean temperatures are increasing and the incidence of egg laying in both the birds and the insects has changed in a very similar manner, so the life cycles remain coordinated. Not all population or species adapt to global warming so well.

EXAM HINT

The rise in temperature is an abiotic factor. It is causing changes in the biotic components of the ecosystem. You learned in **Chapter 5B** that any change in the biotic component of an ecosystem can have widespread effects, some of which are very difficult to predict.

Breeding earlier in the year may mean that some animal species can have more than one breeding cycle in each year, so those populations will increase. Changes in temperature could have dramatic effects on other organisms. For example, the embryos of some reptiles are sensitive to temperature during their development. Male crocodiles develop only if the eggs are incubated at 32–33 °C. If the eggs are cooler or warmer, females develop. If global warming means only female crocodiles develop then it could be the end of a species that has survived virtually unchanged for millions of years (see **fig B**).

▲ **fig B** Crocodiles have existed almost unchanged for millions of years, but global warming could signal the end of the species.

PRACTICAL SKILLS CP12

It is possible to model the effect of increasing temperature on the development of living organisms in the laboratory. There are many different experimental procedures which you can use. You can investigate the effect of different temperatures on plants, by looking at the germination of seeds or the growth rate of young seedlings. You can investigate the effect of temperature on the growth of microorganisms, growing inoculated plates at different temperatures and counting the numbers and size of the bacterial colonies that grow at different temperatures. Finally, you can investigate the effect of temperature on animals, by investigating the hatching rate of brine

shrimp eggs at different temperatures. You will need to plan the temperature differences you use and control the temperatures of your investigations very carefully.

If you use brine shrimps, make sure you take into account the ethical use of animals in practical investigations and treat them well. In **fig C**, you can see brine shrimps are tiny, simple animals which are not complex enough to suffer mental or physical stress when used in an investigation. They obviously cannot give consent to be used, so you must treat them with high standards of animal welfare to make sure they are not badly affected by being used in the laboratory.

eggs

adults

▲ **fig C** You can model the effect of global warming on tiny invertebrates like these brine shrimps.

Safety Note: After handling seeds, plants or microorganisms hands should be washed with soap and water. Disposable gloves should be worn where there is a risk of allergic reaction.

RAINFALL CHANGES

Global warming and its associated climate change seem to be altering the patterns of rainfall across the globe. Some countries, including Pakistan, are facing many difficulties. The monsoon rains are becoming heavier at times and in some places, and in recent years Pakistan has suffered a number of devastating floods. These floods destroy millions of homes, kill thousands of people and damage millions of acres of farmland. On the other hand, there are also increasing periods with no rain and so repeated droughts also affect the country, particularly around Balochistan and Sindh provinces. The lack of rain, combined with raised temperatures, has greatly reduced the ability of people to grow the crops they need to survive and it is also pushing many plants and animals towards local extinction.

Countries such as Jordan are also badly affected by changes in rainfall resulting from global warming. If nothing is done to reduce the global changes, scientists have estimated that the country will receive 30% less rainfall by 2100, and temperatures will be 4–5 °C higher than now. Reservoirs are already dangerously low and the winter rains have become erratic. This level of drought means even the well-adapted animals and plants in the country will struggle, and people will find it almost impossible to grow enough food to survive.

Practical Skills
Practical skills boxes explain techniques or apparatus used in the Core Practicals, and also detail useful skills and knowledge gained in other related investigations.

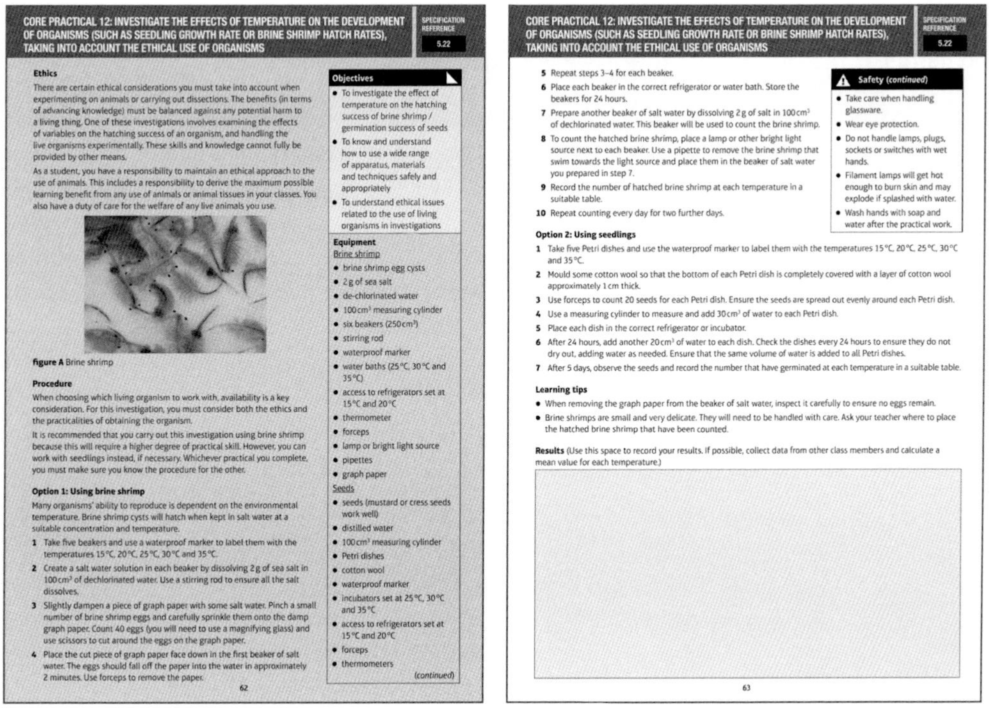

CORE PRACTICAL 12: INVESTIGATE THE EFFECTS OF TEMPERATURE ON THE DEVELOPMENT OF ORGANISMS (SUCH AS SEEDLING GROWTH RATE OR BRINE SHRIMP HATCH RATES), TAKING INTO ACCOUNT THE ETHICAL USE OF ORGANISMS

SPECIFICATION REFERENCE 5.22

Ethics

There are certain ethical considerations you must take into account when experimenting on animals or carrying out dissections. The benefits (in terms of advancing knowledge) must be balanced against any potential harm to a living thing. One of these investigations involves examining the effects of variables on the hatching success of an organism, and handling the live organisms experimentally. These skills and knowledge cannot fully be provided by other means.

As a student, you have a responsibility to maintain an ethical approach to the use of animals. This includes a responsibility to derive the maximum possible learning benefit from any use of animals or animal tissues in your classes. You also have a duty of care for the welfare of any live animals you use.

figure A Brine shrimp

Procedure

When choosing which living organism to work with, availability is a key consideration. For this investigation, you must consider both the ethics and the practicalities of obtaining the organism.

It is recommended that you carry out this investigation using brine shrimp because this will require a higher degree of practical skill. However, you can work with seedlings instead, if necessary. Whichever practical you complete, you must make sure you know the procedure for the other.

Option 1: Using brine shrimp

Many organisms' ability to reproduce is dependent on the environmental temperature. Brine shrimp cysts will hatch when kept in salt water at a suitable concentration and temperature.

1 Take five beakers and use a waterproof marker to label them with the temperatures 15 °C, 20 °C, 25 °C, 30 °C and 35 °C.
2 Create a salt water solution in each beaker by dissolving 2 g of sea salt in 100 cm³ of dechlorinated water. Use a stirring rod to ensure all the salt dissolves.
3 Slightly dampen a piece of graph paper with some salt water. Pinch a small number of brine shrimp eggs and carefully sprinkle them onto the damp graph paper. Count 40 eggs (you will need to use a magnifying glass) and use scissors to cut around the eggs on the graph paper.
4 Place the cut piece of graph paper face down in the first beaker of salt water. The eggs should fall off the paper into the water in approximately 2 minutes. Use forceps to remove the paper.

Objectives

• To investigate the effect of temperature on the hatching success of brine shrimp / germination success of seeds
• To know and understand how to use a wide range of apparatus, materials and techniques safely and appropriately
• To understand ethical issues related to the use of living organisms in investigations

Equipment

Brine shrimps
• brine shrimp egg cysts
• 2 g of sea salt
• de-chlorinated water
• 100 cm³ measuring cylinder
• six beakers (250 cm³)
• stirring rod
• waterproof marker
• water baths (25 °C, 30 °C and 35 °C)
• access to refrigerators set at 15 °C and 20 °C
• thermometer
• forceps
• lamp or bright light source
• pipettes
• graph paper
Seeds
• seeds (mustard or cress seeds work well)
• distilled water
• 100 cm³ measuring cylinder
• Petri dishes
• cotton wool
• waterproof marker
• incubators set at 25 °C, 30 °C and 35 °C
• access to refrigerators set at 15 °C and 20 °C
• forceps
• thermometers

(continued)

62

CORE PRACTICAL 12: INVESTIGATE THE EFFECTS OF TEMPERATURE ON THE DEVELOPMENT OF ORGANISMS (SUCH AS SEEDLING GROWTH RATE OR BRINE SHRIMP HATCH RATES), TAKING INTO ACCOUNT THE ETHICAL USE OF ORGANISMS

SPECIFICATION REFERENCE 5.22

5 Repeat steps 3–4 for each beaker.
6 Place each beaker in the correct refrigerator or water bath. Store the beakers for 24 hours.
7 Prepare another beaker of salt water by dissolving 2 g of salt in 100 cm³ of dechlorinated water. This beaker will be used to count the brine shrimp.
8 To count the hatched brine shrimp, place a lamp or other bright light source next to each beaker. Use a pipette to remove the brine shrimp that swim towards the light source and place them in the beaker of salt water you prepared in step 7.
9 Record the number of hatched brine shrimp at each temperature in a suitable table.
10 Repeat counting every day for two further days.

Option 2: Using seedlings

1 Take four Petri dishes and use the waterproof marker to label them with the temperatures 15 °C, 20 °C, 25 °C, 30 °C and 35 °C.
2 Mould some cotton wool so that the bottom of each Petri dish is completely covered with a layer of cotton wool approximately 1 cm thick.
3 Use forceps to count 20 seeds for each Petri dish. Ensure the seeds are spread out evenly around each Petri dish.
4 Use a measuring cylinder to measure and add 30 cm³ of water to each Petri dish.
5 Place each dish in the correct refrigerator or incubator.
6 After 24 hours, add another 20 cm³ of water to each dish. Check the dishes every 24 hours to ensure they do not dry out, adding water as needed. Ensure that the same volume of water is added to all Petri dishes.
7 After 5 days, observe the seeds and record the number that have germinated at each temperature in a suitable table.

Learning tips

• When removing the graph paper from the beaker of salt water, inspect it carefully to ensure no eggs remain.
• Brine shrimps are small and very delicate. They will need to be handled with care. Ask your teacher where to place the hatched brine shrimp that have been counted.

Results (Use this space to record your results. If possible, collect data from other class members and calculate a mean value for each temperature.)

⚠ Safety (continued)

• Take care when handling glassware.
• Wear eye protection.
• Do not handle lamps, plugs, sockets or switches with wet hands.
• Filament lamps will get hot enough to burn skin and may explode if splashed with water.
• Wash hands with soap and water after the practical work.

63

This Student Book is accompanied by a **Lab Book**, which includes instructions and writing frames for the Core Practicals for students to record their results and reflect on their work.

Practical skills checklists, practice questions and answers are also provided.

The Lab Book records can be used as preparation and revision for the Practical Skills Papers.

ASSESSMENT OVERVIEW

The following tables give an overview of the assessment for Pearson Edexcel International Advanced Level course in Biology. You should study this information closely to help ensure that you are fully prepared for this course and know exactly what to expect in each part of the examination. More information about this qualification, and about the question types in the different papers, can be found on page 314 of this book.

PAPER / UNIT 4	PERCENTAGE OF IA2	PERCENTAGE OF IAL	MARK	TIME	AVAILABILITY
ENERGY, ENVIRONMENT, MICROBIOLOGY AND IMMUNITY Written examination Paper code WBI14/01 Externally set and marked by Pearson Edexcel Single tier of entry	40%	20%	90	1 hour 45 minutes	January, June and October First assessment : January 2020

PAPER / UNIT 5	PERCENTAGE OF IA2	PERCENTAGE OF IAL	MARK	TIME	AVAILABILITY
RESPIRATION, INTERNAL ENVIRONMENT, COORDINATION AND GENE TECHNOLOGY Written examination Paper code WBI15/01 Externally set and marked by Pearson Edexcel Single tier of entry	40%	20%	90	1 hour 45 minutes	January, June and October First assessment : June 2020

PAPER / UNIT 6	PERCENTAGE OF IA2	PERCENTAGE OF IAL	MARK	TIME	AVAILABILITY
PRACTICAL SKILLS IN BIOLOGY II Written examination Paper code WBI16/01 Externally set and marked by Pearson Edexcel Single tier of entry	20%	10%	50	1 hour 20 minutes	January, June and October First assessment : June 2020

ASSESSMENT OBJECTIVES AND WEIGHTINGS

ASSESSMENT OBJECTIVE	DESCRIPTION	% IN IAS	% IN IA2	% IN IAL
A01	Demonstrate knowledge and understanding of science	36–39	31–34	34–37
A02	(a) Application of knowledge and understanding of science in familiar and unfamiliar contexts.	34–36	33–36	33–36
	(b) Analysis and evaluation of scientific information to make judgments and reach conclusions.	9–11	14–16	11–14
A03	Experimental skills in science, including analysis and evaluation of data and methods	17–18	17–18	17–18

RELATIONSHIP OF ASSESSMENT OBJECTIVES TO UNITS

UNIT NUMBER	ASSESSMENT OBJECTIVE (%)			
	A01	A02 (A)	A02 (B)	A03
UNIT 1	17–18	17–18	4.5–5.5	0
UNIT 2	17–18	17–18	4.5–5.5	0
UNIT 3	2–3	0	0	17–18
TOTAL FOR INTERNATIONAL ADVANCED SUBSIDIARY	36–39	34–36	9–11	17–18

UNIT NUMBER	ASSESSMENT OBJECTIVE (%)			
	A01	A02 (A)	A02 (B)	A03
UNIT 1	8.5–9.0	8.5–9.0	2.2–2.8	0
UNIT 2	8.5–9.0	8.5–9.0	2.2–2.8	0
UNIT 3	1–1.5	0	0	17–18
UNIT 4	7.3–7.8	8.4–8.9	3.6–4.0	0
UNIT 5	7.3–7.8	8.4–8.9	3.6–4.0	0
UNIT 6	1–1.5	0	0	8.8–9.2
TOTAL FOR INTERNATIONAL ADVANCED LEVEL	34–37	33–36	11–14	17–18

TOPIC 5 ENERGY FLOW, ECOSYSTEMS AND THE ENVIRONMENT

CHAPTER 5A PHOTOSYNTHESIS

Plants make food by photosynthesis. The more food they make, the bigger they grow. Photosynthesis is the reaction that underpins almost all ecosystems because plants are the first stage in most food chains. The oxygen expelled as a waste product is vital for aerobic respiration for most organisms. Thus, plants are crucial to the health of the planet.

In this chapter, you will find out about photosynthesis. You will discover how chlorophyll pigments absorb light of different wavelengths to maximise their efficiency. You will see how the ultrastructure of the chloroplasts is adapted to their functions, in both the light-dependent and the light-independent reactions of photosynthesis.

You will also discover how carbon dioxide is fixed from the air in the process of photosynthesis and learn about the enzymes and reactions of the Calvin cycle. You will discover the roles of cyclic and non-cyclic photophosphorylation in the production of the NADP and ATP needed in the process, and the oxygen needed by plants and other organisms for cellular respiration. You will investigate the different factors that affect the rate of photosynthesis and consider these effects on plants in their natural habitats.

MATHS SKILLS FOR THIS CHAPTER

- Recognise and make use of appropriate units in calculations (*e.g. rate of photosynthesis in $mm^3 CO_2 cm^{-2} leaf h^{-1}$*)
- Use ratios, fractions and percentages (*e.g. calculating R_f values for photosynthetic pigments*)
- Use appropriate number of significant figures (*e.g. in R_f values for photosynthetic pigments*)
- Translate information between graphical, numerical and algebraic forms (*e.g. absorption spectra of chlorophyll*)
- Plot two variables from experimental or other data (*e.g. practical work on rates of photosynthesis*)
- Determine the intercept of a graph (*e.g. practical work on rates of photosynthesis*)
- Calculate rate of change from a graph showing a linear relationship (*e.g. the effect of temperature on the rate of photosynthesis*)
- Draw and use the slope of a tangent to a curve as a measure of rate of change (*e.g. the effect of carbon dioxide concentration on the rate of photosynthesis*)

What prior knowledge do I need?

Chapter 1A (Book 1: IAS)
- The structure of starch and cellulose

Chapter 2B (Book 1: IAS)
- How enzymes catalyse and control metabolic pathways
- Factors affecting enzyme-controlled reactions

Chapter 4A (Book 1: IAS)
- The ultrastructure of plant cells and chloroplasts
- The transport tissues in plants and how the products of photosynthesis are transported around the plant
- The importance of chlorophyll and cellulose to plants

What will I study in this chapter?
- The absorption and action spectra of photosynthetic pigments
- The effects of different wavelengths of light on the rate of photosynthesis
- The detailed ultrastructure of the chloroplasts
- The light-dependent and the light-independent stages of photosynthesis
- The role of the thylakoid membranes in the light-dependent stage of photosynthesis
- The processes of cyclic and non-cyclic photophosphorylation
- The role of the stroma in the light-independent stage of photosynthesis
- How carbon dioxide is fixed by combination with ribulose bisphosphate by the enzyme RUBISCO
- The factors that limit photosynthesis, including carbon dioxide levels, light intensity and temperature

What will I study later?

Chapter 5C
- Net and gross primary productivity and the efficiency of biomass and energy transfer between trophic levels

Chapter 7A
- The production of ATP during cellular respiration
- The principles of chemiosmosis

Chapter 8B
- Chemical control in plants and how plants have evolved responses that enable them to maximise their opportunities for photosynthesis

Chapter 8C
- Gene technology and ways in which it can be used to make photosynthesis and other plant metabolic pathways more efficient

LEARNING OBJECTIVES

■ Understand how photophosphorylation of ATP requires energy and that hydrolysis of ATP provides an immediate supply of energy for biological processes.

Energy is essential to life. If the supply of energy to the cells of a living organism fails for any reason, the organism will die. Very large amounts of energy continually flow through the biosphere and organisms can be classified according to how they get their energy. **Autotrophic** organisms make organic compounds from carbon dioxide. Most of them do this by **photosynthesis**: they capture energy from the Sun and transfer it into chemical energy in the bonds of organic molecules such as glucose and starch. These compounds are used as an energy source by the organism, and as the building blocks of other important molecules such as proteins. Plants, algae and some bacteria are the main photosynthetic organisms (see **fig A**). There are a few autotrophic bacteria that are not photosynthetic. They use energy from chemical reactions to synthesise their food. **Heterotrophic** organisms generally eat plants or other animals which have eaten plants. They use the products of photosynthesis indirectly for making necessary molecules, and as fuels to supply energy for activities. The Sun is thus the ultimate source of energy for almost all organisms.

LEARNING TIP

Remember that light is the source of energy for photosynthesis. It is, therefore, the original source of energy for almost all living processes.

▲ **fig A** The ability of plants to trap and use energy from the Sun in photosynthesis supports almost all life on Earth.

ATP: THE ENERGY SOURCE FOR THE CELL

Making chemical bonds needs an input of energy. Chemical bonds are constantly being broken in the cells of any living organism. Energy has to be constantly available in an accessible form, ready for use instantly in a multitude of different reactions. One molecule is believed to be the universal energy supplier in cells. It is found in all living organisms in exactly the same form. Anything that interferes with its production or breakdown is fatal to the cell and, ultimately, the organism. This remarkable compound is called **adenosine triphosphate (ATP)**. **Fig B** shows the structure of ATP.

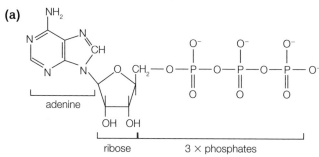

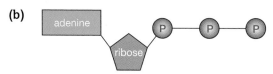

▲ **fig B** ATP is a nucleotide with three phosphate groups attached. It is the energy stored in the phosphate bonds, particularly the end one, which is made available to cells to use to make or break bonds. (a) Structural formula of ATP (some hydrogen atoms have been omitted for simplicity). (b) Simplified diagram of an ATP molecule.

When energy is needed, the third phosphate bond can be broken by a hydrolysis reaction. This is catalysed by the enzyme **ATPase**. The result of this hydrolysis is **adenosine diphosphate (ADP)**, a free inorganic phosphate group (P_i) and energy (see **fig C**). About 34 kJ of energy is released per mole of ATP hydrolysed. Some of this energy is lost as heat and is wasted. The rest is used for any biological activity in the cell which requires energy. Examples include active transport (see **Section 2A.4 (Book 1: IAS)**), anabolic reactions in which large molecules are built up from smaller ones, and muscle contraction (see **Chapter 7B**).

EXAM HINT

Remember that synthesising ATP is a condensation reaction and breakdown of ATP is a hydrolysis reaction. There is a close link to similar reaction studies in **Chapter 1A (Book 1: IAS)**.

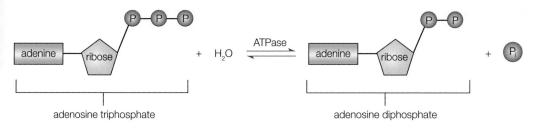

▲ **fig C**   When ATP is hydrolysed (broken down) to ADP + P_i (left → right on the diagram), energy is made available for cellular reactions. When ATP is synthesised (made) from ADP and P_i in the reverse reaction, the same amount of energy is absorbed. This energy comes from cellular respiration.

The breakdown of ATP into ADP and phosphate is a reversible reaction. ATP can be synthesised (made) from ADP and a phosphate group. This synthesis reaction is also catalysed by the enzyme ATPase and it requires an input of energy (34 kJ per mole of ATP produced). The energy needed to drive the synthesis of ATP usually comes from catabolic (breakdown) reactions or **reduction/oxidation reactions (redox reactions)**. As a result, an ATP molecule provides an immediate supply of energy for your cells, ready for use when needed (see **fig D**).

coupled to catabolism,
e.g. respiration

condensation

P_i

ADP

ATP

hydrolysis

P_i

energy coupled
to anabolism

▲ **fig D**   The energy released in catabolic reactions is used to drive the production of ATP. ATP acts as a store of energy which is released when it is needed for cell functions, including anabolic reactions, when large molecules are built up from smaller ones. This results in the formation of ADP and inorganic phosphate which can be resynthesised into ATP.

MAKING ATP: THE ELECTRON TRANSPORT CHAIN

There are two main ways in which ATP is formed from ADP and inorganic phosphate in the cell. One is using energy released from the catabolic reactions which take place, for example, in cellular respiration. However, the main way in which ATP is synthesised is by the removal of hydrogen atoms from several of the intermediate compounds in a metabolic pathway.

When two hydrogen atoms are removed from a compound, they are collected by a hydrogen carrier or acceptor. The acceptor is reduced. Electrons from the hydrogen atoms are then transferred along a series of carriers known as an **electron transport chain**. The components of the chain are reduced when they receive the electrons, and oxidised again when they transfer the electrons to the next part of the chain. These redox reactions each release a small amount of energy which is used to drive the synthesis of a molecule of ATP. In this way, the energy is readily available for use when it is needed in the cell. You will learn more about the production of ATP later (see **Chapter 7A**).

CHECKPOINT

1. (a) The breakdown reaction of ATP to ADP + P_i is described as reversible. Explain what this means.

    (b) Both the formation and the breakdown of ATP are controlled by the same enzyme. State the name of this enzyme.

SKILLS   CREATIVITY

2. ATP is the universal energy supply molecule in living things. Explain how ATP is suited for this role in the cells.

SKILLS   CRITICAL THINKING

3. Some people describe photosynthesis as the most important reaction in living organisms. Give one reason why it might be, and one reason why not.

SUBJECT VOCABULARY

autotrophic organisms that make complex organic compounds from simple compounds in the environment

photosynthesis the process by which living organisms, particularly plants and algae, capture the energy of the Sun using chlorophyll and use it to convert carbon dioxide and water into simple sugars

heterotrophic organisms that obtain complex organic molecules by feeding on other living organisms or their dead remains

adenosine triphosphate (ATP) a nucleotide that acts as the universal energy supply in cells. It is made up of the base adenine, the pentose sugar ribose and three phosphate groups

ATPase the enzyme which catalyses the formation and breakdown of ATP, depending on the conditions

adenosine diphosphate (ADP) a nucleotide formed when a phosphate group is removed from ATP, releasing energy to drive reactions in the cell

reduction/oxidation reactions (redox reactions) reactions in which one reactant loses electrons (is oxidised) and another gains electrons (is reduced)

electron transport chain a series of electron-carrying compounds along which electrons are transferred in a series of oxidation/reduction reactions, driving the production of ATP

LEARNING OBJECTIVES

■ Understand the overall reaction of photosynthesis as requiring energy from light to split apart the strong bonds in water molecules, storing the hydrogen in a fuel (glucose) by combining it with carbon dioxide and releasing oxygen into the atmosphere.
■ Understand the structure of chloroplasts in relation to their role in photosynthesis.
■ Understand what is meant by the terms *absorption spectrum* and *action spectrum*.
■ Understand that chloroplast pigments can be separated using chromatography and the pigments identified using R_f values.

Photosynthesis is the process used by living organisms, particularly plants, to capture the energy of the Sun using chlorophyll and use it to convert carbon dioxide and water into simple sugars. This equation summarises photosynthesis:

$$\text{carbon dioxide} + \text{water} \xrightarrow[\text{chlorophyll}]{\text{light energy}} \text{glucose} + \text{oxygen} \qquad \Delta H \approx +2880\,\text{kJ}$$

$$6CO_2 + 6H_2O \xrightarrow[\text{chlorophyll}]{\text{light energy}} C_6H_{12}O_6 + 6O_2 \qquad \Delta H \approx +2880\,\text{kJ}$$

The energy from light is used to break the strong H–O bonds in the water molecules. The hydrogen which is released is combined with carbon dioxide to form a fuel for the cells (glucose). Oxygen is released into the atmosphere as a waste product of this process.

Simple models of photosynthesis such as the equation above show a one-step process. They include the most important points of the process and make it relatively easy to understand. But the whole process is extremely complex. You will learn some of the details of the process in this section.

The structure of plants has evolved around the process of photosynthesis. The different parts of the plant are adapted for efficiently obtaining the raw materials carbon dioxide and water, and for trapping as much sunlight as possible.

THE IMPORTANCE OF CHLOROPLASTS

Chloroplasts are relatively large organelles found in the cells of the green parts of plants (see **fig A**). An average green plant cell contains 10–50 chloroplasts which are uniquely adapted for the process of photosynthesis (see **Section 4A.2 (Book 1: IAS)**). Each chloroplast is surrounded by an outer and an inner membrane with a space between the two, known as the **chloroplast envelope**. Inside the chloroplast is a system of membranes that are arranged in layers called **grana**. A single granum is made up of layers of membrane discs known as **thylakoids**. This is where the green pigment chlorophyll is found. The pigment molecules are arranged on the membranes in the best possible position for capturing light energy. Electron micrographs show that the granal membranes are covered in particles. Scientists think these particles are involved in ATP synthesis. The grana are joined together by **lamellae**, which are extensions of the thylakoid membranes. These lamellae connect two or more grana. The lamellae act as a skeleton inside the chloroplast, maintaining a distance between the grana so that they receive the maximum light and function as efficiently as possible.

The membrane layers are surrounded by a matrix called the **stroma**. The stroma contains all the enzymes needed to complete the process of photosynthesis and produce glucose. Glucose can then be used in cellular respiration, converted to starch for storage or used as an intermediate for the synthesis of other organic compounds such as amino acids and lipids.

LEARNING TIP

The processes of photosynthesis and cellular respiration are closely linked.

Photosynthesis takes in energy and uses carbon dioxide and water to synthesise glucose and oxygen. It is endothermic.

Aerobic respiration uses glucose and oxygen to produce energy in the form of ATP for the cell, with carbon dioxide and water as the waste products. It is exothermic. You will learn more about aerobic respiration in **Chapter 7A**.

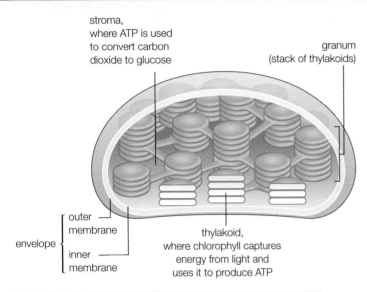

stroma, where ATP is used to convert carbon dioxide to glucose

granum (stack of thylakoids)

envelope { outer membrane, inner membrane

thylakoid, where chlorophyll captures energy from light and uses it to produce ATP

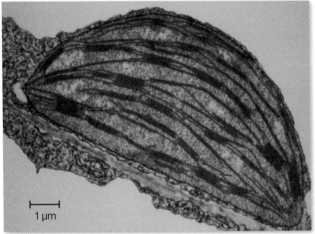

1 μm

▲ **fig A** Evidence from electron micrographs has helped scientists to produce a realistic and complex model demonstrating how the structure of a chloroplast is adapted to its functions in photosynthesis.

EXAM HINT

Make sure you spell *stroma* correctly. Do not confuse it with *stoma*.

LEARNING TIP

A good way to recall and record the details of chloroplast structure is to draw a diagram and annotate it fully. You need to write brief but detailed notes around the diagram, not just labels.

CHLOROPHYLL

Chlorophyll is the other major adaptation of the chloroplasts. It is a light-capturing, photosynthetic pigment. Chlorophyll is not a single molecule. It is a mixture of closely related pigments. These include **chlorophyll *a*** (blue-green), **chlorophyll *b*** (yellow-green), the chlorophyll **carotenoids** (orange carotene and yellow xanthophyll) and also a grey pigment **phaeophytin**, which is a breakdown product of the others. Chlorophyll *a* is found in all photosynthesising plants and in the highest quantity of the five pigments. The other pigments are found in varying proportions in different plants. These differences give the leaves of plants their

great variety of different greens. Each of the pigments absorbs and captures light from particular areas of the light spectrum. As a result, much more of the energy from the light falling on the plant can be used than if only one pigment was involved.

The different photosynthetic pigments can be demonstrated in a number of ways.

ABSORPTION SPECTRA AND ACTION SPECTRA

The **absorption spectrum** describes the different amounts of light of different wavelengths that a photosynthetic pigment absorbs. It is usually represented as a graph. We can find the absorption spectra of the different photosynthetic pigments by measuring their absorption of light of differing wavelengths (see **fig B**). It is also possible to produce an absorption spectrum for whole chloroplasts, with all the photosynthetic pigments combined.

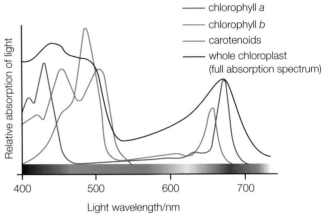

—— chlorophyll *a*
—— chlorophyll *b*
—— carotenoids
—— whole chloroplast (full absorption spectrum)

▲ **fig B** The different photosynthetic pigments absorb light of different wavelengths. This enables the plant to use more of the available light.

You can compare the rate of photosynthesis with the wavelength of light. The first person to do this was T.W. Engelmann in the late 1800s. He placed a strand of a filamentous alga in light of different wavelengths. He used bacteria that move towards oxygen to show where most oxygen was released, because this is directly related to the amount of photosynthesis occurring. In this way, he achieved an **action spectrum**, which is a way of demonstrating the rate of photosynthesis according to the wavelength of light (see **fig C**).

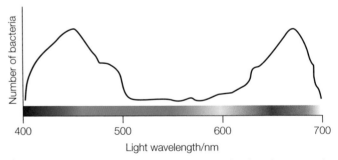

▲ **fig C** Engelmann's first action spectrum was developed using numbers of bacteria to show where photosynthesis was occurring.

Modern action spectra use electronic data logging instead of bacterial movements to measure the rate of photosynthesis at different wavelengths of light. However, the action spectra they produce still compare the rate of photosynthesis to the wavelength of light. Action spectra show us that the rate of photosynthesis is very closely related to the combined absorption spectrum of all the photosynthetic pigments in a plant, as you can see in **fig D**. This demonstrates that having different photosynthetic pigments makes a much bigger portion of light available to plants and therefore gives them an adaptive advantage.

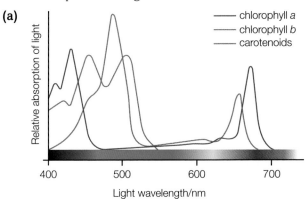

(a)

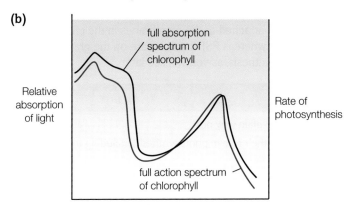

(b)

▲ **fig D** Here you can see (a) the absorption spectrum of the individual photosynthetic pigments compared with (b) the absorption spectrum and action spectrum of chlorophyll in a chloroplast. This shows the advantage of having more than one pigment available to absorb the light.

EXAM HINT

Remember that different proportions of the photosynthetic pigments will produce a different colour of leaf; this can be a major adaptation to habitat. For example, many aquatic plants are red or brown. These colours absorb the blue light that penetrates water easily.

CHROMATOGRAPHY

Plants look green. If you extract the pigments from a plant by grinding up leaves with propanone and then filtering, the filtrate looks green. So how can you show that there are several different pigments? The answer is by chromatography using paper or silica gel. The pigments travel up the solid medium at different speeds and are readily separated using a suitable solvent (see **fig E**).

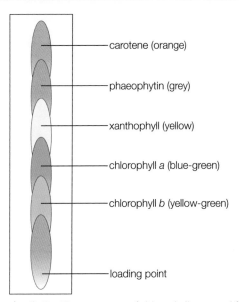

▲ **fig E** Chromatogram of chlorophyll extracted from a plant showing five photosynthetic pigments.

Once you have conducted chromatography on the photosynthetic pigments you can determine their R_f **values** and compare them to the R_f values of known pigments in the same solvent. It is important to compare R_f values using the same solvent because the pigments can have very different values with different solvents. The R_f value is the ratio of the distance travelled by the pigment to the distance travelled by the solvent alone (see **fig F**). The R_f value is always between 0 and 1 and this is how you can calculate it:

$$R_f \text{ value} = \frac{\text{distance travelled by solute (photosynthetic pigment)}}{\text{distance travelled by solvent}}$$

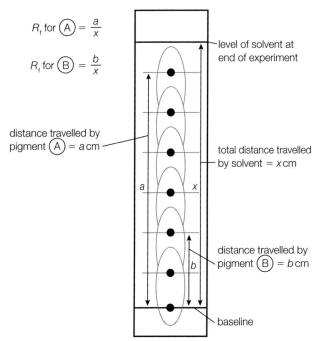

▲ **fig F** This example shows a method for calculating the R_f values for two photosynthetic pigments on a silica gel chromatogram.

PHOTOSYNTHETIC PIGMENT	R_f VALUE FOR SPINACH LEAVES EXTRACTED WITH HEXANE AND CHROMATOGRAPHED WITH 3:1:1 PETROLEUM ETHER-PROPANONE-CHLOROFORM SOLVENT ON SILICA GEL
carotene	0.98
chlorophyll a	0.59
chlorophyll b	0.42
phaeophytin	0.81
xanthophyll 1	0.28
xanthophyll 2	0.15

table A Photosynthetic pigments and their R_f values

PHOTOSYSTEMS

The photosynthetic pigments absorb light in two distinct chlorophyll complexes known as **photosystem I (PSI)** and **photosystem II (PSII)**. Each system contains a different combination of chlorophyll pigments and therefore absorbs light in a slightly different area of the spectrum (wavelength 700 nm for PSI and 680 nm for PSII). Electron micrographs have revealed that the different photosystems are differently sized particles attached to the membranes in the chloroplasts. PSI particles are mainly on the intergranal lamellae, whereas PSII particles are on the grana themselves. They have different functions in photosynthesis, as you will see later.

SKILLS CRITICAL THINKING

SKILLS ANALYSIS

CHECKPOINT

1. (a) Chloroplasts are not present in all plant cells. Explain why not.
    ▶ (b) Summarise the adaptations of chloroplasts for their role in photosynthesis, including the role of membranes.
2. ▶ Explain, using the data in **fig B**, why plant leaves usually appear green.
3. Measure the distances travelled by the solute and solvent and calculate the R_f values for the pigments labelled A and B in **fig F**. Identify these two pigments, assuming they were extracted with hexane and chromatographed with 3:1:1 petroleum ether–propanone–chloroform solvent on silica gel.

SUBJECT VOCABULARY

chloroplast envelope the outer and inner membranes of a chloroplast including the intermembrane space
grana layers of thylakoid membranes within a chloroplast
thylakoids membrane discs found in the grana of a chloroplast
lamellae extensions of the thylakoid membranes which connect two or more grana and act as a supporting skeleton in the chloroplast; they maintain a working distance between the grana so that these receive the maximum light and function as efficiently as possible
stroma the matrix which surrounds the grana and contains all the enzymes needed to complete the process of photosynthesis and produce glucose
chlorophyll a a blue-green photosynthetic pigment, found in all green plants
chlorophyll b a yellow-green photosynthetic pigment
carotenoids photosynthetic pigments consisting of orange carotene and yellow xanthophyll
phaeophytin a grey pigment which is produced by the breakdown of the other photosynthetic pigments
absorption spectrum a graph showing the amount of light absorbed by a pigment against the wavelength of the light
action spectrum a graph demonstrating the rate of photosynthesis against the wavelength of light
R_f value the ratio of the distance travelled by the pigment to the distance travelled by the solvent alone when pigments are separated by chromatography
photosystem I (PSI) a combination of chlorophyll pigments which absorbs light of wavelength 700 nm and is involved in cyclic and non-cyclic photophosphorylation
photosystem II (PSII) a combination of chlorophyll pigments which absorbs light of wavelength 680 nm and is involved only in non-cyclic photophosphorylation

LEARNING OBJECTIVES

- Understand the light-dependent reactions of photosynthesis, including how light energy is trapped by exciting electrons in chlorophyll and the role of these electrons in generating ATP, reducing NADP in cyclic and non-cyclic photophosphorylation and producing oxygen through photolysis of water.
- Understand the light-independent reactions as reduction of carbon dioxide using the products of the light-dependent reactions (carbon fixation in the Calvin cycle, the role of GP, GALP, RuBP and RUBISCO).
- Know that the products of photosynthesis are simple sugars that are used by plants, animals and other organisms in respiration and the synthesis of new biological molecules (polysaccharides, amino acids, proteins, lipids and nucleic acids).

Photosynthesis is a two-stage process involving a complex series of reactions (see **fig A**). The reactions in the first stage only occur in light. The reactions of the second stage occur independently of light. The **light-dependent reactions** produce materials to be used in the **light-independent reactions**. The whole process occurs all the time during the hours of daylight. The light-independent reactions can continue when it is dark.

EXAM HINT

Do not refer to these as the light stage and the dark stage. The light-dependent reactions occur only when light is available. The light-independent reactions occur in both light and dark conditions.

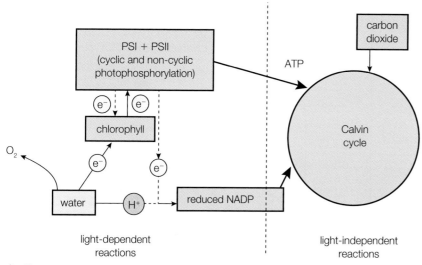

▲ **fig A** A simplified summary of photosynthesis.

THE LIGHT-DEPENDENT STAGE OF PHOTOSYNTHESIS

The light-dependent stage of photosynthesis occurs on the thylakoid membranes of the chloroplasts. It has two main functions:

- to break up water molecules in a **photochemical reaction**, providing hydrogen ions to reduce carbon dioxide and produce carbohydrates in the light-independent stage
- to produce ATP, which supplies the energy to build carbohydrates (see **Section 5A.1**).

Light is a form of electromagnetic radiation and the smallest unit of light is a photon. When a photon of light hits a chlorophyll molecule, the energy is transferred to the electrons of the chlorophyll molecule. The electrons are excited and are raised to higher energy levels. If an electron is raised to a sufficiently high energy level, it leaves the chlorophyll molecule completely. The excited electron is collected by a carrier molecule called an electron acceptor and this results in the synthesis of ATP by one of two processes: **cyclic photophosphorylation** and **non-cyclic photophosphorylation**. Both these processes occur at the same time and, in both cases, ATP is formed as the excited electron is transferred along an electron transport chain. In non-cyclic photophosphorylation, reduced NADP is also produced.

LEARNING TIP

Here, phosphorylation means adding a phosphate group to ADP. There are a number of different ways that phosphorylation can be achieved. Photophosphorylation occurs during photosynthesis; it can be cyclic or non-cyclic. Oxidative phosphorylation occurs in respiration; it involves electron transport chains and oxygen as the final electron acceptor. Oxidative phosphorylation can also occur by direct substrate-level reactions which occur mostly in anaerobic respiration.

DID YOU KNOW?

A two-stage process

How do we know that photosynthesis is a two-stage process? There are several pieces of evidence for this current model.

1 Photochemical reactions obtain the energy they need from light, so temperature should not affect the rate of the reaction. However, when the rate of photosynthesis is investigated experimentally, temperature can be shown to have a clear effect (see **fig B**). Initially, photochemical (light-dependent) reactions limit the rate of the overall process and so temperature has no effect. However, once there is plenty of available light, the process seems to be limited by temperature-sensitive reactions. This shows there are two distinct phases to photosynthesis, one dependent on light and the other controlled by temperature-sensitive enzymes.

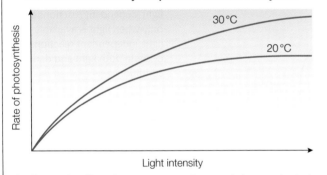

▲ **fig B** The effect of temperature on the rate of photosynthesis shows that two different processes are involved.

2 A plant that is exposed to alternating periods of dark and light forms more carbohydrate than a plant in continuous light. How can we explain this? The light-dependent reactions produce a chemical that feeds into the light-independent stage. In continuous light, this chemical accumulates, because the light-independent stage is not as fast as the light-dependent stage. The increasing concentration of this chemical inhibits the enzymes which control the light-independent reactions that make carbohydrates. A period of darkness allows all of the products of the light stage to be converted into carbohydrate without their concentration becoming too high. This system is very efficient in a natural environment with periods of light and dark (day and night). Scientists have been able to isolate regions of the chloroplast with newly developed techniques. The reactions occurring on the grana have been shown to depend on the presence of light, whereas those in the stroma do not.

CYCLIC PHOTOPHOSPHORYLATION

Cyclic photophosphorylation involves only photosystem I (PSI) and drives the production of ATP. When light hits a chlorophyll molecule in PSI, a light-excited electron leaves the molecule. It is collected by an electron acceptor and transferred directly along an electron transport chain to produce ATP. When an electron returns to the chlorophyll molecule in PSI, it can then be excited in the same way again (see **fig C**).

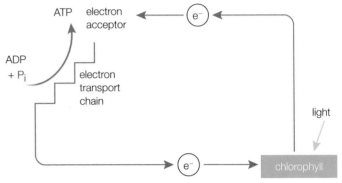

▲ **fig C** Cyclic photophosphorylation

NON-CYCLIC PHOTOPHOSPHORYLATION

During non-cyclic photophosphorylation, water molecules are broken down, providing hydrogen ions to reduce NADP. ATP is also produced. The process involves both photosystem I and photosystem II (see **fig D**).

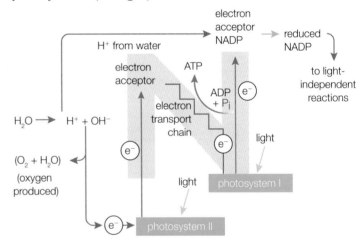

▲ **fig D** Non-cyclic photophosphorylation: one electron leaves a chlorophyll molecule in PSI and moves into the light-independent stage of the process. A different electron is returned to PSI from PSII by an electron transport chain thus driving the production of more ATP.

In the light, photons constantly hit chlorophyll molecules in both PSI and PSII. This excites the electrons to a higher level. They are, therefore, lost from the chlorophyll molecule and collected by electron acceptors. An excited electron from PSII is collected by an electron acceptor and transferred along an electron transport chain to PSI, driving the synthesis of one molecule of ATP. PSI receives an electron to replace one that was lost to the light-independent reactions. Now the chlorophyll molecule in PSII is missing one electron and so it is unstable. The original electron cannot be returned to the chlorophyll because it has continued on to PSI. So another electron is needed to restore the chlorophyll to its original state. This electron comes from the breaking down of water molecules, a process that is known as **photolysis** because it depends on light. Water molecules dissociate (break down) spontaneously into hydrogen (H^+) ions and hydroxide (OH^-) ions. As a result, there are many hydrogen and hydroxide ions in every part of the cell, including in the chloroplasts. These ions are used to replace the lost electrons from chlorophyll. Once the chlorophyll molecule in PSII has received an electron it is restored to its original state, ready to be excited again when hit by another photon of light.

At the same time, electrons in PSI are also being excited by light and collected by an electron acceptor. Electrons are transferred along an electron transport chain and collected by the electron acceptor, nicotinamide adenine dinucleotide phosphate (NADP). The NADP also collects a hydrogen ion from the dissociated water to form reduced NADP.

The reduced NADP and ATP which are produced during non-cyclic photophosphorylation provide the source of reducing power and energy respectively in the light-independent reactions of photosynthesis to make glucose.

Photosynthesis is a reaction that occurs millions of times in every chloroplast. This means that many hydrogen ions are removed by NADP, and many hydroxide ions remain. The hydroxide ions react together to form oxygen and water. Electrons are freed as a result of the reaction and are absorbed by chlorophyll. Four chlorophyll molecules regain electrons in the production of one molecule of oxygen:

$$4OH^- - 4e^- \text{ (lost to chlorophyll)} \rightarrow O_2 + 2H_2O$$

EXAM HINT

Make sure you are clear about the differences between PSI and PSII.

Remember that both photosystems need light. Do not confuse PSI and PSII with the light-dependent and light-independent stages of photosynthesis.

THE LIGHT-INDEPENDENT STAGE OF PHOTOSYNTHESIS

The light-independent stage of photosynthesis uses the reducing power (reduced NADP) and ATP produced by the light-dependent stage to build carbohydrates. This stage consists of a series of reactions known as the **Calvin cycle** and occurs in the stroma of the chloroplast. A series of small steps results in the reduction of carbon dioxide from the air leading to the synthesis of carbohydrates (see **fig E**). Each stage of the cycle is controlled by enzymes.

THE CALVIN CYCLE

In the first step of the Calvin cycle, carbon dioxide from the air combines with the 5-carbon compound **ribulose bisphosphate (RuBP)** in the chloroplasts. The carbon dioxide is said to be fixed, so this process is known as carbon fixation. This vital step needs

the enzyme **ribulose bisphosphate carboxylase/oxygenase (RUBISCO)**. Research has shown that RUBISCO is the rate-limiting enzyme in the process of photosynthesis.

Theoretically, the result of the reaction between RuBP and carbon dioxide is a 6-carbon compound. Scientists are convinced that this theoretical compound exists but it is highly unstable and it has never been isolated. It immediately separates into two molecules of **glycerate 3-phosphate (GP)**, a 3-carbon compound. GP is then reduced (hydrogen is added) to form **glyceraldehyde 3-phosphate (GALP)**, a 3-carbon sugar. The hydrogen for this reduction comes from reduced NADP, and the energy required comes from ATP; both of these are produced in the light-dependent stage.

Much of the 3-carbon GALP follows a series of steps to replace the RuBP needed in the first step of the cycle. However, some 3-carbon GALP is synthesised into the 6-carbon sugar glucose or transferred directly into the glycolysis pathway. In this pathway, it may be used for the synthesis of other molecules needed by the plant (see next page). The reactions of the Calvin cycle occur in both the light and the dark. These reactions only stop in the dark when no products of the light reaction remain, that is to say, no reduced NADP or ATP is available in the chloroplasts. The reactions of the Calvin cycle are summarised in **fig E**, and the whole process of photosynthesis is put together in **fig F**.

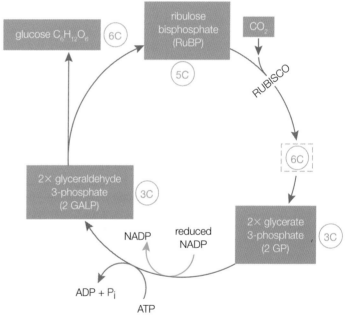

▲ **fig E** The Calvin cycle: the products of the light-dependent stage of photosynthesis are used in a continuous cycle to fix carbon dioxide. The end result is new carbohydrates.

EXAM HINT

The Calvin cycle is very complex. This is a simplified version with all the detail you need. Each step is controlled by enzymes. Therefore, the rate of the whole cycle is affected by the factors that affect enzyme activity. Remember that you need to be ready to make use of the knowledge you learned at IAS to help answer questions.

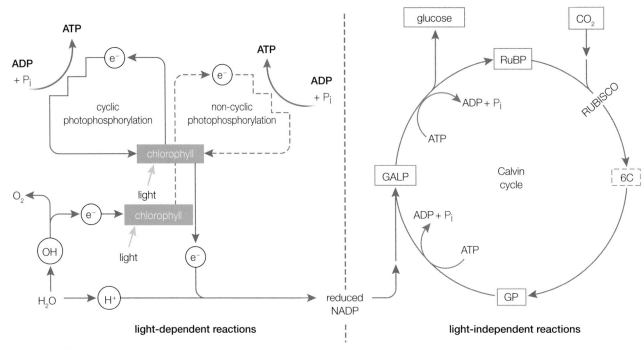

▲ **fig F**   The full process of photosynthesis occurs continuously in plants when they are exposed to light.

DID YOU KNOW?

RUBISCO and photorespiration

RUBISCO makes up about 30% of the total protein of a leaf, so it is probably the most common protein on Earth. It is also possibly the most important enzyme because of its role in fixing carbon dioxide during photosynthesis. But RUBISCO is very inefficient. The active site cannot distinguish between the carbon–oxygen double bonds of carbon dioxide molecules and the oxygen–oxygen double bonds of oxygen molecules. As a result, both molecules compete for the active site of the enzyme (see **Section 2B.2 (Book 1: IAS)**). This is called competitive inhibition.

RUBISCO functions as a carboxylase in high levels of carbon dioxide/relatively low oxygen. RUBISCO binds to the carbon dioxide and combines it with RuBP, producing two molecules of 3C GP which enter into the Calvin cycle.

RUBISCO functions as an oxygenase in low levels of carbon dioxide/relatively high oxygen. RUBISCO binds to the oxygen and combines it with RuBP to form one molecule of GP and one molecule of glycolate-2-phosphate. The glycolate-2-phosphate is converted into GP in a reaction that uses products of the Calvin cycle and ATP and releases carbon dioxide. This process is known as **photorespiration** because it uses oxygen and releases carbon dioxide.

Photorespiration wastes both carbon and energy. Fortunately for human beings, RUBISCO is attracted to carbon dioxide 80 times more than it is attracted to oxygen. However, in plants about 25% of the products of the Calvin cycle are lost in photorespiration. This means that in many plants, the efficiency of photosynthesis is reduced by 25%.

Why is RUBISCO so inefficient? All the evidence suggests that when RUBISCO evolved the atmosphere was high in carbon dioxide with very little oxygen. Because of this, photorespiration never occurred and there was no selection pressure against it. Even today, photorespiration is not a problem for plants because of our high oxygen/low carbon dioxide atmosphere. There is no selection pressure for the enzyme to evolve to become more specific to carbon dioxide. However, if RUBISCO became more efficient, our crop plants could become 25% more productive, which would be extremely useful to people.

USING THE PRODUCTS OF PHOTOSYNTHESIS

GALP is the primary end-product of the process of photosynthesis. It is the most important molecule for the synthesis of everything else needed in the plant. Some of the GALP is used directly in cellular respiration. You will learn more about this process in **Chapter 7A**. Some of the GALP produced in the Calvin cycle is used to produce glucose in a process called **gluconeogenesis**.

This glucose may be converted into disaccharides such as sucrose for transport round the plant; into polysaccharides such as starch for energy storage; and into cellulose for structural support (see **Sections 1A.2** and **1A.3 (Book 1: IAS))**.

The GALP that enters cellular respiration is used to provide energy in the form of ATP for the functions of the cell. Compounds from these pathways are also used as the building blocks of amino acids. The molecules combine with nitrates from the soil. GALP can also continue round the Calvin cycle and, in that case, it can combine with phosphates from the soil to produce nucleic acids.

Some of the GALP that enters the cellular respiration pathways is converted into a chemical called acetyl coenzyme A. This compound is then used to synthesise the fatty acids needed for the production of phospholipids for membranes, and lipids needed for storage and other functions within the plant.

GP is also part of this process, but GALP is regarded as the main molecule leading to the synthesis of all the other molecules needed by the plant (see **fig G**).

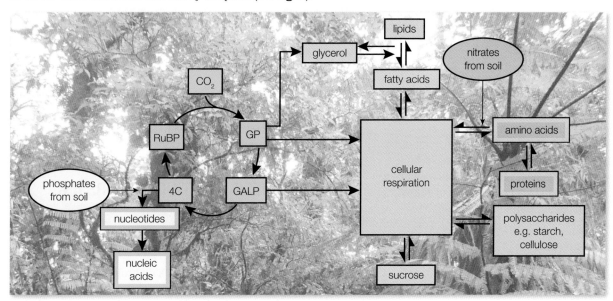

▲ **fig G**   The body of a plant is composed of many different chemicals. Most of these chemicals are formed from the products of photosynthesis.

LIMITING FACTORS IN PHOTOSYNTHESIS

When you understand the process of photosynthesis, you can see why certain factors affect the ability of a plant to photosynthesise. Photosynthesis is limited by the factor that is nearest to its minimum value.

LIGHT

Light intensity affects the amount of chlorophyll which is excited and, therefore, the amount of reduced NADP and ATP produced in the light-dependent stage of the process. If there is a low level of light, insufficient NADP and ATP will be produced to allow the reactions of the light-independent stage to progress at their maximum rate. Light is, therefore, the **limiting factor** for the process in this situation. Both the light intensity and the wavelength of the light falling on a plant will affect the rate of photosynthesis.

CARBON DIOXIDE

Carbon dioxide concentration is very important in photosynthesis. If there is not enough carbon dioxide available for fixing in the Calvin cycle, the reactions of photosynthesis cannot proceed at the maximum rate. Carbon dioxide concentration is, therefore, the limiting factor in this case. In natural environments, carbon dioxide concentration is the most common limiting factor of photosynthesis in plants. Changes in the concentration of carbon dioxide have a clear effect on the rate of photosynthesis (see **fig H**). When commercial growers of some fruits and vegetables grow their crops in greenhouses, they often supply extra carbon dioxide to the atmosphere to increase their production.

EXAM HINT

Always remember to refer to *light intensity* rather than *the amount of light*. When you describe the effect of light remember to state 'as light intensity increases, the rate of photosynthesis increases'. It is not enough to say 'light increases the rate of photosynthesis'.

EXAM HINT

Again, remember to refer to *the concentration of carbon dioxide*. If you are describing or explaining a relationship, remember to say 'as carbon dioxide concentration increases ... '.

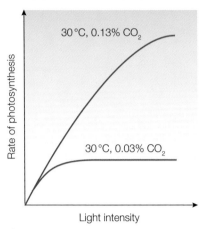

▲ **fig H** The effect of carbon dioxide concentration as a limiting factor in photosynthesis.

TEMPERATURE

The other main factor which limits the rate of photosynthesis is temperature. All of the Calvin cycle reactions and many of the light-dependent reactions of photosynthesis are controlled by enzymes and are, therefore, sensitive to temperature (see **Section 2B.2 (Book 1: IAS)**). A plant can only photosynthesise rapidly when the temperature is in the right range, even if light and carbon dioxide levels are abundant. The rate of photosynthesis in a wild plant is often determined by a combination of these factors, some or all of them having a limiting effect on the process (see **fig I**).

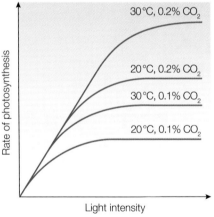

▲ **fig I** Different factors interact to limit the rate of photosynthesis.

EXAM HINT

As part of your study of this topic, you will conduct **Core Practical 10: Investigate the effects of light intensity, light wavelengths, temperature and availability of carbon dioxide on the rate of photosynthesis**. Make sure you understand this practical well because your understanding of the experimental method may be assessed in your examination.

PRACTICAL SKILLS **CP10**

You will investigate the effect of different factors on the rate of photosynthesis in plants. You will have to use an aquatic plant to do this. It is very hard to measure the rate of photosynthesis in land plants, but aquatic plants release bubbles of oxygen gas which you can see in the water. You can:

- count the rate at which the bubbles are released in a given time interval and calculate the rate of photosynthesis in bubbles per minute

- collect the gas produced over a given time interval and calculate the rate of photosynthesis from the volume of gas produced per minute.

With both of these methods, you can then change the conditions for your plant. For example, you can change the wavelength of the light shining on the plant or you can change the temperature of the water the plant is in. By measuring any change in the rate of photosynthesis, you can observe the effect of different factors on how your water plant makes food.

Safety Note: Lights must not be splashed with water nor electric cables, plugs or sockets handled with wet hands. After handling plant materials, hands should be washed with soap and water.

PHOTOSYNTHESIS, LIMITING FACTORS AND REAL PLANTS

Photosynthesis and its limiting factors are relatively easy to investigate in the laboratory, but what happens in the real world?

Plants do not usually grow in pots in a controlled environment. They are found in woods, gardens, ponds, mountains, swamps, savannah and desert. How plants grow and the ecosystems that develop are controlled by competition between plants for the factors that can limit photosynthesis and growth. For example, carbon dioxide concentration does not generally vary much in the air, but plants compete for light and warmth. They also compete for the nutrients they need to convert carbohydrate into proteins and fats. Plants are adapted to get as much light as possible so that photosynthesis is not limited. For example, they can grow tall, spread their leaves, climb or develop large leaves. Methods of seed dispersal have also evolved to reduce competition by ensuring that seedlings do not develop in the shade of their parents. The biochemistry investigated in the artificial situation of the laboratory is extremely important to help us understand the lives of plants in their natural habitats.

CHECKPOINT

1. Make a table to compare what happens in cyclic and non-cyclic photophosphorylation.

2. Calvin cycle reactions are also known as the light-independent reactions of photosynthesis. Explain why this name is appropriate as well as being, in some ways, inaccurate.

3. Explain why GALP is sometimes referred to as *the primary product of photosynthesis.*

4. ▶ Greenhouses and polytunnels are used in many countries to change the conditions so farmers can grow high-value food crops and flowers as economically and quickly as possible. In many of these artificial growing environments, the light levels, temperature and carbon dioxide levels are carefully monitored and controlled. Explain this in terms of limiting factors.

SKILLS ▶ CREATIVITY

SUBJECT VOCABULARY

light-dependent reactions the reactions that take place in the light on the thylakoid membranes of the chloroplasts; the reactions produce ATP and break down water molecules in a photochemical reaction, providing hydrogen ions to reduce carbon dioxide and produce carbohydrates

light-independent reactions the reactions that use the reduced NADP and ATP produced by the light-dependent stage of photosynthesis in a pathway known as the Calvin cycle; this occurs in the stroma of the chloroplast and results in the reduction of carbon dioxide from the air to cause the synthesis of carbohydrates

photochemical reaction a reaction initiated by light

cyclic photophosphorylation a process that drives the production of ATP; light-excited electrons from PSI are taken up by an electron acceptor and transferred directly along an electron transport chain to produce ATP, with the electron returning to PSI

non-cyclic photophosphorylation a process involving both PSI and PSII in which water molecules are broken into smaller units using light energy to provide reducing power to make carbohydrates and at the same time produce more ATP

photolysis the breaking down of a molecule into smaller units using light

Calvin cycle a series of enzyme-controlled reactions that take place in the stroma of chloroplasts and result in the reduction of carbon dioxide from the air to bring about the synthesis of carbohydrate

ribulose bisphosphate (RuBP) a 5-carbon compound that combines with carbon dioxide from the air in the Calvin cycle to fix the carbon dioxide and form a 6-carbon compound

ribulose bisphosphate carboxylase/oxygenase (RUBISCO) a rate-controlling enzyme that catalyses the reaction between carbon dioxide/oxygen and ribulose bisphosphate

glycerate 3-phosphate (GP) a 3-carbon compound thought to be the result of breakdown of a theoretical highly unstable 6-carbon compound formed as a result of the reaction between RuBP and carbon dioxide in the Calvin cycle

glyceraldehyde 3-phosphate (GALP) a 3-carbon sugar produced in the Calvin cycle using reduced NADP and ATP from the light-dependent stage; GALP is the key product of photosynthesis and is used to replace the RuBP needed in the first step of the cycle, in glycolysis and the Krebs cycle, and in the synthesis of amino acids, lipids, etc. for the plant cells

photorespiration the alternative reaction catalysed by RUBISCO in a low carbon dioxide environment which uses oxygen and releases carbon dioxide, making photosynthesis less efficient

gluconeogenesis the synthesis of glucose from non-carbohydrates

limiting factor the factor needed for a reaction to progress that is closest to its minimum value

C4 PHOTOSYNTHESIS

There are different types of photosynthesis, each with differing biochemistry. Some types of photosynthesis are more efficient than others. Scientists are looking for ways to use this to make crop plants more productive.

INFOGRAPHIC

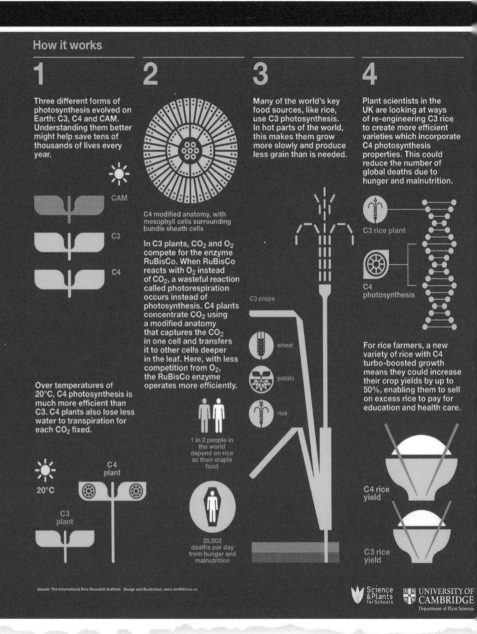

PlantPower

Can understanding photosynthesis help save thousands of lives a year?

How C4 photosynthesis can turbo-charge crop growth and help tackle starvation.
www.saps.org.uk/c4rice

Rice is among the three most important crops in the world. It is the staple food source for more than half the world's population. Like many crops rice uses an inefficient method of photosynthesis known as C3.

But some other plants have evolved a more efficient method, called C4 photosynthesis. UK plant scientists are looking at ways of re-engineering rice to incorporate C4 photosynthesis properties. For farmers in the developing world, this could mean much higher crop yields, helping them to feed their families and pay for health care and education.

How it works

1 Three different forms of photosynthesis evolved on Earth: C3, C4 and CAM. Understanding them better might help save tens of thousands of lives every year.

CAM
C3
C4

Over temperatures of 20°C, C4 photosynthesis is much more efficient than C3. C4 plants also lose less water to transpiration for each CO$_2$ fixed.

20°C
C4 plant
C3 plant

2 C4 modified anatomy, with mesophyll cells surrounding bundle sheath cells

In C3 plants, CO$_2$ and O$_2$ compete for the enzyme RuBisCo. When RuBisCo reacts with O$_2$ instead of CO$_2$, a wasteful reaction called photorespiration occurs instead of photosynthesis. C4 plants concentrate CO$_2$ using a modified anatomy that captures the CO$_2$ in one cell and transfers it to other cells deeper in the leaf. Here, with less competition from O$_2$, the RuBisCo enzyme operates more efficiently.

1 in 2 people in the world depend on rice as their staple food

25,000 deaths per day from hunger and malnutrition

3 Many of the world's key food sources, like rice, use C3 photosynthesis. In hot parts of the world, this makes them grow more slowly and produce less grain than is needed.

C3 crops
wheat
potato
rice

4 Plant scientists in the UK are looking at ways of re-engineering C3 rice to create more efficient varieties which incorporate C4 photosynthesis properties. This could reduce the number of global deaths due to hunger and malnutrition.

C3 rice plant
C4 photosynthesis

For rice farmers, a new variety of rice with C4 turbo-boosted growth means they could increase their crop yields by up to 50%, enabling them to sell on excess rice to pay for education and health care.

C4 rice yield
C3 rice yield

Source: The International Rice Research Institute Design and illustration: www.smithltd.co.uk

Science & Plants for Schools

UNIVERSITY OF CAMBRIDGE
Department of Plant Sciences

▲ **fig A** Based on part of a poster produced by Science and Plants for Schools (SAPS) to introduce different types of photosynthesis to students.

SCIENCE COMMUNICATION

Fig A is part of an infographic, which is a very popular way of providing information.

Visit www.saps.org.uk/attachments/article/1266/C4%20Rice%20-%20poster.pdf to see the whole poster.

1   What is an infographic?

2   In an infographic, there is not a lot of space to explain ideas. Every word is important. The colour and size of the font used helps indicate which information is most important. Look carefully at the infographic on C4 photosynthesis. It would normally be the size of a wall poster. Choose an example of:

   (a) a statement explaining what the infographic is about

   (b) a piece of important information on the infographic

   (c) a piece of information which is interesting but not key to understanding C4 photosynthesis on the infographic.

   In each case, describe how well the wording explains an idea, and describe the information that is given to the reader by the size and colour of the lettering.

3   Does this poster help you understand why C4 photosynthesis is an exciting alternative to the C3 photosynthesis that you have studied in **Chapter 5A**? Explain your answer.

BIOLOGY IN DETAIL

You have looked at C3 photosynthesis in some detail. This poster gives you some information about another form of photosynthesis that uses a rather different biochemical pathway.

4   Make a flow diagram to summarise the process of C3 photosynthesis.

5   Using only the information on the poster:

   (a) Summarise the process of C4 photosynthesis.

   (b) Explain why scientists are attempting to genetically modify rice plants so they use C4 photosynthesis instead of C3.

ACTIVITY

Look into the science behind the poster to find out more about C4 photosynthesis and the way it might be used to help feed the growing population of the world. The best place to start is with the resources produced by the team at SAPS, who work from the Botanic Gardens at the University of Cambridge in the UK, and study plants and ecosystems all over the world. Visit http://www.saps.org.uk/students/further-reading/1266 to start. You can look at other sources as well. Now choose one of the following activities.

   • Produce your own poster to inform other International A Level students who are not studying biology about C4 photosynthesis and its potential to improve the yield of crops, including rice.

   • Find out more about C4 and CAM photosynthesis and make a table to compare the three types of photosynthesis.

   • Make your own infographic to compare C3 and C4 photosynthesis.

The resources produced by Science and Plants for Schools (SAPS) are a great place to start when looking into plants and photosynthesis. Remember to always reference your sources.

5A EXAM PRACTICE

1 The following substances are all involved in photosynthesis.
   1 glyceraldehyde phosphate
   2 glycerate 3-phosphate
   3 carbon dioxide
   4 glucose

   In which order of substances will a carbon atom be transferred?   [1]
   A 2–1–4–3   **B** 3–1–2–4
   C 3–2–1–4   **D** 1–2–4–3

   (Total for Question 1 = 1 mark)

2 What are the products of the light-dependent stage of photosynthesis?   [1]
   A ATP and reduced NAD
   B ATP and oxidised NADP
   C ATP and oxidised NAD
   D ATP and reduced NADP

   (Total for Question 2 = 1 mark)

3 How many molecules of ribulose bisphosphate are broken down to produce two molecules of glucose?   [1]
   A 10   **B** 12   **C** 14   **D** 18

   (Total for Question 3 = 1 mark)

4 One of the reactions of photosynthesis can be summarised as shown below.

   water → hydrogen ions + oxygen gas + electrons

   (a) Name the reaction shown.   [1]
   (b) State **one** other factor, not shown above, that is required for this reaction to occur in a chloroplast.   [1]
   (c) Describe the role of the electrons in the light-dependent stage of photosynthesis.   [4]
   (d) Explain how the products of the light-dependent reaction are involved in the production of glyceraldehyde 3-phosphate (GALP).   [5]
   (e) GALP does not accumulate in a chloroplast during photosynthesis. State how GALP is used following its production.   [2]

   (Total for Question 4 = 13 marks)

5 The diagram below summarises the light-dependent reactions of photosynthesis.

   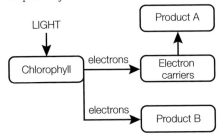

(a) State the precise location within a chloroplast where this sequence of reactions occurs.   [1]
(b) (i) What is the name of product A?   [1]
        A glucose   **B** ADP
        C reduced NADP   **D** ATP
    (ii) What is the name of product B?   [1]
        A glucose   **B** ADP
        C reduced NADP   **D** ATP
(c) Explain why the chlorophyll in photosystem II loses an electron.   [3]
(d) A chemical called atrazine prevents the flow of electrons to the electron carriers. Describe the likely effect of atrazine on the production of carbohydrate in a chloroplast. Explain your answer.   [4]

   (Total for Question 5 = 10 marks)

6 The diagram below shows what happens to electrons during part of the light-dependent stage of photosynthesis. Any excited electrons that are not taken up by electron carriers follow pathway A and release energy as light in a process called fluorescence. The excited electrons that are taken up by electron carriers follow pathway B.

   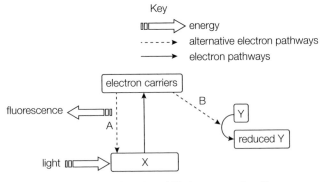

(a) Name the molecules X and Y shown on the diagram.   [2]
(b) Explain the importance of reduced Y in the process of photosynthesis.   [2]
(c) A light was shone on a leaf and left switched on.
    The graph below shows changes in the amount of light given off as fluorescence by the leaf.

   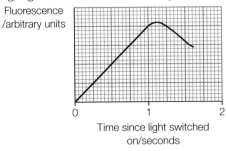

(i) Suggest an explanation for the increase in fluorescence. [2]

(ii) Suggest a reason for the fall in fluorescence. [2]

(d) Explain why an inhibitor of carbon dioxide fixation would lead to an increase in fluorescence. [4]

(Total for Question 6 = 12 marks)

7 The diagram below shows structures found in a chloroplast.

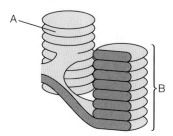

(a) What are the correct names of structures A and B? [1]

A granum and thylakoid

B thylakoid and granum

C thylakoid and stroma

D lamella and thylakoid

(b) State precisely where chlorophyll is found. [2]

(c) A student investigated the effect of different coloured lights on the rate of photosynthesis. She placed coloured filters between a lamp and a beaker containing a piece of pondweed (*Cabomba*) and counted the number of bubbles released from the pondweed in two minutes. Her results are shown in the table.

Colour of filter	Number of bubbles counted in two minutes
black	10
blue	82
green	21
yellow	32
red	97

(i) State the colour of light that enabled the most rapid rate of photosynthesis. [1]

(ii) Explain the result shown for the black filter. [2]

(iii) Explain the result shown with the green filter. [2]

(iv) The student expected there to be no oxygen released when using the black filter. Suggest **two** reasons why some bubbles were observed when the black filter was in place. [2]

(v) Suggest **two** ways in which the student could have improved this investigation. [2]

(Total for Question 7 = 12 marks)

8 The carbohydrates in green plants are formed during the light-independent stage of photosynthesis. They are synthesised from glycerate 3-phosphate (GP).

(a) State precisely where the synthesis of carbohydrates takes place during the light-independent stage of photosynthesis. [1]

(b) (i) Name the products of the light-dependent stage of photosynthesis used during the synthesis of carbohydrates. [2]

(ii) Name the source of carbon used to manufacture sugars. [1]

(c) What is the role of ribulose bisphosphate (RuBP) in the light-independent stage of photosynthesis? [1]

A to convert carbon dioxide to GP

B to fix carbon dioxide

C to accept electrons from the light-dependent stage

D to release energy for the light-independent stage

(d) An investigation of photosynthesis in cells taken from a green alga was carried out. Samples of the algal cells were taken at 1 minute intervals over a period of 6 minutes. The quantities of GP and RuBP in these cell samples were measured.

At the start of the investigation, the algal cells were kept in an atmosphere with 1% carbon dioxide. After 3 minutes, the concentration of carbon dioxide was decreased to 0.003%.

The graph below shows the results of this investigation.

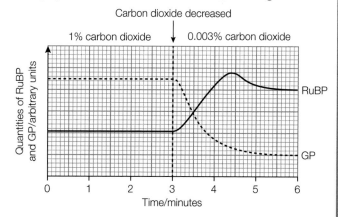

Describe the effects on the quantities of GP and RuBP with the decrease in the carbon dioxide concentration and deduce explanations for the effects you have described. [4]

(Total for Question 8 = 9 marks)

TOPIC 5 ENERGY FLOW, ECOSYSTEMS AND THE ENVIRONMENT

CHAPTER 5B ECOLOGY

Around the world, different species are becoming rare. For example, in the UAE there are only around 100 nesting hawksbill turtles left, and there have only been about 10 reported sightings of the beautiful and elusive sand cat in the last decade or so. Globally, scientists and citizens alike are working hard to identify organisms that are at risk, and to conserve them. But if the data is poor, the identification of the species is uncertain or the habitat is damaged or destroyed during the survey then, however well it is presented, the information is of no value to anyone.

In this chapter, you will learn about the nature of ecosystems. You will define the term and look at the variations in size. You will learn how rock is colonised and how ecosystems progress to maturity. You will look at how living and non-living factors affect the organisms which can survive in an ecosystem, and consider the importance of the niche to the numbers of organisms in any habitat.

You will discover a variety of techniques that can be used to measure the abundance and distribution of organisms in an ecosystem, and gain practical experience of as many of them as possible. You will learn to select the best ecological technique to use according to the ecosystem and organisms you are studying. Slow-moving animals and plants require different techniques to faster moving, flying or swimming organisms. Finally, you will learn several statistical techniques that can be used to analyse the data you collect.

MATHS SKILLS FOR THIS CHAPTER

- **Construct and interpret frequency tables and diagrams, bar charts and histograms** (*e.g. interpret diagrams to show secondary succession*)
- **Translate information between graphical, numerical and algebraic forms** (*e.g. interpret data on increase in biodiversity during colonisation*)
- **Determine the intercept of a graph** (*e.g. when looking at data on two different populations*)
- **Understand the principles of sampling as applied to scientific data** (*e.g. when analysing data from field investigations*)
- **Select and use a statistical test** (*e.g. the Student's t-test, the Spearman's rank correlation and the chi squared test*)
- **Substitute numerical values into algebraic equations using appropriate units for physical quantities** (*e.g. the Student's t-test, the Spearman's rank correlation and the chi squared test*)

What prior knowledge do I need?

Chapter 4C (Book 1: IAS)

- That organisms occupy niches according to physiological, behavioural and anatomical adaptations
- That biodiversity is important and can be assessed at species and genetic levels
- That the maintenance of biodiversity is important for both ethical and economic reasons
- That conservation may be in situ or ex situ

Chapter 5A

- The process of photosynthesis
- Factors that affect or limit the process of photosynthesis

What will I study in this chapter?

- How ecosystems develop over time, including using the terms *colonisation* and *succession*, and types of climax communities
- The effects of biotic and abiotic factors on population size
- The nature of ecosystems
- How to select appropriate ecological techniques according to the ecosystem and organisms to be studied
- How to use statistical tests to analyse data, including Student's *t*-test, Spearman's rank correlation and chi squared test

What will I study later?

Chapter 5C

- How energy is transferred between trophic levels using the terms net primary productivity and gross primary productivity
- How to calculate the efficiency of energy transfer between different trophic levels and account for the loss of energy at each level
- The role of microorganisms in the recycling of nutrients within an ecosystem
- Human effects on ecosystems

LEARNING OBJECTIVES

■ Understand what is meant by the terms *population*, *community*, *habitat* and *ecosystem*.
■ Understand that the numbers and distribution of organisms in a habitat are controlled by biotic and abiotic factors.
■ Understand how the concept of niche accounts for the distribution and abundance of organisms in a habitat.

You studied adaptation, biodiversity and endemism in **Chapter 4C (Book 1: IAS)**. As a result, you know that real ecology is often very different from the emotive picture of the environment shown in the media. Ecology is the study of the interactions between organisms and their environment. These interrelationships determine the distribution and **abundance** of organisms within a particular environment. The word ecology comes from the Greek *oikos*, meaning 'house'. Put simply, ecology is the study of living things in their home environment.

WHAT IS AN ECOSYSTEM?

An **ecosystem** is a life-supporting environment. It includes all of the living organisms, which interact together, the nutrients that cycle throughout the system, and the physical and chemical environment in which the organisms are living. An ecosystem consists of a network of habitats and the communities of organisms associated with them. Wherever we live, our ecosystem is one of thousands on the surface of the earth (see **fig A**).

▲ **fig A** There are many ecosystems on Earth, from the Arctic wastes to tropical forests, arid deserts, oceans and caves.

USEFUL TERMS

Ecology uses a number of very specific terms. Some of these are reminders from your **Book 1: IAS** studies. Others will be new in this part of your course.

- A **habitat** is the place where an organism lives, such as a stream, a tropical rainforest or a sand dune. You can think of the habitat of an organism as its address. Many organisms live only in a small part of a habitat; a single fig on a tree may be home to a fig wasp for example. Such habitats are referred to as **microhabitats** (see **fig B** for another example).

- A **population** is a group of organisms of the same species, living and breeding together in a habitat. The tilapia fish in a pond or the dust mites in your mattress are examples.

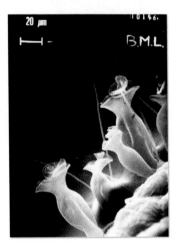

▲ **fig B** These tiny protista live in a microhabitat on the palps of a freshwater shrimp.

- A **community** is all the populations of all the different species of organisms living in a habitat at any one time. For example, in a habitat such as a rock pool, the community consists of populations of different seaweeds, sea anemones, small fish such as gobies, shrimps and crabs and other crustacea, molluscs such as mussels and barnacles, and many other species. More details about a particular community may be given in the name, such as the soil community, or the animal community in the soil.

- The **niche** of an organism can be described as the role of the organism in the community, or its way of life. If the habitat is the address of the organism, the niche describes its profession. Several organisms can share the same habitat, occupying different niches. So for example, in a Central American rainforest the jaguar fills the top **predator**, carnivore niche; different species of monkey occupy the tree-dwelling omnivore niche; and agouti fill the large, ground-living herbivore niche.

- **Abiotic factors** are the non-living elements of the habitat of an organism. They include those related to the climate, such as the amount of sunlight, temperature and rainfall, and those related to the soil (edaphic factors), including the drainage and the pH. In aquatic habitats, the oxygen availability in the water is very important. Abiotic factors have a big effect on the success of an organism in a particular habitat.

- **Biotic factors** are the living elements of a habitat that affect the ability of a group of organisms to survive there. For example, the presence of suitable **prey** species will affect the numbers of predators in a habitat.

BIOMES: THE MAJOR ECOSYSTEMS

The **biosphere** is the largest ecosystem on Earth. However, it is so large that it is very difficult to study as a whole. So it is divided into smaller parts, distinguished by their similar climates and plant communities. These major ecosystems or **biomes** are shown in **fig C**. Biomes are generally subdivided into smaller ecosystems to make them easier to study.

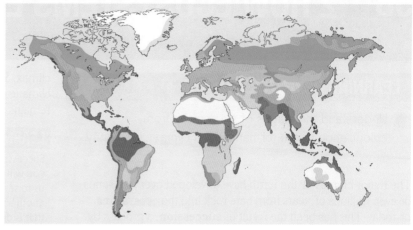

▲ **fig C**   Map of the world showing the major land biomes (key in **table A**).

COLOUR	NAME OF BIOME	DESCRIPTION OF BIOME	LEVEL OF BIODIVERSITY
	tropical rainforest	high humidity (rain all year), warm and plenty of sunlight	very high
	tropical seasonal forest	drier than tropical rainforest, warm, sunny	high
	savannah	dry tropical grassland	medium
	tropical woodland	wetter than savannah, grassland with bushes and trees	more than savannah
	desert	very little rainfall, often extremes of temperature between day and night	very low
	temperate grassland	warm dry temperate areas, e.g. prairies, steppes and pampas	medium
	temperate shrublands	hot dry summers and cool wet winters	medium
	temperate forests	warm moist regions, including deciduous and conifers	less than tropical rainforest
	taiga	evergreen forests in cold subarctic and subalpine regions	low
	tundra	very cold, arctic and high mountain regions	very low
	high mountain	very cold, high altitude	very low
	polar ice	very cold, little available water	very low

table A Major land biomes

CHECKPOINT

1. How does the habitat of an organism differ from its niche? Give examples to illustrate your answer.

2. Choose **three** of the major biomes of the Earth.
    (a) Find out about the climate of the biome, including the range of temperatures experienced and the level of rainfall.
    (b) Find out about **two** plants and **two** animals found in each biome you have chosen and their adaptations to the conditions.

    SKILLS ADAPTIVE LEARNING

    ▶ (c) Link the water availability and temperatures in the biomes you have chosen to the level of biodiversity found using **fig C**.

SUBJECT VOCABULARY

abundance the relative representation of a species in a particular ecosystem

ecosystem an environment including all the living organisms interacting within it, the cycling of nutrients and the physical and chemical environment in which the organisms are living

habitat the place where an organism lives

microhabitat a small area of a habitat

population a group of organisms of the same species, living and breeding together in a habitat

community all the populations of all the different species of organisms living in a habitat at any one time

niche the role of an organism within the habitat in which it lives

predator an organism which hunts and eats other organisms (prey)

abiotic factors the non-living elements of the habitat of an organism

biotic factors the living elements of a habitat that affect the ability of a group of organisms to survive there

prey an organism which is hunted and eaten by other organisms (predators)

biosphere all of the areas of the surface of the Earth where living organisms survive

biomes the major ecosystems of the world

LEARNING OBJECTIVES

■ Understand the stages of succession from colonisation to the formation of a climax community.

The major biomes of the Earth have developed over thousands or even millions of years from bare rock into the ecosystems of today. This has been the result of **succession**, a process by which communities of animals and plants colonise an area and then, over periods of time, are replaced by other, usually more varied, communities.

PRIMARY SUCCESSION

Primary succession starts with an empty inorganic surface, such as bare rock or a sand dune. You can observe this type of succession after a volcanic eruption or landslide or after the emergence of a new volcanic island. The first stage of the succession is **colonisation**, and the first organisms to appear are **opportunists** or **pioneer species** such as algae, mosses and fungi. These organisms can penetrate the rock surface in a number of ways, including dissolving the rock with acids they secrete, and penetrating tiny cracks with root hairs and hyphae. This helps to break the rock into small grains (particles), and trap organic material that will break down to form humus. The inorganic rock grains and the organic humus are the start of the formation of soil.

Once there is soil, other species such as grasses and ferns can establish root systems. The action of their roots, and the humus they form when they die and decay, add to the soil. As the soil layer develops, more water and nutrients are kept in the soil and become available for plant roots; this means less resistant species can survive. Gradually, larger plants can be supported and the diversity of species increases. This increase in plant biodiversity results in an increase in the diversity of animals that can be supported. Eventually, a **climax community** is reached, in which the biodiversity and range of species are generally constant. A climax community is self-sustaining and is usually the most productive group of organisms that the environment can support. Primary succession must have been of key importance in the formation of the biosphere, but today it is found in only a few places, such as the island of Surtsey off the coast of Iceland (see **Worked example 1, Section 5B.7**).

In 1916, F.E. Clements proposed the idea that climate is the major factor in determining the composition of the climax community in a particular place. He said that for any given climate, there was only one possible climax community, and that this should be known as the **climatic climax community**. Scientists have modified this view over time as they learn more about ecosystems and recognise that many factors interact to determine any given climax. You will learn more about this later in the chapter. So a climatic climax community is now defined as a community that remains generally constant over time.

LEARNING TIP

Succession happens over time; it can take hundreds of years. Usually, you will not see previous stages in succession. Some ecosystems allow us to see a series of stages occurring as we move from one place to another; for example, on a sand dune or at the edges of an old pond that is drying up.

In the modern landscape, there are many examples of another type of climax community: a **plagioclimax**. These are constant and self-sustaining, but they are not truly natural. They are a final community that is partly the result of human intervention. Humans have changed the landscape, for example by clearing trees, using water for irrigation and introducing grazing animals (those that eat grass). These changes affect the ultimate climax community. Examples include miombo (*Brachystegia*) woodlands and lowland heaths. Much of the countryside in many countries around the world consists of plagioclimax communities rather than climatic climax communities, because people have affected the ecosystems. If the limiting factors are removed, for example if people move away from the area, a climatic climax community will eventually develop.

EXAM HINT

A plagioclimax is a sub-climax community where succession has been held back or deflected by human activity.

SECONDARY SUCCESSION

Secondary succession is the development of an ecosystem from existing soil that is clear of vegetation. It occurs as rivers change their courses, after fires and floods, and after disturbances caused by humans. The sequence of events is very similar to that seen in primary succession, but because the soil is already formed and contains seeds, roots and soil organisms, the numbers of plants and animals present right from the beginning of the succession are much higher. Simply digging a patch of earth and leaving it is sufficient to observe the beginnings of a secondary succession. This process is illustrated in **fig A**.

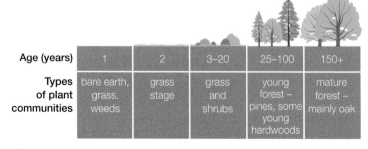

Age (years)	1	2	3–20	25–100	150+
Types of plant communities	bare earth, grass, weeds	grass stage	grass and shrubs	young forest – pines, some young hardwoods	mature forest – mainly oak

▲ **fig A** Oak woodlands are found in many temperate areas of the world. This diagram shows the stages of a secondary succession from bare earth to oak woodland. The timescale is very approximate.

The time it takes to go from an area of bare earth to a climax community varies enormously. It depends on many different factors, including temperature, rainfall and the fertility of the soil. However, a succession of different types of plants and animals is always seen. The climax community that is formed will depend on the climatic factors, and also on the plants, animals and microorganisms that are able to colonise the area. This can mean that a secondary climax community differs from the original primary climax community. You can see this difference when virgin rainforest is cut down and the area is left to regenerate naturally.

Observing succession is not always easy, because it is a process that occurs over a long time. In hot, arid desert countries, little or nothing can grow on the sand dunes. However, in some countries, such as the US and the UK, sand dunes can show a complete record of the stages of the succession. The oldest dunes (those furthest from the sea) are in the late stages of the succession. Nearest to the sea are the youngest dunes, in the very earliest stages of colonisation (see **fig B**).

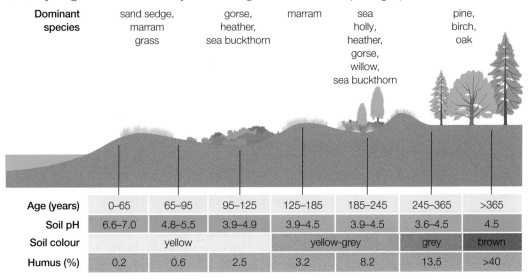

Dominant species	sand sedge, marram grass	gorse, heather, sea buckthorn	marram	sea holly, heather, gorse, willow, sea buckthorn		pine, birch, oak
Age (years)	0–65	65–95	95–125	125–185	185–245	245–365	>365
Soil pH	6.6–7.0	4.8–5.5	3.9–4.9	3.9–4.5	3.9–4.5	3.6–4.5	4.5
Soil colour	yellow			yellow-grey		grey	brown
Humus (%)	0.2	0.6	2.5	3.2	8.2	13.5	>40

▲ **fig B** You can see very clearly on this transect through some sand dunes on the east coast of the UK, the gradual change from sand to mature soil, the changing shape of the landscape and the differences in the plants (and therefore the animals) making up the communities.

CHECKPOINT

1. Define the term *ecological succession*.

2. Define *climax community*, and comment on whether such communities always look the same.

3. Describe the difference between primary and secondary succession.

SUBJECT VOCABULARY

succession the process by which the communities of organisms colonising an area change over time
colonisation the process by which new species spread to new areas
opportunists/pioneer species species which are the first to colonise new or disturbed ecosystems
climax community a self-sustaining community with relatively constant biodiversity and species range. It is the most productive group of organisms that a given environment can support long term
climatic climax community the only climax community possible in a given climate
plagioclimax a climax community that is at least in part the result of human intervention

3 THE EFFECT OF ABIOTIC FACTORS ON POPULATIONS

LEARNING OBJECTIVES

■ Understand that the numbers and distribution of organisms in a habitat are controlled by abiotic and biotic factors.

It is important to understand how different factors affect living things and determine the distribution and population size of organisms in a particular habitat. The community of organisms in a habitat is controlled by both abiotic (non-living) and biotic (living) factors, whether in the early stages of colonisation or in a climax community. The interaction of these factors results in different ecological niches, which change as the factors change. Understanding the abiotic and biotic factors that affect living organisms can help us predict the effect of changes on an ecosystem.

Abiotic factors can vary a lot within a habitat to produce **microclimates**. These provide different niches and so determine the distribution and abundance of different populations within the habitat. For example, logs may be placed in an open area to sit on. The area will be largely dry and well lit, except under a log where it will be damp and shady, allowing very different organisms to grow and become strong and healthy compared with those in the open areas.

LIGHT

The amount of light (energy input) in a habitat has a direct effect on the numbers of organisms found there (see **Section 5B.6**). Plants are dependent on light for photosynthesis, so any plant populations that will grow well in habitats with low light levels must be able to function without much light. For example, in temperate countries where trees shed their leaves in the relatively cold and dark winter season, some plants reproduce very early in the year (see **fig A**). This is to avoid the shade caused by larger plants in the warmer, lighter summer months.

▲ **fig A** Bluebells are a European woodland plant. The carpet of blue flowers appears early in the year before the leaves on the trees are fully open, so the bluebells still get the sunlight they need.

Plants that grow in the understory (lower levels) of a rainforest must be able to photosynthesise and reproduce successfully in low light levels. They have adaptations such as extra chlorophyll or a different ratio of photosynthetic pigments that are sensitive to lower light levels. This enables these plants to be successful in a niche where other plants would die. Another strategy for plants that grow in low light levels is to have very large leaves to absorb light. The data in **fig B** illustrates how plants growing in shady locations may have a bigger leaf area than the same species of plant growing in a lighter habitat. This enables the plants in shade to carry out enough photosynthesis to survive.

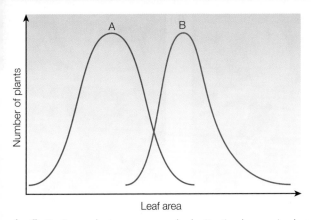

▲ **fig B**   Some plants overcome a shady situation by growing larger leaves. Curve A shows the range of leaf area of a plant species grown in bright light. Curve B shows the range of leaf areas found in a population of the same species of plant growing in the shade. On average they have a much bigger leaf area and they also show less variation in size.

Animals are affected by light levels indirectly, as a result of the distribution of food plants. Seasonal light changes also affect reproductive patterns in many animals. All animals are affected by the 24-hour cycle of light and dark. This cycle influences the natural rhythms that control much of animal physiology and behaviour.

TEMPERATURE

Every organism has a range of temperatures within which it can grow and successfully reproduce. Reproduction does not occur above or below that range, even if the individual organism survives. It is the extremes of temperature that determine where an organism can live, not the average.

The temperature of the environment particularly affects the rate of enzyme-controlled reactions in plants and ectothermic animals such as reptiles. In some areas of the world, the daytime temperatures can exceed the temperature at which endotherms can normally control their body temperatures. Many animals have evolved behaviours and physiological adaptations that enable them to deal with this, for example the Namaqua chameleon (see **fig C**). Organisms without these adaptations do not survive.

(a)

(b)

▲ **fig C**   The Namaqua chameleon lives in the Namibian desert, where temperatures can range from around 45 °C in the day to below freezing at night. (a) Early in the morning, the chameleon is dark in colour. It absorbs radiation from the Sun and warms up so it can move fast to catch prey. (b) In the heat of the day it turns white, to reflect light and keep as cool as possible.

WIND AND WATER CURRENTS

Wind has a direct effect on organisms in a habitat. Wind increases cooling and water loss from the body and so adds to the environmental stress on an organism. Fewer species can survive in areas with strong prevailing (dominant) winds, while occasional gales and hurricanes can cause extreme damage to populations. Whole woodlands may be destroyed and the communities of plant and animal life within them lost.

In water where there are currents, organisms have to flow with the current, be strong swimmers or be able to attach to a surface and resist the force of the water. Currents are most damaging to populations when the strength increases suddenly, such as when flooding occurs.

WATER AVAILABILITY

In a terrestrial environment, the availability of water is affected by several factors such as the amount of precipitation (rain, sleet, snow or hail), the rate of evaporation, and the rate of loss by drainage through the soil. Water is vital for living organisms, so where water supply is limited, it will cause severe problems.

If the water stress becomes too severe, organisms will die unless (like camels, kangaroo rats and cacti) they have special adaptations to enable them to survive and reproduce in very dry conditions. An increase in the availability of water can lead to a huge change in a habitat and to a huge increase in population size of some organisms. For example, in deserts after a little rain has fallen, the seeds of many desert plants germinate, grow and flower in a very short time period in the phenomenon known as the flowering of the desert (see **fig D**). This provides optimum food conditions for insects and other animals that normally just manage to survive in the very difficult conditions, so there is a population explosion.

▲ **fig D**   The arrival of rain in the desert will cause the germination of many seeds. The seeds may have survived in the ground for many years since the last time it rained.

LEARNING TIP

Don't forget that a desert is a place with low rainfall. Deserts are often hot but there are also cold deserts.

OXYGEN AVAILABILITY

Oxygen can be in short supply in both water and soil. When water is cold or fast flowing, there will be enough oxygen dissolved in it to support life. If the temperature of the water rises, or it becomes still and no longer flows, then the oxygen content will drop. This affects the survival of populations within it.

Soil is usually a well-aerated habitat. The spaces between the soil particles contain air, so there is plenty of oxygen for the respiration of plant roots. However, in waterlogged (extremely wet) soil, the air spaces are filled with water and the plant roots cannot obtain oxygen. The plants may die as a result. Some plants, like mangroves, have special adaptations such as aerial roots that allow them to grow successfully in waterlogged conditions.

EDAPHIC FACTORS: SOIL STRUCTURE AND MINERAL CONTENT

Edaphic factors relate to the structure of the soil and can affect the various populations associated with it. Sand has a loose, shifting structure that allows very little to grow on it. Plant populations that are linked by massive root and rhizome networks, such as dune grass, can and do survive on sand. They can reproduce successfully and also bind the sand together, which makes it more suited for colonisation by other species. Dune grass and similar species are also well adapted to survive the physiological drought (very dry) conditions that occur on the seashore. The leaves of some plants that grow in deserts or on sand dunes by the coast curl round on themselves with the stomata on the inside, creating a microenvironment that reduces water loss.

Soils that contain a high proportion of sand are light, easily worked and easily warmed. However, they are also very easily drained. Water passes through them rapidly, carrying with it minerals that may be needed by plants. This **leaching** of minerals reduces the population density of plants that can grow in the soil. In comparison, it is difficult for water to drain through soils that are made up predominantly of very small clay particles. They are heavy, take longer to warm up, are hard to work and are easily waterlogged. Mineral leaching is not a problem in soil of this type, but the populations which it will support are still limited. The ideal soil is loam, and it has particles in a wide range of sizes. It is heavier and leaching is less likely than in sandy soils, yet it is easier to warm and work than clay. Different types of plant have evolved to grow well in different, specific, soil types; they will not grow as well in other soils. **Fig E** shows you two plants specialised for very different types of soil.

▲ **fig E**　(a) This beautiful orchid is adapted to grow on the thin soil of the mountains of Crete. (b) The torch ginger needs the wet warmth of a rainforest.

CHECKPOINT

1. (a) What are abiotic factors? Give some examples.

   (b) Why do abiotic factors have such a major impact on the distribution of all organisms in a habitat?

2. Abiotic factors interact to make up the conditions of a particular habitat. Describe an example of the way in which the impact of one abiotic factor may be influenced by another.

3. Choose one abiotic factor and investigate organisms that are adapted to survive in the extremes of these conditions.

SUBJECT VOCABULARY

microclimate a small area with a distinct climate that is different from the surrounding areas
edaphic factors factors that relate to the structure of the soil
leaching the loss of minerals from soil as water passes through rapidly

4 THE EFFECT OF BIOTIC FACTORS ON POPULATIONS

LEARNING OBJECTIVES

■ Understand that the numbers and distribution of organisms in a habitat are controlled by abiotic and biotic factors.

Biotic factors are all of the living elements in a habitat, including plants, algae, fungi, herbivores, predators, parasites and disease-causing organisms.

PREDATION

It is easy to see how numbers of one species can affect the abundance of another species. Horses grazing a field reduce the reproduction of the grass by eating the potentially flower-forming parts. In the same way, a predator such as a fox will reduce the numbers of a prey species, reducing the population numbers of the local insects, rodents, reptiles and birds.

A mathematical model that describes the relationships between predator and prey populations predicts that the populations will oscillate (change between two extremes) in a repeating cycle. The reasoning behind this model is simple. As a prey population increases there is more food for the predators and so, after an interval, the predator population grows too. The predators will increase to the point at which they are eating more prey than are replaced by reproduction, so the numbers of prey will fall. This will reduce the food supply of the predators, so they will not produce as many offspring, and their numbers will fall, allowing the abundance of prey to increase again and so on (see **fig A**).

> **LEARNING TIP**
>
> The interaction of biotic factors is very complex. It is often impossible to consider one factor in isolation and the effects of changing one biotic factor can have huge implications on the rest of the community.

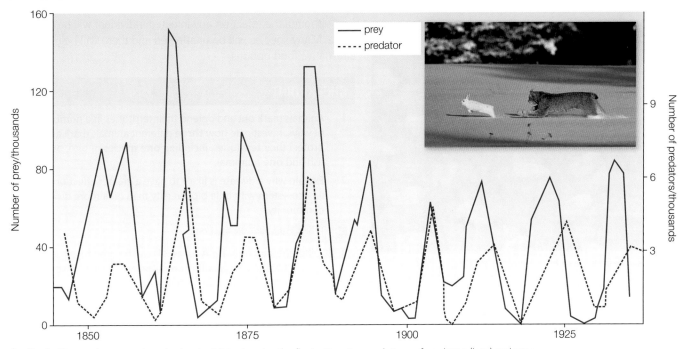

▲ **fig A** The graph represents a classic set of data showing the fluctuations in populations of predators (lynx) and prey (snowshoe hares) in Canada.

The situation in a natural habitat is always more complex than a model. Other research shows that the prey population follows a similar pattern, even in areas where there are no predators. The prey are responding to cycles in their food availability that appear to be related to climatic variations and changes in insect pest populations. This is why it is important to study all the factors in an ecosystem. You will be looking at this in more detail in this section.

FINDING A MATE

Reproduction is a powerful driving force and the probability of finding a mate, or achieving pollination for plants, affects which organisms are found in any habitat. For example, if a single seed is transported to a new area it may germinate, grow and survive. However, that species of plant is unlikely to become a permanent resident unless other plants of the same species also grow in the area, or the plant can reproduce successfully asexually. In the same way, a single individual of any animal species in an area does not mean that the species lives in the habitat. For many larger species, there must be males and females, so mates can be found. Therefore, the abundance of any type of animal that reproduces sexually found in an area is greatly affected by the availability of mates.

TERRITORY

Many species of animals show clear territorial behaviour. A territory is an area held and defended by an animal or group of animals against other organisms, which may be of the same or different species. Territories have different functions for different animals, but they are almost always used in some way to make sure that a breeding pair has sufficient resources to raise young. The type and size of territory will help to determine which species live in a particular community (see **fig B**). In **Section 5B.5**, you will look at territories in more detail in relation to the role of competition in determining the distribution of animals living in a habitat.

▲ **fig B**   Gannets are birds which have very small breeding territories because they feed out at sea. Without one of these small territories, a bird cannot nest and reproduce.

PARASITISM AND DISEASE

Parasitism and disease are biotic factors that can have a devastating effect on individuals. Diseased animals will be weakened and often do not reproduce successfully. Sick predators cannot hunt well, and diseased prey animals are more likely to be caught. Some diseases are very infectious and can be spread without direct contact, such as avian (bird) flu, which can be spread in the faeces of an infected bird.

Parasites affect their hosts, usually by feeding off the living body of their host and so weakening it. Occasionally they can remove whole populations. Examples include a parasitic fungal disease called Panama disease which, in the 1950s, destroyed banana crops around the world. Scientists developed new resistant strains of banana but now these are threatened by another parasitic fungal disease called black sigatoka. In Europe, the elm tree population has been almost eliminated by a fungal disease carried from tree to tree by beetles which dig into the bark, carrying the fungus with them.

Parasites and other infectious diseases spread more rapidly in situations of high population density, because individuals are close to each other. The problem is worst when almost all of the organisms in an area are of the same species. In these circumstances, an infectious disease or parasite can have a devastating impact, as it will affect most of the individuals directly or indirectly. In a community with greater biodiversity, the effect of a disease or parasite on the whole community will be much less, although the effect on any infected individual will be the same. Many species will be unaffected and there will be plenty of alternative food options.

CHECKPOINT

1. Animals mark out and defend their territories in a number of ways. Investigate how **three** different animals mark and protect their territories, including **one** species of bird, **one** fish and **one** mammal.

2. Explain why a disease is likely to have a greater effect on an ecosystem with little biodiversity than on a more diverse community.

   SKILLS   INNOVATION, ANALYSIS

3. ▶ 'Different regions of **fig A** (page 31) can be used to argue for or against the idea of repeating predator/prey cycles.' Discuss this statement with reference to the data.

LEARNING OBJECTIVES

■ Understand that the numbers and distribution of organisms in a habitat are controlled by abiotic and biotic factors.

■ Understand how the concept of niche accounts for the distribution and abundance of organisms in a habitat.

It is difficult to distinguish clearly between the biotic and abiotic factors that shape an ecosystem when we consider the ecology of an area and the impact of changes on the populations of living organisms. This is because they are often interlinked, and it is rare that one factor works alone. In a natural habitat, the factors determining the distribution and abundance of organisms are always complex. Here are several examples that demonstrate the importance of studying all the factors in an ecosystem, as well as the niche occupied by each different organism. You need to take them all into account when considering the numbers of organisms in a population and their distribution in a habitat.

CASE STUDY 1: SNOWSHOE HARES AND LYNXES

In **Section 5B.4 fig A**, you saw a classic graph showing the relationship between numbers of prey and predators based on data which was collected many years ago. More recently, a team from the University of British Columbia have investigated snowshoe hare (prey) populations in Yukon territory, Canada. Their hypothesis was that both food supply and predation interact to affect the hare population and result in the pattern of the lynx (predator)/hare abundance. They chose nine $1\,km^2$ areas of natural forest and measured the populations of hares over 11 years (the average length of the population cycle). In two of the areas, extra food was supplied to the hares all year round. Another two areas were fenced off so that hares could pass through but lynxes could not, and extra food was provided in one of these areas, but not the other. In two more areas, the soil was fertilised to increase the food quality. The remaining three areas were untouched and left as controls. Snowshoe hares occupy the niche of ground plant eaters predated by lynxes and other animals. The data collected (see **fig A**) show that both aspects of this niche (i.e. availability of food and predation) affect the abundance of hares.

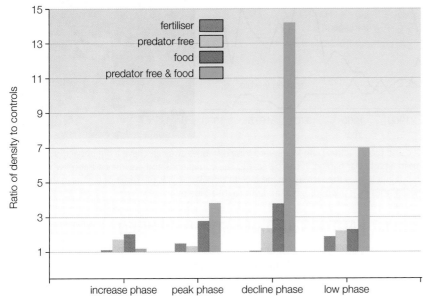

▲ **fig A** Results of investigations into factors affecting snowshoe hare populations in Canada. Population censuses were conducted at intervals through the 11-year cycle. The results of four of these censuses are shown here, when the population was in (a) an increase phase, (b) a peak phase, (c) a decline phase and (d) a low phase.

Scientists often have to think differently to explain their observations. The obvious answer is not always the right one, as you will see in the second case study.

CASE STUDY 2: WOODLAND BIRDS

A regular census of breeding birds in a woodland in southern England showed major changes in the populations of breeding birds between 1950 and 1980 (see **fig B**). The woodpigeon population almost doubled quite suddenly between 1965 and 1970. Garden warblers, which are small songbirds, disappeared altogether in 1971. Blue tits are small birds which live successfully in woodlands and in gardens. The breeding blue tit population increased fairly steadily from 1950 to 1980.

The biggest change in the habitat of the wood itself was that regular tree cutting stopped, so there were increasingly more mature and dead trees. Had the increase in pigeons caused the drop in garden warblers, or had the end of felling caused the changes in bird populations? There appeared to be correlation between these events, but it did not seem enough to explain the changes in the bird populations. Scientists felt they needed to look again to find the real causes of the changes.

It turned out that changes in biotic and abiotic factors a long distance from the wood itself were affecting the bird populations. During the late 1960s, many farmers across the south of England started growing a crop called oilseed rape, and oilseed rape fields provide woodpigeons with plenty of food through the winter. Many more woodpigeons survived to breed in the woodland due to a biotic factor outside that habitat. They expanded their seed eating niche to include oilseed rape seed.

Garden warblers were found to be affected by abiotic factors thousands of miles away in West Africa. These small birds migrate to Africa for the winter, but lack of rain meant severe drought in their overwintering grounds, so the numbers surviving and returning to the UK to breed fell dramatically.

Only the blue tits were affected by changes in the woodland habitat itself. Blue tits occupy a niche which includes breeding in small holes in trees (or in nest boxes provided by people). The increase in mature and dead trees in the wood meant an increase in the small holes that blue tits nest in, so the breeding population increased because more nest sites were available.

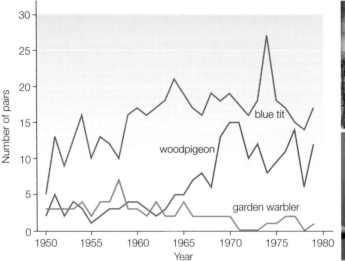

▲ **fig B**   Changes in abiotic and biotic factors both locally and long distances away can have a big impact on the distribution of organisms in a particular habitat. It can be easy to see correlations and assume causation where there is none. For example, in this graph, only the increase in the blue tit population was the result of factors in the habitat where they were observed. Photo (a) is a woodpigeon and photo (b) a blue tit.

DENSITY-INDEPENDENT FACTORS AND DENSITY-DEPENDENT FACTORS

The abiotic and biotic factors that affect the number of organisms occupying a particular niche may be **density-independent factors** or **density-dependent factors**.

The effect of density-independent factors is the same regardless of population size. For example, extremes of temperature will have the same effect on all individuals, however many individuals live there. This type of factor will limit the distribution of individuals and therefore of species. One of the effects of climate change as a result of global warming is a change in the distribution of species as extremes of temperature change.

The impact of a density-dependent factor will depend on how many organisms there are in a specific area. For example, disease and parasitism are factors that are strongly density dependent. The more individuals there are in a given area, the more likely it is that a disease or parasite can be transmitted between individuals. Breeding success in territorial animals and birds is also usually density dependent, because individuals without territories, or with reduced territories, are less able to breed. Only the strongest individuals are able to hold territories if space is a limited resource. Density-dependent factors are important in limiting the abundance of species.

COMPETITION

Individual organisms often have to compete for density-dependent factors, and this can determine the size and density of a population. Competition occurs when two organisms compete for a resource that is in limited supply. The competition may be for abiotic resources such as sunlight, minerals or water, or for biotic resources such as territories, nest sites or mates. For example, competition for mates has led to sexual selection, and to the evolution of sexual dimorphism, where the males and females of a species look very different (see **fig C**).

▲ **fig C** Here you can see sexual dimorphism in mammals and birds. (a) The male nyala is larger than the female, he is a different colour and has large horns. (b) The peacock and peahen are well known; only the male has the magnificent tail feathers used to display and impress the females and intimidate other males.

INTRASPECIFIC COMPETITION

Intraspecific competition is competition between members of the same species within the same niche for a limited resource. For example, meerkats are small mammals that live in family groups, and both males and females will defend their territory against

other groups of meerkats wanting to search for food there. The territory provides the food they need for their growing young, and so it is a limiting resource.

Sometimes a territory is small or there is relatively little food available as a result of abiotic factors such as rainfall or temperature. Therefore, as a result of intraspecific competition, some individuals may not survive, or may not reproduce, and so population growth slows. In contrast, if there are large quantities of resources, there is little or no competition and the numbers of individuals will increase.

Another example of intraspecific competition involves *Eleutherodactylus coqui*, a species of frog that lives in the tropical rainforests of Puerto Rico (see **fig D**). These frogs feed on insects and are active at night, hiding during the day to avoid predators. They lay eggs in damp places and the male guards them until tiny froglets hatch out. There is more than enough food in the forest to support a larger population of frogs than scientists have observed. The scientists wanted to discover the limiting resource. They investigated whether competition for shelter could be a factor limiting the population size, by dividing the study area into plots of 100 m². In some areas, the researchers provided the frogs with many small bamboo shelters, which were ideal for nests. In other areas, the habitat was left unchanged. Frogs rapidly moved in to all of the shelters in the test plots and reproduced. The population density therefore increased, whereas the population density of the frogs in the control plots remained the same. This confirmed the hypothesis that it was the availability of nest sites which was limiting the population. This is a similar situation to the blue tits described in the UK woodland in **fig B**.

▲ **fig D** Coqui frogs got their common name from the noise the males make: ko-KEE. The 'ko' sound marks the territory and the 'kee' attracts females. When more frogs had territories, the forest became noisier as more frogs called.

EXAM HINT

Be careful to spell the terms *intraspecific* and *interspecific* correctly. Don't confuse these two similar terms. Remember that *inter-* means 'between' (in this case, between species).

INTERSPECIFIC COMPETITION

Interspecific competition occurs when different species within a community compete for the same resources. The niches of the species overlap. Competition will reduce the abundance of the competing species. If there is a greater density of one species, or it has a faster reproduction rate than its competitor, then one of the species may become extinct in that area. For example, years ago sailors released goats on Pinta Island, which is part of the Galápagos archipelago famous for the work of Charles Darwin almost 200 years ago. Each island had its own species of giant tortoise: huge, slow moving, slow growing, slow reproducing, plant-eating reptiles which can live for more than 100 years. Goats are relatively large, fast-breeding mammals, which eat a lot. Since their introduction, the growing goat population has consumed huge numbers of the plants on the island, including the ones which the giant tortoises ate. The reptiles could not compete effectively because they reproduce much more slowly, and in the 1960s they became extinct on Pinta Island. Isabela Island, which supports a higher proportion of endemic Galápagos species than any of the other islands, faces the same problem. In **fig E** you can see both the tortoises and the goats in the same habitat. Local scientists are working to eliminate the goats and remove the competition which people have introduced.

▲ **fig E**   On Isabela Island many endemic species, including the giant tortoises, are threatened by competition from the fast-breeding goats.

CHECKPOINT

1.   Explain the difference between intraspecific and interspecific competition and how they affect the distribution and abundance of organisms in a habitat.

2.   Use data from **fig A** to answer these questions.

  (a) Describe the impact of the different experimental conditions on the density of the hare population.

  ▶ (b) Suggest an explanation for the difference between the effect of simply adding food or removing predators, compared with the enclosures where both food was added and predators excluded.

  ▶ (c) Suggest an explanation for the hare population declining towards the end of the population cycle and back this up with evidence if possible. Why do you think the population where the grass was fertilised does not show this dip?

3. ▶ Explain how both abiotic and biotic factors can be density independent and density dependent, and how they affect the population numbers and distribution of organisms.

SUBJECT VOCABULARY

density-independent factors factors affecting the number of organisms occupying a niche which are the same regardless of population size

density-dependent factors factors affecting the number of organisms occupying a niche which are dependent on the number of organisms in a specific area

intraspecific competition competition between members of the same species for a limited resource

interspecific competition competition between different species within a community for the same resources

LEARNING OBJECTIVES

■ Describe the ecological techniques used to assess the abundance and distribution of organisms.

When scientists are considering the ecology of an area of natural habitat they will measure both the **abundance** and the **distribution** of the organisms in the area. We can collect this type of data using a number of different ecological techniques. The technique we use depends on the type of ecosystem and the organisms that we want to study. For example, we need different techniques to study the fish in an ocean from the techniques needed to study the palm trees at an oasis.

ABUNDANCE AND DISTRIBUTION

The abundance of an organism refers to the relative representation of a species in a particular ecosystem. It is more than simply the numbers of an organism. It is the numbers of an organism relative to the numbers of other organisms in the same habitat.

The distribution of an organism describes where a species of organism is found in the environment and how it is arranged. The distribution of organisms can change. Animals that migrate, for example, are found in different countries or even on different continents at different times of the year. Some organisms are always found in very high densities, for example, nematode worms in soil, ants in an ants' nest, people in a city or grass in a lawn. Other species are found spread throughout a habitat. These are often the top predators. The distribution of an organism in a habitat usually enters into one of three main patterns (see **fig A**).

- Uniform distribution. This occurs when resources are thinly but evenly spread or when individuals of a species are hostile to each other. The territories may be very large or very small, for example nesting penguins, polar bears or hawks.

- Clumped distribution. This is the most common distribution seen with herds of animals or groups of plants and animals that have specific resource requirements and therefore group together (clump) in areas where those resources are found, for example a caravan of camels, a pod of dolphins or a stand of palms.

- Random distribution. This is the result of plentiful resources and no hostility. It is seen, for example, in weeds in a grass lawn.

EXAM HINT

Remember that abundance and distribution may vary through the day and also through the year. It may be important to repeat measurements at different times.

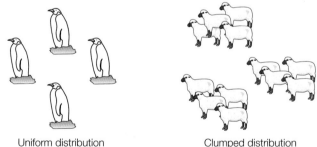 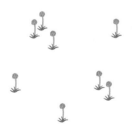

Uniform distribution   Clumped distribution   Random distribution

▲ **fig A** Distribution patterns of different types of organism.

ECOLOGICAL TECHNIQUES

In **Chapter 4C (Book 1: IAS)**, you saw the importance of biodiversity within a habitat. Scientists often want to measure the abundance of an organism or its distribution or both, in order to calculate the biodiversity of an ecosystem. You need to assess both the number of species in the area and the size of their populations. You also need to identify them correctly.

EXAM HINT

As part of your study of this topic, you will undertake **Core Practical 11: Carry out a study of the ecology of a habitat**. This will involve using quadrats and transects to determine the distribution and abundance of organisms, and measuring abiotic factors appropriate to the habitat. Make sure you understand the practical techniques you might use; your understanding of the experimental method may be assessed in your examination.

QUADRATS

The simplest way to sample an area is to use a **quadrat** (see **fig B**). This method is particularly useful for plants and for animals that do not move much. A frame quadrat is usually a square frame divided into sections that you lay on the ground to identify the sample area. Quadrats come in different sizes, from very large to very small, but we often use ones with sides of 50 cm (with an area of 0.25 m^2) or 25 cm (with an area of 0.0625 m^2), because they are easy to handle. This is known as a quantitative sampling technique as it allows us to quantify and measure such things as the number of individual organisms in the area (**individual counts**) or, using the divided grid, the area covered by the above-ground parts of a particular species (the **percentage cover**). A number of results are obtained to get a mean abundance or distribution.

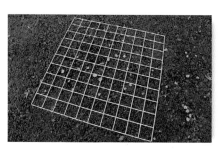

Everything in or touching the quadrat = 8 plants

Only organisms completely within the quadrat = 5 plants

▲ **fig B** When you are using quadrats to measure the abundance or distribution of organisms, you have to decide before you start whether organisms partly covered by the quadrat count as 'in' or 'out'. It does not matter which, as long as you decide and use it consistently.

EXAM HINT

A quadrat is simply a tool to help take a sample. Placing quadrats correctly is a vital part of the procedure. Simple random placing may mean that important organisms are missed and it may be necessary to modify your approach to ensure all organisms are sampled appropriately.

Frame quadrats are very important tools for the ecologist, but there are limitations, including:

- limitations to the area you can sample
- the randomness (or not) of the sampling sites
- decisions made about whether to include or exclude organisms partly covered by the quadrat.

One way of measuring the abundance of organisms in a quadrat, or any other given area, is to use the **ACFOR scale**. This is a simple and quite subjective scale, but it is nevertheless useful in the field. It describes the abundance of a species in a given area as:

A = abundant

C = common

F = frequent

O = occasional

R = rare.

Limitations of the ACFOR scale include the following.

- The measurements given on the scale are subjective; two people would probably never give exactly the same scores.
- There are no set definitions of the terms; for example, how common is common?
- Species can easily be rated according to how obvious they are rather than how abundant they are.

A point quadrat is a horizontal bar supported by two or three legs. At set intervals along the bar are holes through which a long pin is dropped. The species that the pins touch are recorded (see **fig C**). Point quadrats are often used to estimate the percentage cover by recording how many of the 10 pins touch a particular species of plant. For example, if 5 out of 10 pins touch plants of species A, the percentage cover is 50%. You can obtain a number of results and calculate a mean percentage cover.

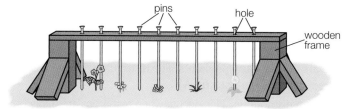

▲ **fig C** A point quadrat is often used to calculate percentage cover by an organism.

Permanent quadrats are left in place all the time. In this way, you can collect data through the seasons or from year to year reliably from the same place. For example, in a study of more than 40 years' duration at penguin breeding sites at Punta Tombo in Patagonia, researchers return to exactly the same (100 m^2) quadrat areas every year to count the number of active nests. This makes their data reliable, because it always refers to the same regions of the breeding colony.

Aerial quadrats are a high-tech version of this technique. Aerial photographs of large areas of land are divided into measured areas and surveyed for numbers of species. The technique is used, for example, when surveying large areas of desert or savannah.

You can use simple quadrats to randomly sample areas by placing them at random coordinates, or simply throwing them in a given area. You can also use them to systematically sample a site using different types of transect.

EXAM HINT

When using a quadrat it is important to state its size. The number of times the quadrat is used will depend on the area being studied; it is important to ensure that enough samples are taken to achieve a representative impression of the area.

TRANSECTS

Line transects are a widely used way of gathering data about an area in a systematic way. A transect is not random. You stretch a tape between two points, for example in a wildflower meadow, across sand dunes (see **Section 5B.2**) or down a mountainside. You then record every individual plant (or animal) that touches the tape. An alternative is a **belt transect**. In this case you stretch two tapes and the ground between them is surveyed. You can use quadrats for systematic sampling if you place them along a tape to form a belt transect. You can use transects to investigate whether a change in the distribution of organisms is linked to a change in an abiotic factor. You can investigate correlations between organisms and abiotic factors by sampling the abundance of organisms at regular intervals along the transect, and also sampling and recording abiotic factors such as soil pH or light intensity. **Fig D** summarises the difference between the two main types of transect.

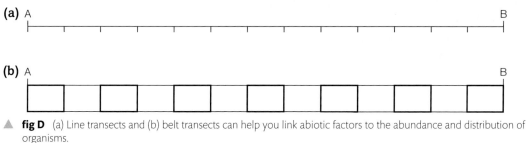

▲ **fig D**　(a) Line transects and (b) belt transects can help you link abiotic factors to the abundance and distribution of organisms.

EXAM HINT

As with other sampling techniques, multiple transects should be carried out so that mean values can be calculated. While a transect may be a systematic method of collecting data, the starting point of the quadrat may be random.

MEASURING ABIOTIC FACTORS

It is important to be able to measure abiotic factors which may affect the abundance and distribution of organisms. You will use some of them when you carry out practical investigations into the abundance and/or distribution of organisms in a specific ecosystem. **Table A** summarises some of the most common instruments used to measure abiotic factors.

ABIOTIC FACTOR	METHOD OF MEASUREMENT
temperature of air, soil, water, etc.	glass or digital thermometers; thermistors (temperature probes, particularly useful for measuring water temperature)
light intensity	light meters
pH	chemical pH tests based on pH indicators; pH meters
wind speed	anemometers
slope of an incline	clinometers
oxygen levels (in water)	digital oxygen probe

table A　Some common instruments for measuring abiotic factors

Data loggers combine ways of measuring several different aspects of a habitat or ecosystem. They are relatively cheap and portable and are commonly used.

PRACTICAL SKILLS　　　　CP11

Core Practical 11: Carry out a study of the ecology of a habitat involves field work. You will investigate organisms in a local habitat and discover how their distribution and abundance relates to abiotic factors such as the amount of light, availability of water, the pH of the soil, the position on a seashore or the temperature. You will use a number of different techniques, and the method you choose will depend on what you are investigating.

Whether you use random quadrats, line transects or belt transects, make sure you understand the theory behind the method and can explain fully why you used it. The student in **fig E** is investigating the organisms found on a rocky seashore. When the tide goes out, the organisms which do not swim in the water are left on the beach, exposed to the air and the sun. The student wants to determine if the abundance and distribution of different organisms is affected by how long they are exposed to the air. He uses a belt transect from the low water mark to the high tide line to record the types of organisms found at 1 m intervals moving up the beach. The student records the length of time the different areas are exposed to the air, and the temperature on the beach compared to the sea temperature. All of this data can be put together to develop a model of which organisms are adapted to cope with relatively long periods of exposure, and which need to be covered with sea water most of the day.

▲ **fig E**   A belt transect on a rocky shore.

Safety Note: Wear clothing and footwear that is appropriate for the area being studied. Gloves may be needed where there are stinging or thorny plants.

LEARNING TIP

A good way to display data from a transect is a kite graph or a series of kite graphs plotted together on the same axes. This makes it easy to spot how abundance changes as an abiotic factor changes.

SKILLS   DECISION MAKING

CHECKPOINT

1.   Why is a quadrat more useful for measuring the abundance and distribution of plants than of animals?
2. ▶ Compare the use of frame quadrats, point quadrats and permanent quadrats in a field investigation.

SUBJECT VOCABULARY

abundance the relative representation of a species in a particular ecosystem
distribution where a species of organism is found in the environment and how it is organised
quadrat a sample area used in practical ecology, often measured using a square frame divided into sections that you lay on the ground
individual counts a measure of the number of individual organisms in an area
percentage cover the area covered by the above-ground parts of a particular species
ACFOR scale a simple scale used to describe the abundance of a species in a given area
line transect a way of collecting data more systematically; a tape is stretched between two points and every individual plant (or animal) that touches the tape is recorded
belt transect when two tapes are stretched out and the ground between them surveyed or a tape stretched out and quadrats are taken at regular intervals

LEARNING OBJECTIVES

- Understand how the concept of niche accounts for the distribution and abundance of organisms in a habitat.
- Describe the ecological techniques used to assess the abundance and distribution of organisms.

In any field study, you will collect a lot of data. It is not always easy to understand the data. It is hard to know whether patterns that appear are the result of a biological relationship or merely due to chance. This is why we use statistical tests to help us decide. Ecologists widely use a number of statistical tests that are particularly useful. These include **Spearman's rank correlation coefficient**, the **Student's *t*-test** and the **chi squared test**.

LEARNING TIP

One of the most difficult steps in statistical testing is deciding what test to use. It is best to decide on the test before you start and then ensure you collect data suitable for that test.

WHAT ARE YOU TESTING FOR?

Before you start to collect data in the field or lab, it is important to have a clear idea of what you are trying to prove or disprove. In other words, you need a clearly stated **null hypothesis** (which assumes any differences between data sets are the result of chance). This allows you to choose the right statistical test to help you analyse and evaluate your data.

After running your chosen statistical test, you will obtain a number known as the observed value. You need to look up this value in a published table of critical values to see which probability (p) value this corresponds to, so you can decide whether or not your results are **significant** or whether you should accept your null hypothesis. In other words, if your results are significant they are not due to chance. If your observed result fits with the critical value corresponding to the $p < 0.05$ level of significance given in the table, you can be 95% certain that the relationship between the factors you are studying is due to something other than chance.

Scientists use statistical computer programs to run the tests and give them a p value, without the need to use tables. However, it is important that you learn how to use these tests and do the calculations yourself.

HYPOTHESIS TESTING

In statistical tests, by definition, you need something to test. In each situation you investigate, you will make biological observations and collect sets of data. You need to produce a null hypothesis in order to conduct a statistical analysis of your results. In general terms, the null hypothesis assumes that your current model explains your results.

When you have two or more sets of data, your null hypothesis will usually state that any differences between the data sets are simply due to chance. As scientists, we always want to find out something new. If statistical analysis shows a high probability that the null hypothesis is due to chance, it usually means we have not found something new and have confirmed an existing idea. For example: You want to investigate the effect of goat grazing on the biodiversity of an area of scrub. The null hypothesis is that differences in biodiversity between grazed and similar ungrazed areas are due to chance.

A statistical analysis will show you the probability that any differences between the groups you observe are the result of chance. If it is chance, you accept the null hypothesis and you may need to study more data to create a new hypothesis to investigate. If the probability is that the differences are significant and so are not due to chance, then you reject the null hypothesis, but may need to do a lot more analysis to discover exactly what is causing those differences.

SPEARMAN'S RANK CORRELATION COEFFICIENT

You can plot a scatter diagram to determine whether there is correlation between two variables. This is a useful technique but it does not always give a clear result. You can measure more precisely whether there is significant correlation between two variables using a correlation coefficient. We calculate Spearman's rank correlation coefficient using a formula that gives a correlation coefficient r_s between -1 and $+1$. The size of the correlation coefficient indicates how strong the correlation is and the sign indicates whether the correlation is positive or negative:

- A value of $+1$ indicates that there is perfect positive correlation.
- A value of -1 indicates that there is perfect negative correlation.
- A value of 0 indicates that there is no correlation between the variables.

USING THE TEST

Step 1: State the null hypothesis

The null hypothesis is that there is no correlation between the two variables, in which case r_s is equal to 0.

Step 2: Calculate the correlation coefficient, r_s (the observed value)

The formula for the Spearman's rank correlation coefficient is:

$$r_s = 1 - \frac{6 \sum d^2}{n(n^2 - 1)}$$

where d is the difference in rank between each pair of variables and n is the number of pairs.

To use it, find the difference in rank between each pair, square these differences and add them all together. Now substitute this value into the formula in place of $\sum d^2$.

Step 3: Decide whether to accept the null hypothesis

If your value for r_s is anything except 0, then it suggests a correlation and you need to see if the correlation is statistically significant by looking up your value in a table of critical values.

Find the probability (p) value that relates to your observed value and the number of pairs (n). Ignore whether r_s is positive or negative. If $p < 0.05$, the correlation is considered to be statistically significant, so you can reject your null hypothesis that there is no correlation.

WORKED EXAMPLE 1

After the new volcanic island of Surtsey was formed in the 1960s, scientists observed the colonisation of the bare lava. Now you will analyse some of the data scientists collected about the abundance of species there over time.

Table A shows you how the numbers of species on Surtsey changed over time. This is an example of a data set where Spearman's rank correlation coefficient can be used.

YEARS AFTER FORMATION	0	5	10	15	20	25	30	35	40
SPECIES TOTAL	0	4	10	13	11	18	34	46	54

table A The total number of species on Surtsey over time

The null hypothesis is that there is no correlation between the time in years after Surtsey was first formed and the number of species present. Rank the variables and calculate d and d^2 for each pair of data (see **table B**).

x VALUES (NUMBER OF SPECIES)	RANKING OF x VALUES FROM LOWEST TO HIGHEST	y VALUES (YEARS AFTER FORMATION IN 1965)	RANKING OF y VALUES FROM LOWEST TO HIGHEST	DIFFERENCE IN RANK BETWEEN PAIR OF VARIABLES (d)	d^2
0	1	0	1	0	0
4	2	5	2	0	0
10	3	10	3	0	0
13	5	15	4	−1	1
11	4	20	5	1	1
18	6	25	6	0	0
34	7	30	7	0	0
46	8	35	8	0	0
54	9	40	9	0	0
Sum of	$n = 9$				$\sum d^2 = 2$

table B Ranking number of species against years of formation of Surtsey

Now substitute your data into the formula:

$$r_s = 1 - \frac{6\sum d^2}{n(n^2 - 1)} = 1 - \frac{6 \times 2}{9 \times 80} = 1 - \frac{12}{720} = 1 - 0.0167 = \mathbf{0.983}$$

You need to use a critical values table (see **table C**) to see if you can accept your null hypothesis.

NUMBER OF PAIRS OF VALUES (n)	4	5	6	7	8	9	10	11	12
CRITICAL VALUES	1.00	0.90	0.83	0.79	0.74	0.68	0.65	0.61	0.59

table C Extract from a table of critical values for Spearman's rank correlation coefficient at $p = 0.05$

The critical values table shows that for $n = 9$ you need a number greater than 0.68 for $p < 0.05$. Your figure of 0.983 is considerably higher than this, relating to a p value of much less than 0.05, so the correlation is considered statistically significant. Therefore, you can reject the null hypothesis that there is no correlation between the time since Surtsey was first formed and the number of species present, and confirm that the longer the island of Surtsey has been in existence, the more species of plants are found growing on it.

THE STUDENT'S *t*-TEST

The Student's *t*-test is used to determine whether the mean of a variable in one group differs significantly from the mean of the same variable in a different group; for example, comparing the blood pressure of males and females. Stated simply, it tests whether two sets of data are significantly different from each other. The variable being considered must follow a normal distribution. The Student's *t*-test considers the degree of overlap between the two sets of data and allows you to judge whether any difference between the mean values of the two groups is statistically significant or just due to chance.

USING THE TEST

Step 1: State the null hypothesis

The null hypothesis is that there is no significant difference between the mean of the variable for the two categories.

Step 2: Calculate your observed value, t

You will need to calculate t, but first you need to find the mean and the variance s^2 for both groups. The sample size is n (for example, the number of people in the group) and x is a value of the measured variable (for example, one person's blood pressure reading).

$$\bar{x} = \frac{\sum x}{n}$$

$$s^2 = \frac{\sum x^2 - \frac{(\sum x)^2}{n}}{n - 1}$$

You can now use the following formula to calculate the value of t:

$$t = \frac{\bar{x}_1 - \bar{x}_2}{\sqrt{\frac{s_1^2}{n_1} + \frac{s_2^2}{n_2}}}$$

where

$\bar{x}_1$ = mean of the first set of data

$\bar{x}_2$ = mean of the second set of data

s_1^2 = variance of the first set of data

s_2^2 = variance of the second set of data

n_1 = number of data values in first set of data

n_2 = number of data values in second set of data

Step 3: Decide whether to accept the null hypothesis

To decide whether this value of t indicates a significant difference in variable between the two groups, you need to compare your observed value of t with a critical values table. You will need to know the degree of freedom; for the Student's *t*-test, it is calculated as:

$df = n_1 + n_2 - 2$ (the total number of data values minus 2).

Find the probability (p) value that relates to your observed value and degrees of freedom; if $p < 0.05$, the difference between the means of the two groups is considered to be statistically significant and not due to chance.

WORKED EXAMPLE 2

The Student's t-test is very useful in field work for helping you decide if there are genuine differences between populations, for example the size of shellfish such as mussels (see **fig A**) on different beaches or different areas of the same beach.

In this example, a student compared the shell length (the interval variable) of the largest mussel in each of 10 quadrats on the rocks at the low water level on two different beaches (the nominal variable). The results are shown in **table D**.

▲ **fig A** Mussels on a rocky shore.

The null hypothesis is that there is no significant difference between the mean shell length of the largest mussels on the rocks at the low water mark on the different beaches.

	x_1 VALUES (LENGTH OF LARGEST MUSSELS FROM BEACH 1) (mm)	x_1^2	x_2 VALUES (LENGTH OF LARGEST MUSSELS FROM BEACH 2) (mm)	x_2^2
	76	5776	77	5929
	82	6724	71	5041
	65	4225	73	5329
	90	8100	69	4761
	71	5041	79	6241
	69	4761	75	5625
	75	5625	82	6724
	59	3481	76	5776
	84	7056	81	6561
	78	6084	85	7225
SUM OF (Σ)	749	56 873	768	59 212

table D Length of largest mussels from beach 1 and beach 2

So

$$\bar{x}_1 = \frac{749}{10} = 74.9$$

$$\bar{x}_2 = \frac{768}{10} = 76.8$$

$$s_1^2 = \frac{56\,873.0 - (749.0)^2/10.0}{9.0} = \frac{56\,873.0 - 56\,100.1}{9.0} = \frac{772.9}{9.0} = 85.88$$

$$s_2^2 = \frac{59\,212.0 - (768.0)^2/10.0}{9.0} = \frac{59\,212.0 - 58\,982.4}{9.0} = \frac{229.6}{9.0} = 25.51$$

$$n_1 = 10$$

$$n_2 = 10$$

Now you can substitute these numbers to calculate the value of t:

$$t = \frac{\bar{x}_1 - \bar{x}_2}{\sqrt{\frac{s_1^2}{n_1} + \frac{s_2^2}{n_2}}} = \frac{74.9 - 76.8}{\sqrt{\frac{85.88}{10} + \frac{25.51}{10}}} = \frac{-1.9}{\sqrt{11.14}}$$

$$= \frac{-1.9}{3.34} = -0.57 \text{ (to 2 decimal places)}$$

For the Student's t-test, the sign of the observed value does not matter – if you swapped the two data sets around we would get +0.57 – so you just need to take the absolute value and ignore the sign.

You now need to use a critical values table (see **table E**) to see whether you can reject the null hypothesis that there is no significant difference between the maximum shell lengths of mussels on the two beaches. The degrees of freedom is 20 – 2 = 18 (the total number of data values minus 2).

DEGREES OF FREEDOM (*df*)	LEVELS OF SIGNIFICANCE (*p*)				
	0.05	0.02	0.01	0.002	0.001
18	2.101	2.552	2.878	3.610	3.922

table E Extract from a table of critical values

The critical value table shows that you need a figure greater than 2.101 to give $p < 0.05$ level of significance with 18 degrees of freedom.

The number we calculated was –0.57. This absolute value 0.57 is far less than the critical value of 2.101, indicating a *p* value much greater than 0.05. This means that there is no significant difference between the means of shell length on the two beaches. Therefore we accept the null hypothesis. There is no significant difference in the maximum shell sizes of the mussels measured at the low water mark on the two beaches.

THE CHI SQUARED (χ^2) TEST

The chi squared (χ^2) test is a relatively simple statistical method used to estimate the probability that any differences between the observed and expected results are due to chance. It is particularly useful when you are counting things that you can put into categories. For example, the distribution of blue and yellow flowers or the distribution of insects on different plants. The chi squared test will tell you whether or not your results differ significantly from the expected outcome. The test involves some fairly simple arithmetic followed by the use of a chi squared table.

The formula for the chi squared (χ^2) test is:

$$\chi^2 = \Sigma \frac{(O - E)^2}{E}$$

In this equation, *O* is your observed result, *E* is the expected result and the symbol Σ means the sum of.

USING THE TEST

Step 1: State the null hypothesis

The basic null hypothesis for the chi squared test is that there is no significant difference between the observed and expected frequencies. This is modified to be specific to the data you have collected.

Step 2: Calculate chi squared using the formula given above

To calculate χ^2 (chi squared) you will need to generate a table similar to **table F**.

PHENOTYPE	HYPOTHESIS (RATIO OF ROUND TO WRINKLED SEEDS)	OBSERVED RESULTS (*O*)	EXPECTED RESULTS (*E*)	*O* – *E*	$(O - E)^2$	$\frac{(O - E)^2}{E}$
plants with round pea seeds	3	640	660	640 – 660 = –20	–20² = 400	400/660 = 0.61
plants with wrinkled pea seeds	1	240	220	240 – 20 = 20	20² = 400	400/220 = 1.82

table F The χ^2 test applied to the results of a simple genetic cross

$$\chi^2 = 0.61 + 1.82 = 2.43$$

Step 3: Decide whether to accept the null hypothesis

First calculate the degrees of freedom in the system. This measures the spread of the data. In the chi squared test, you find the degrees of freedom by taking the categories of observed information and then subtracting 1. The degree of freedom is always one less than the number of categories of observed information you are working with.

degrees of freedom = $n - 1$

Next find the probability (*p*) value that relates to your observed value and degree of freedom.

Once you have the χ^2 value and the degree of freedom, you can use them with a χ^2 probability table to find out whether there is any significant difference between your predicted and observed results. This will tell you whether your null hypothesis is correct or not, and whether any differences in the results you get are due to chance. If $p < 0.05$, the difference between the two groups is considered to be statistically significant. They are not due to chance, but to some undetermined factor which you then need to investigate.

WORKED EXAMPLE 3

The chi squared test is used in field work to determine if there is a statistically significant association between the distribution of two species such as the clams and pearl oysters shown in **fig B**. In this example, divers recorded the species found in 50 quadrats of 1 m^2 on the sea bed.

▲ **fig B** Clams and pearl oysters are both found in the Arabian Gulf.

The divers found 7 quadrats contained both species, 15 contained only pearl oysters, 20 contained only clams and 8 contained neither species. This can be represented as in the chart below.

PEARL OYSTERS	CLAMS PRESENT	ABSENT	TOTAL
PRESENT	7	15	22
ABSENT	20	8	28
TOTAL	27	23	50

The null hypothesis is that there is no significant difference between the distribution of the two species (in other words, distribution is random). The alternative hypothesis is that there is a significant difference between the distributions of the species (in other words, the species are associated).

Now you have the observed frequencies of the shellfish, but to use the chi squared formula you also need the expected frequencies if the null hypothesis is correct. You can calculate these by multiplying the row total by the column total and dividing the answer by the total number of quadrats.

PEARL OYSTERS	CLAMS PRESENT	ABSENT	TOTAL
PRESENT	11.88 (22 × 27/50)	10.12 (22 × 23/50)	22
ABSENT	15.12 (28 × 27/50)	12.88 (28 × 23/50)	28
TOTAL	27	23	50

Now carry out the calculations needed for the chi squared formula.

PEARL OYSTERS		CLAMS PRESENT	ABSENT
PRESENT	O	7	15
	E	11.88	10.12
	$(O - E)^2$	23.81	23.81
	$\dfrac{(O - E)^2}{E}$	2.0	2.35
ABSENT	O	20	8
	E	15.12	12.88
	$(O - E)^2$	23.81	23.81
	$\dfrac{(O - E)^2}{E}$	1.58	1.85

Based on these results, $\chi^2 = 2.0 + 2.35 + 1.58 + 1.85 = 7.78$

Now you must determine the degrees of freedom. When you are comparing the distribution of two species, as in this example, the degree of freedom is 2 – 1 = 1.

This allows you to identify the p value.

DEGREES OF FREEDOM	p VALUES FOR χ^2 DISTRIBUTION							
	0.90	0.75	0.50	0.25	0.10	0.05	0.025	0.01
1	0.016	0.102	0.455	1.320	2.706	3.841	5.024	6.635

With 1 degree of freedom, you need a value greater than 3.841 for the results to be significant and so for the null hypothesis to be disproved. In this example the data gives you a value of 7.78. This means the difference is significant and that the distribution of clams and pearl oysters on the sea bed is not random. As they are relatively rarely found together, this suggests a negative association: where you find clams, you don't find oysters. This gives scientists a basis to further investigate which conditions suit the different species.

EXAM HINT

Many students find it difficult to report the results of a statistical test. They are confused by what the test shows in relation to the null hypothesis. It is important to remember that you trying to disprove a null hypothesis. If your test shows that the null hypothesis is incorrect it can be rejected and another hypothesis can be accepted.

CHECKPOINT

1. In an area where some bog sites had been restored, ecologists wanted to evaluate if the process had been successful. They looked at the percentage cover of two different bog plants, *Deschampsia flexuosa* and *Agrostis* sp., in 26 quadrats on the same site and ranked the relative cover of the two species. They wanted to discover if there was any relationship between the populations of the two plants.

   (a) Suggest a null hypothesis for this investigation.

   (b) Suggest a suitable statistical test to use to analyse the data they collected and explain why you have chosen it.

   (c) State how many degrees of freedom there will be for this data set.

2. (a) State what is meant by the term *Student's* t-*test*.

   (b) Explain why the Student's *t*-test is not the right test to apply to the data described in Q1 above.

   (c) Suggest data that might be collected from that site and analysed using the Student's *t*-test.

3. The ghaf tree (*Prosopis cineraria*) is regarded as the national tree in the UAE. However, this native tree is now threatened by, among other things, an invading species, the mesquite (*Prosopis juliflora*). Scientists wondered if the two trees could live together in harmony. Here are the results of one survey, based on 150 quadrats.

MESQUITE	GHAF		
	PRESENT	ABSENT	TOTAL
PRESENT	25	85	110
ABSENT	15	25	40
TOTAL	40	110	150

   (a) State your null hypothesis.

   (b) Deduce if the two tree species show an association by using a chi squared test on these results.

SUBJECT VOCABULARY

Spearman's rank correlation coefficient a statistical tool used to test whether two variables are significantly correlated

Student's *t*-test a statistical test that allows you to judge whether any difference between the means of the two sets of data is statistically significant

chi squared test a statistical test that enables you to determine whether there is a statistically significant association between the distribution of two species

null hypothesis the hypothesis that any differences between data sets are the result of chance

significant not due to chance

1 (a) Which statistical test, **A**, **B**, **C** or **D**, is most suitable for each of the investigations below (i–iv)?

    A Spearman's rank test

    B Student's *t*-test

    C chi squared (χ^2) test

    D standard deviation

  (i) Investigate whether the height of limpet (*Patella*) shells is linked to the distance up a rocky shore. [1]

  (ii) Investigate whether the height of limpet (*Patella*) shells depends on whether they grow on the exposed side or the sheltered side of rocks. [1]

  (iii) Compare the number of worms in two different fields. [1]

  (iv) Investigate how the percentage cover of marram grass changes with the increasing salt concentration in the sand. [1]

(b) Explain how competition leads to succession. [5]

(Total for Question 1 = 9 marks)

2 Salt marshes develop in sheltered areas of coastline where deposits of mud build up. The sea frequently floods these marshes with salt water and only plants with a tolerance to salt can survive. The frequency of flooding becomes less as the height of the land above low tide increases. As a result the highest tides flood some areas only a few times a year.

The distribution of plant species in a salt marsh was investigated. The presence of plant species was recorded at 10 sites along a transect running from low water (site 1) to higher ground 50 metres inland (site 10).

The diagram below shows a profile of this salt marsh and the plant species recorded at each site.

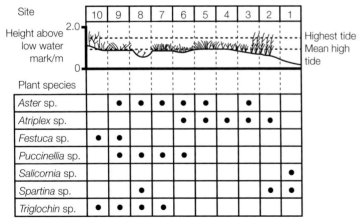

Plant species	10	9	8	7	6	5	4	3	2	1
Aster sp.		•	•	•	•	•		•		
Atriplex sp.				•	•	•	•	•		
Festuca sp.	•	•								
Puccinellia sp.		•	•	•	•					
Salicornia sp.									•	
Spartina sp.			•					•	•	
Triglochin sp.	•	•	•	•						

• Species present

(a) Describe a suitable sampling method to obtain the plant species data recorded. [2]

(b) Compare and contrast the data for sites 1 and 9. [2]

(c) (i) Which species is most tolerant to salt? [1]

    A *Spartina* sp.

    B *Festuca* sp.

    C *Salicornia* sp.

    D *Atriplex* sp.

  (ii) Different species are found at each site due to competition between species which leads to succession. What term is used for competition between species? [1]

    A species competition

    B interspecific competition

    C intraspecific competition

    D extraspecific competition

(d) Site 8 includes a hole which fills with seawater when the area floods. This makes the soil waterlogged for longer periods than the surrounding area. As the water evaporates, it leaves high concentrations of salt in the soil. The photograph below shows *Spartina* sp. growing in the hole filled with seawater.

  (i) Using all the information given, deduce why *Spartina* sp. is found at site 8. [3]

  (ii) Suggest how the hole at site 8 may change over a long period of time. [5]

(Total for Question 2 = 14 marks)

3 For many centuries, sheep have grazed on the grasslands on Scottish islands. These grassland communities are examples of a plagioclimax.

As the demand for wool from sheep reduced, sheep farming on many of the islands stopped. Within a few years, the grasslands on most of these islands developed into shrub communities.

(a) Describe what is meant by the term *climax community*. [2]

(b) (i) Explain why the grassland on the islands developed into shrub communities after sheep farming stopped. [4]

(ii) Suggest what is likely to happen to these communities in the future. [1]

(c) On some of the islands where sheep farming stopped, the grasslands remained unchanged. Suggest **two** possible reasons for this. [2]

(Total for Question 3 = 9 marks)

4 An area of abandoned grassland was studied over a period of more than 100 years. During that time, there were changes to the plant communities which resulted in changes to the number of species and population density of small birds. The table below shows these changes.

Time since abandoned/years	1–10	10–25	25–100	100+
Plant community	Grass	Shrubs	Pine trees	Mixed woodland
Number of species of small birds	2	8	15	19
Population density of small birds/number of birds per 40 hectares	54	246	226	466

(a) Name the process by which communities change over time. [1]

(b) Describe the effect of the change in the plant community on the number of species of small birds. Suggest an explanation for these changes. [5]

(c) The population density of birds drops when shrubs are replaced by pine trees, but then increases with the change to mixed woodland. Suggest explanations for these changes in population density. [3]

(Total for Question 4 = 9 marks)

5 (a) What is the best definition of a community? [1]

A a group of living things in its environment

B a group of organisms of the same species living together

C the place where animals live

D a group of organisms of different species living in the same place

(b) What is the correct term for the non-living factors in the environment? [1]

A edaphic

B abiotic

C biotic

D non-biological

(c) Interspecific competition is one way in which species can interact.

(i) Define the term *interspecific competition*. [2]

(ii) Name and describe **two** other ways in which species can interact. [4]

(Total for Question 5 = 8 marks)

ENERGY FLOW, ECOSYSTEMS AND THE ENVIRONMENT

ENVIRONMENT AND CLIMATE CHANGE

The collection of leaf litter is a widely used if somewhat surprising way of measuring the productivity of an ecosystem. Leaf fall measurements range from around $0.6\,t\,ha^{-1}\,y^{-1}$ in the Arctic tundra to $25.4\,t\,ha^{-1}\,y^{-1}$ in lowland tropical rainforests. Rainforests, with their dense, leafy canopies, are very productive plant communities, so they have a high level of leaf fall. But not all the leaves end up as leaf litter. Much plant material is eaten by animals including birds, monkeys and insects before it reaches the ground. In the warm moist climate of a rainforest, decomposers work fast. In spite of the high productivity of the plants, the forest floor does not disappear under a metres-deep layer of old leaves and fruits because many fungi, bacteria, arthropods and worms break down the plant material and return the nutrients to the soil to sustain the continual growth.

In this chapter, you will learn about the efficiency of ecosystems. Light from the Sun catalyses the production of plant material which is then passed through different trophic levels. Biologists try to calculate the efficiency of these processes and calculate where and how biomass is used at every level. Biomass represents the energy stored within the chemical bonds of the biological molecules. New biomass is made constantly through photosynthesis but the inorganic components of biological molecules are finite. You will consider the vital role of microorganisms and other decomposers in the recycling of nutrients within an ecosystem.

You will be considering the effect humans have on ecosystems. You will look at some of the evidence that the climate of the world is changing, and the apparent correlations with human activities that increase the emissions of greenhouse gases. You will also look at the human impact on biodiversity and population sizes on the land and in the oceans. You will discuss the idea that the sustainability of the world's resources depends on effective management of the conflict between human needs and the conservation of biodiversity.

MATHS SKILLS FOR THIS CHAPTER

- **Recognise and make use of appropriate units in calculations** (*e.g. the amount of carbon stored in different carbon sinks*)
- **Estimate results** (*e.g. to see if biomass calculations through trophic levels are appropriate*)
- **Use appropriate number of significant figures** (*e.g. when calculating NPP*)
- **Construct and interpret frequency tables and diagrams, bar charts and histograms** (*e.g. pyramids of numbers, biomass, link between lactation numbers and methane production in cows*)
- **Translate information between graphical, numerical and algebraic forms** (*e.g. biomass transfers, evidence on carbon dioxide levels from different sources*)
- **Use ratios, fractions and percentages** (*e.g. percentage contributions of different greenhouse gases to the greenhouse effect*)

What prior knowledge do I need?

Chapter 1A (Book 1: IAS)

   Biological molecules

Chapter 5A

* The process of photosynthesis
* Factors that affect or limit the process of photosynthesis

Chapter 5B

* The nature of ecosystems
* Succession in ecosystems and the interactions of abiotic and biotic factors in determining population numbers
* The ecological techniques used to assess abundance and distribution of organisms in a natural habitat, including quadrats, transects, ACFOR scale, percentage cover and individual counts

What will I study in this chapter?

* How energy is transferred between trophic levels using the terms net primary productivity and gross primary productivity
* How to calculate the efficiency of energy transfer between different trophic levels and account for the loss of energy at each level
* The carbon cycle in nature
* Human effects on ecosystems
* Some of the evidence relating to human effects on ecosystems, including climate change and depletion of biological resources
* The effect of climate change on changes in allele frequencies and speciation
* How sustainability of resources depends on effective management of the conflict between human needs and conservation, including reforestation and the use of sustainable alternative fuels such as biofuels, to reduce the possible causes of climate change

What will I study later?

Chapter 6C

* The role of microorganisms in the recycling of nutrients within an ecosystem
* The process of decay

LEARNING OBJECTIVES

■ Know how to calculate the efficiency of biomass and energy transfers between trophic levels.

In the 1920s, Charles Elton, a young British biologist, began to study the relationships between the animals on Bear Island (off the northern coast of Norway) and their scarce food resources. The island had limited numbers of plant and animal species, and this made his studies easier and more effective than if he had studied a more diverse habitat. The main animals he observed were arctic foxes and the birds that they ate, such as sandpipers and ptarmigan. The birds ate the leaves and berries of plants or, in some cases, ate the insects that fed on the plants. Elton called these feeding interactions a **food chain** and proposed a general model to explain the flow of resources through a community. Each link in the food chain represents a specific **trophic level** (see **fig A**).

EXAM HINT

Do not confuse trophic levels with tropisms, which are the directional growth movements of plants in response to environmental changes.

A MODEL FOR A FOOD CHAIN

Elton proposed a general model for the food chain based on the following trophic levels.

- **Producers** make food. In photosynthesis, plants and algae trap light from the Sun and this drives the production of ATP, which they then use to make glucose from carbon dioxide and water (see **Chapter 5A** to remind you of the process of photosynthesis).
- **Primary consumers** are the organisms, mainly animals, that eat producers. They are herbivores. They use the molecules in plants to supply the raw materials needed for their metabolic reactions.
- **Secondary consumers** are the animals that feed on herbivores. They are carnivores. They use the molecules in the herbivores to supply the raw materials needed for their metabolic reactions.
- **Tertiary consumers** are animals that feed on other carnivores. They are usually the top predators in an area unless there is a quaternary consumer. They use the molecules in the carnivores to supply the raw materials needed for their metabolic reactions.
- **Decomposers** are the final trophic level in any set of feeding relationships. They are the microorganisms, such as bacteria and fungi, that break down the remains of animals and plants and return the mineral nutrients to the soil. Decomposers are rarely included in a food chain.

producers
(plants)

primary consumers
(arctic insects)

secondary consumers
(ptarmigan)

tertiary consumers
(arctic foxes)

▲ **fig A** A food chain like this is the simplest representation of the feeding relationships and energy flows within an ecosystem.

LEARNING TIP

Remember that light from the Sun is the source of energy for almost all food chains. A food chain always starts with a producer which produces the food eaten by all the other levels.

FOOD WEBS AND BEYOND

The description of a food chain makes sense and is relatively simple to understand. However, it is too simplified to be very useful to scientists. Few animals eat a single food except, for example, giant pandas eating bamboo and koala bears eating eucalyptus. Most animals have a variety of food sources. They do not exist in simple food chains but as part of complex interconnected food webs. This is exactly what Elton later observed on Bear Island, and ecologists have constructed similar models from ecosystems all over the world (see **fig B**).

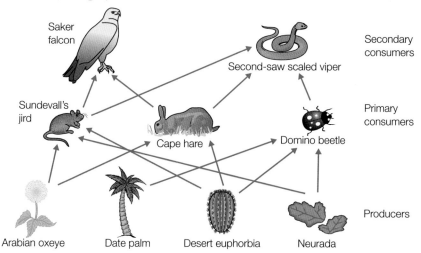

▲ **fig B** A simplified food web from the United Arab Emirates. In a food web like this, if the numbers of one of the organisms drop, it will not be a disaster for the other organisms as they all have more than one food source.

The ecosystem is easily disrupted in situations involving a single food chain, such as the giant panda with its diet exclusively of bamboo. Any event that reduces the availability of bamboo will also threaten the panda. This makes pandas very vulnerable to habitat destruction, whether as a result of human actions or natural disasters. An organism which is part of a complex food web is in a more stable situation. A change in any one component can potentially affect the balance of the ecosystem but it is not likely to have catastrophic effects (see **fig B**).

ECOLOGICAL PYRAMIDS

Although a food chain is a highly simplified model of the trophic levels within an ecosystem, it has helped us to understand the energy economy of ecosystems. We can use pyramids to illustrate what is happening in an ecosystem.

PYRAMIDS OF NUMBERS

In many food chains, the number of organisms decreases at each trophic level. There are more producers than primary consumers, more primary consumers than secondary consumers, and so on. These simple observations are represented by a **pyramid of numbers** (see **fig C**).

PYRAMIDS OF BIOMASS

In many situations, a pyramid of numbers does not accurately reflect what is happening in an ecosystem. For example, a single rosebush will support a very large population of aphids, which will be eaten by a smaller population of ladybirds and hoverfly larvae. These larvae are the prey of relatively few birds. This gives a very inaccurate picture. You get a much more realistic model by using a **pyramid of biomass**. This shows the combined mass of all the organisms in a particular habitat.

It can be difficult to count the plants and animals for pyramids of numbers. Measuring biomass is even harder. You can use either wet or dry biomass; however, wet biomass is very inaccurate. It is affected by water uptake in the soil, transpiration in plants, and drinking, urinating, defaecating and, in some cases, sweating in animals. If you use dry mass, it eliminates the inaccuracy of variable water content in organisms, but involves destroying the material. You can take a small sample of all the organisms involved and obtain their dry mass to avoid destroying the habitat. The total biomass of the population is then calculated from this sample. It takes more time to produce a pyramid of biomass than a pyramid of numbers but it gives you much more accurate information about what is happening in an ecosystem (see **fig D**).

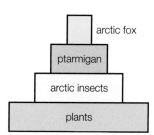

▲ **fig C** A pyramid of numbers for the simple Bear Island food chain shown in **fig A**. The number of individuals decreases as you move up the chain.

LEARNING TIP

If a pyramid of numbers is not the normal pyramid shape, it is probably because the organisms in the different levels are very different sizes.

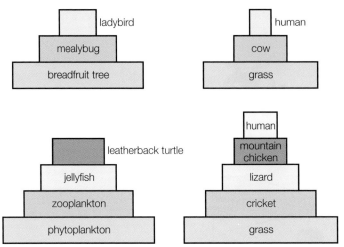

▲ **fig D** Pyramids of biomass give a more accurate picture of what is happening in a food chain than pyramids of numbers do.

PYRAMIDS OF ENERGY

Pyramids of biomass also have their limitations. For example, if the biomass of the organisms in a sample of water from the sea is analysed, the biomass of photosynthetic phytoplankton appears less than the biomass of the zooplankton that eat it. This is because the sample only shows the ecosystem at a single moment. It cannot show that the phytoplankton reproduces much more rapidly than the zooplankton. The biomass of the total phytoplankton population at any one time is smaller than that of the zooplankton, but the turnover of the phytoplankton is much higher and so the biomass over a period of time is much greater. A pyramid of observations over time gives the most accurate model of what is happening in an ecosystem. This is what we try to do with **pyramids of energy**.

The energy in an ecosystem remains the same at every level. It is the size of the different energy stores that changes. Less energy is stored in the organisms and more is stored in the surrounding atmosphere as you move along a food chain. The most accurate way to represent a food chain is to use a pyramid of energy (see **fig E**). The amount of energy stored in the organisms decreases at every trophic level along a food chain and, at the same time, the energy store of the surroundings increases. However, pyramids of energy are extremely difficult to measure and often involve an old model of energy, so pyramids of biomass are still widely used.

You can see the different models for the same food chains in **fig E**.

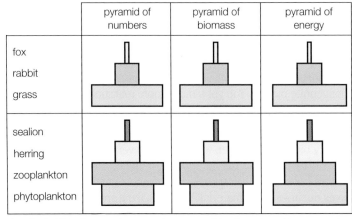

▲ **fig E** This diagram shows you how pyramids of numbers, biomass and energy compare for two different food chains.

LOSSES ALONG A FOOD CHAIN

In **Chapter 5A**, you saw that plants (producers) use light energy in photosynthesis and make a chemical energy store in their cells that is available to herbivores. But only a relatively small proportion of this becomes new animal material. What happens to the rest of it?

- The animal cannot break down and use everything it eats, so some energy is lost to the animal in undigested food and is expelled as unused material in the faeces.
- Much of the material that is digested is used to drive respiration, a series of reactions that result in the production of ATP. This is an exothermic process which heats the tissues of the animal and the surrounding atmosphere.
- Some of the plant material is lost in metabolic waste products such as urea (see **fig F**).

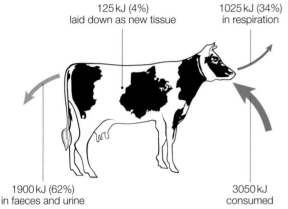

125 kJ (4%)
laid down as new tissue

1025 kJ (34%)
in respiration

1900 kJ (62%)
in faeces and urine

3050 kJ
consumed

▲ **fig F**   The fate of energy in the food eaten by a primary consumer. Similar losses occur in other consumer trophic levels up the food chain.

Only a small amount of the chemical energy store of the plant becomes new animal material and, therefore, part of the energy store of the animal. The rest is lost to the surroundings, increasing the internal energy store of the universe. The process of making new animal biomass is known as **secondary production**.

Similar energy transfers to the surroundings occur between animal trophic levels, from herbivores to primary consumers, and so on up the food chain. The energy store in the biomass of one organism compared with the energy store that ends up in an organism in the next trophic level is a measure of the efficiency of energy transfer. This measure is often quoted as being around 10%, but it varies greatly. Studies have shown it can be as low as 0.1% for some small herbivorous mammals and as great as 80% for some microorganisms. It is most commonly between about 2 and 24%. This wide variation depends on many factors, including the effort required to find food, the digestibility of the food and the metabolic rate of the organism.

EXAM HINT

Remember that biomass contains energy. Therefore, biomass and energy are closely linked. When biomass is respired, energy is released.

DID YOU KNOW?

Endotherms and exotherms

Endotherms (mammals and birds) use food resources to maintain their body temperatures at a constant level, regardless of the external temperature. They increase the internal energy store of their surroundings through radiation, convection and conduction. Consequently, endotherms need to eat a lot more food than exotherms to achieve the same increase in biomass. This is an important consideration in sustainable food production. It takes a lot less plant material to produce fish or insect protein than it does to produce beef or lamb. However, many people eat more meat than fish, particularly in more economically developed countries, and few people want to eat insects.

CHECKPOINT

1. Name the trophic levels in an ecosystem and explain why they are useful.
2. Summarise the advantages and disadvantages of using pyramids of numbers, pyramids of biomass and pyramids of energy as representations of ecosystem structure.

SUBJECT VOCABULARY

food chain a simple way of modelling the feeding relationships between a series of organisms in an ecosystem

trophic level a term which describes the position of an organism in a food chain or web and its feeding relationship with other organisms

producers organisms that make food by photosynthesis or chemosynthesis

primary consumers organisms that eat producers, either plants or algae

secondary consumers animals that feed on primary consumers

tertiary consumers animals that feed on secondary consumers (i.e. they eat other carnivores); they are usually the top predators in an area

decomposers the final trophic level in any set of feeding relationships; these are the microorganisms such as bacteria and fungi that break down the remains of animals and plants and return the mineral nutrients to the soil

pyramid of numbers a model of feeding relationships that represents the numbers of organisms at each trophic level in a food chain

pyramid of biomass a model of feeding relationships that represents the biomass of the organisms at each trophic level in a food chain

pyramid of energy a model of feeding relationships that represents the total energy store of the organisms at each trophic level in a food chain

secondary production the process of making new animal biomass from plant material that has been eaten

LEARNING OBJECTIVES

■ Understand the relationship between gross primary productivity (GPP), net primary productivity (NPP) and plant respiration (R).
■ Be able to calculate net primary productivity.

In **Section 5C.1**, you saw that the starting point in most ecosystems is the effect of light from the Sun on the reactions of photosynthesis. This determines the rate at which producers make organic material and this consequently determines the production of biomass within those ecosystems. Only a very small percentage of the light from the Sun actually results in the production of plant material (see **fig A**). Figures vary, but between 1 and 3% is generally accepted.

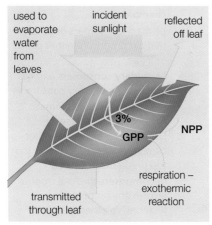

▲ **fig A** Most of the light that reaches a leaf is not available for photosynthesis to produce new plant tissue.

GROSS AND NET PRIMARY PRODUCTIVITY

Gross primary productivity (GPP) in plants is the rate at which plants make new material by photosynthesis. We can measure it in units of biomass area time, such as simple $g\,m^{-2}\,year^{-1}$ or gC (grams of carbon assimilated) $m^{-2}\,year^{-1}$. We can also measure it in units of energy, for example, $kJ\,m^{-2}\,year^{-1}$. Measuring productivity is very difficult. It usually involves finding the mass of representative samples of biomass and then multiplying them to represent a whole ecosystem. Theoretical calculations can be used to convert biomass to energy rather than using empirical measurements, so biomass is the better measure to use.

Plants use at least 25% of the material they produce for their own metabolic needs. Most importantly they respire, breaking down glucose to produce ATP. This is an exothermic process, so it increases the internal energy store of the surroundings (warms the surroundings). The rest of the material is stored as new tissue in plants. This energy store is known as the **net primary productivity (NPP)**.

$$NPP = GPP - R \text{ (where R = losses due to respiration)}$$

EXAM HINT

Remember that net primary productivity is the gross primary productivity minus the losses to respiration.

The NPP of different ecosystems has been estimated. It depends on all the abiotic factors and biotic factors that affect plant growth within the ecosystem. Latitude is important, because the light levels within a specific area are lower at latitudes nearer the pole than closer to the equator (see **fig B**).

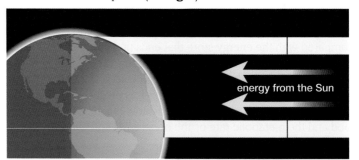

▲ **fig B** Light intensity on a square metre of the Earth's surface at the equator is greater than nearer the Poles because the curvature of the Earth means that the same amount of light is spread out over a larger area of the surface at the Poles.

In **fig C** you can clearly see the importance of factors such as the ambient temperature and availability of water within an ecosystem on the NPP of an area. If you combine the NPP of each type of ecosystem with the area of the Earth's surface it covers, you can see how much each ecosystem type contributes to the overall NPP of the Earth. For example, tropical rainforests cover only about 5% of the Earth's surface, but produce over 30% of the global NPP. The oceans contribute a similar proportion to the global NPP, but only because they are so large (see **fig C**).

The human population of the world is growing at a rapid rate, but even at its current level humans are already consuming up to 40% of the NPP of the planet.

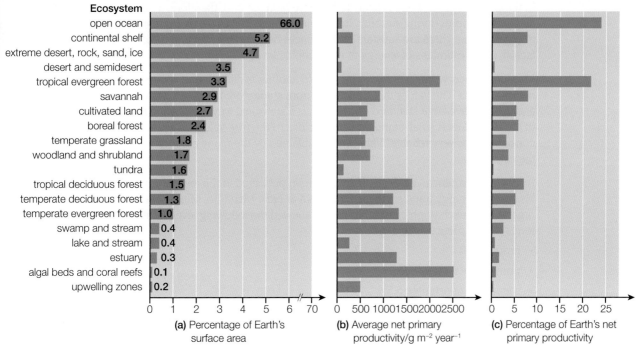

▲ **fig C**   Different ecosystems contribute different amounts to the NPP of the Earth.

ENERGY TRANSFERS TO HIGHER LEVELS

When we look at the efficiency of energy transfer between trophic levels, it is important to include the decomposers as well as the more obvious herbivores and carnivores. Decomposers break down and accumulate the bodies of dead animals and plants, and the faeces of animals. So the calculations look different again if we take into account the transfer of biomass or energy to decomposers. The rate of decomposition and, therefore, the rate of transfer of biomass and energy varies, depending on factors such as temperature, water availability and the season of the year, especially in temperate zones. These considerations add to the difficulty of calculating the energy transfer through an ecosystem.

MEASURING ENERGY TRANSFERS

How is the percentage energy or biomass transfer calculated? This involves a simple percentage calculation.

WORKED EXAMPLE

$96\,680\,kJ\,m^{-2}\,year^{-1}$ of energy is available in a plant source per annum.

$6780\,kJ\,m^{-2}\,year^{-1}$ per annum is transferred into new herbivore material at the next trophic level; the remainder is transferred to the surroundings.

What percentage of energy is transferred from the plants to the herbivore per year?

$$\frac{6780}{96680} \times 100 = 7\%$$

EXAM HINT

Many students find even simple calculations difficult under examination conditions. In this case, the most common error is not knowing what number to use as the denominator (the number below the line in the fraction). Remember that you need to divide the energy in the upper trophic level by the energy in the lower trophic level.

THE LIMITATIONS OF EFFICIENCY CALCULATIONS

It is very difficult to measure energy transfers for whole ecosystems because many assumptions need to be made in calculating the energy in all the organisms, so it is not done very often. First, you need to be certain that you have identified all the most abundant species in the ecosystem. Then, you also need to know how many there are, what an average body size is, how much energy that body size represents, how much of the biomass is transferred into decomposers at any stage of the life of the animal or plant, and so on.

In temperate regions, there is likely to be more biomass and, therefore, a bigger energy store in plant and animal bodies during summer. In the winter, there is likely to be more in the decomposers. In tropical regions, there are substantial differences between the wet and the dry seasons. All of these factors affect estimates of energy at each trophic level.

Measuring energy transfers from one level to another is difficult and is often based on biomass measurements. The values that are quoted from studies are the mean values from the calculations and usually have large standard errors. This is a statistical way of showing how close the mean is to the real value. A small standard error is close to the true value, whereas a large standard error is considered not very reliable. This can be explained by the many assumptions that are made during the calculations.

In 1942, the American ecologist R.L. Lindeman published in the journal *Ecology* the results of his study of energy transfers in an ecosystem at Lake Mendota, Minnesota. He looked at all the organisms at each trophic level. His results showed the following levels of efficiency in the trophic levels of that particular system:

- producers 0.4%
- primary consumers 8.7%
- secondary consumers 5.5%
- tertiary consumers 13%.

A famous study on energy transfers within an ecosystem was carried out at Silver Springs, Florida, by H.T. Odum in 1957 (see **fig D**).

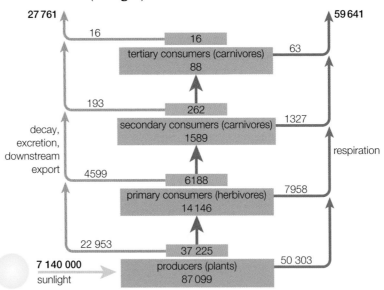

fig D The complex energy interactions between different trophic levels in the river at Silver Springs, Florida.

EXAM HINT

After any calculation remember to use the sanity check. Does your answer seem about right? In many cases, the percentage transfer between trophic levels is around 10%.

For some years, scientists considered that the efficiency of transfer between trophic levels followed specific patterns. Many different studies on energy transfer efficiency between trophic levels have provided very different and wide-ranging values. Scientists now recognise that the efficiency of energy transfer varies greatly and depends on the physiology and behaviour of the organisms involved and the climatic conditions, rather than specific positions in a food chain.

ENERGY TRANSFER AND FOOD CHAIN LENGTH

You have seen that relatively little of the light falling on the Earth is used to produce plant material, and increasingly small amounts of that material are passed from one trophic level to the next in a food chain. One of the main effects of the relatively inefficient transfers through food chains and food webs is to limit the number of trophic levels. At higher trophic levels, the organisms usually need to move over larger distances so, by the fourth or fifth trophic level, it could take more energy to find food and a mate than is needed for growth and reproduction. This would make survival impossible. Some scientists suggest this explains why food chains in ecosystems in tropical regions are generally longer than food chains in polar regions because tropical regions receive more light. However, it is difficult to isolate individual food chains in tropical systems because of the complexity of food webs, so it is hard to confirm or show this idea is wrong.

CHECKPOINT

1. (a) Looking at the data in **fig C**, state which type of ecosystem is most productive by percentage of the area of the Earth that it covers.

    (b) State which type of ecosystem is least productive, based on the same comparison.

    (c) Explain why primary production in the open oceans is so important.

2. (a) Calculate (to 1 decimal place) the efficiency of the net energy transfers between the trophic levels shown in the ecosystem at Silver Springs (see **fig D**).

    (b) Calculate the average energy transfer for this system.

3. ▶ (a) Look at **fig D**. Suggest why Odum's diagram shows only trophic levels and not the individual species in the food web. Give as many reasons as you can.

    (b) State what assumptions Odum would have made in order to create this diagram.

SKILLS   CREATIVITY

SUBJECT VOCABULARY

gross primary productivity (GPP) in plants, the rate at which light from the Sun catalyses the production of new plant material, measured as $g\,m^{-2}\,year^{-1}$, $g\,C\,m^{-2}\,year^{-1}$ or $kJ\,m^{-2}\,year^{-1}$

net primary productivity (NPP) the material produced by photosynthesis and stored as new plant body tissues; that is, NPP = GPP – R (where R = losses due to respiration)

LEARNING OBJECTIVES

■ Understand how knowledge of the carbon cycle can be applied to methods to reduce atmospheric levels of carbon dioxide.

The inefficient transfer of energy between organisms is not a problem because there is a constant supply of fresh energy from the Sun. However, the same is not true for the other ingredients of life such as water, carbon and nitrogen. Complex cycles have evolved which ensure that the chemical components of life are continually cycled through ecosystems. These cycles involve:

- a biotic phase, during which the inorganic ions are incorporated in the tissues of living things
- an abiotic phase, during which the inorganic ions are returned to the non-living part of the ecosystem.

THE CARBON CYCLE

In **Chapter 1A (Book 1: IAS)**, you saw that carbon is fundamental to the formation of the complex organic molecules such as carbohydrates, proteins, fats and nucleic acids, which are the building blocks of life. There is a massive pool of carbon in carbon dioxide which is found in the atmosphere and dissolved in the water of rivers, lakes and oceans. This carbon dioxide is absorbed and the carbon is incorporated into complex compounds in plants during photosynthesis. The carbon then passes to animals through food chains. Carbon dioxide is continually returned to the atmosphere or water through the process of respiration.

We can best summarise the interactions of the **carbon cycle** in a diagram (see **fig A**). You will look at the microorganisms involved in the carbon cycle in more detail in **Chapter 6C**. If you understand the carbon cycle and how it is regulated in the natural world, it will help you recognise the impact of changes in carbon dioxide levels in the atmosphere and help you understand how we might be able to reduce the increasing levels of atmospheric carbon dioxide.

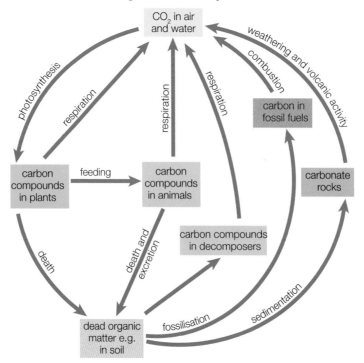

LEARNING TIP

It is often easiest to draw a cycle from memory if you use one colour for the basic cycle involving photosynthesis and respiration and then use other colours to add on other parts of the cycle.

▲ **fig A** The carbon cycle in nature.

CARBON SINKS

Fig A shows you that there are massive abiotic and biotic **carbon sinks** in nature. These are reservoirs where carbon is removed from the atmosphere and locked into organic and inorganic compounds. Carbon is removed from the atmosphere by photosynthesis and stored in the bodies of living organisms in the biotic part of the system. Soil also contains large quantities of organic, carbon-rich material in the form of humus. Rocks such as limestone and chalk, and fossil fuels such as coal, oil and natural gas, hold enormous stores of carbon in the abiotic part of the system.

The oceans also act as massive reservoirs of carbon dioxide and contain around 50 times more dissolved inorganic carbon than is present in the atmosphere. The carbon dioxide is in continual exchange at the surface between the air and water. The phytoplankton that live in the surface waters of the oceans need to absorb carbon dioxide from the water for photosynthesis (see **fig B**). The calcium carbonate shells produced by many different marine organisms and also found in coral reefs store large amounts of carbon. By lowering the concentration of dissolved carbon dioxide, these organisms make it possible for more carbon dioxide from the air to be absorbed by the water.

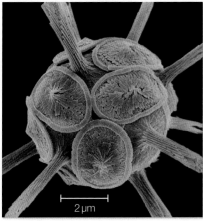

2 μm

▲ **fig B**　Each year, over half of all photosynthesis occurs in microscopic phytoplankton like this, found in the oceans across the world.

The Atlantic Ocean is a particularly important carbon sink and it absorbs up to 23% of carbon produced by humans each year. The Southern Ocean covers a much bigger area but only contains 9% of the total carbon. The differences are due to a variety of factors, including water temperature and ocean currents in the Northern (Atlantic) Ocean which move carbon-rich water downwards and bring water up from the depths to absorb more carbon.

We measure the quantity of carbon stored in the different carbon sinks in petagrams. One petagram is 10^{15} g or 1 billion tonnes. So, for example, photosynthesis removes around 110 petagrams of carbon each year from the atmosphere into the bodies of living organisms. Respiration of living organisms returns about 50 petagrams of carbon dioxide to the atmosphere, and the process of decomposition involves another 60 petagrams. The carbon becomes part of the organic material in the soil (the soil sink) before eventually being released back to the atmosphere as carbon dioxide. **Table A** shows you estimates of the main carbon stores on Earth.

The carbon cycle regulates itself. In other words, there is a balance between the amounts of carbon released in respiration and other natural processes and those absorbed in photosynthesis so that atmospheric carbon dioxide levels remain relatively steady.

SINK	AMOUNT IN PETAGRAMS
atmosphere	from 578 (in 1700) to 766 (in 1999)
soil organic matter	1500–1600
ocean	38 000–40 000
marine sediments and sedimentary rocks	66 000 000–100 000 000
terrestrial plants	540–610
fossil fuel deposits	4000

table A　Estimated major stores of carbon on the Earth

THE HUMAN INFLUENCE

Humans were probably fairly carbon-neutral until the last few hundred years. There has been an enormous increase in the production of carbon dioxide by people since the Industrial Revolution in the 18th and 19th centuries. The industrially related output of carbon dioxide, combined with the development of the internal combustion engine for cars and other vehicles, is now threatening the balance of the carbon cycle. Scientists are collecting evidence that the level of carbon dioxide in the atmosphere is increasing and many predict that this could have major effects on climate, geology and the distribution of organisms.

We will look at some of these effects in the following sections. Understanding the carbon cycle will help you consider models of ways in which atmospheric carbon dioxide levels may be reduced.

CHECKPOINT

SKILLS　CRITICAL THINKING

1. Given that combustion is a natural process, explain how humans are destabilising the natural carbon cycle.

2. In **table A**, the amounts in many of the different carbon sinks are shown as ranges. Suggest reasons why the quantities in the different sinks have changed over time.

SUBJECT VOCABULARY

carbon cycle a series of reactions by which carbon is constantly recycled between living things and the environment
carbon sink a reservoir where carbon dioxide is removed from the atmosphere and locked into organic or inorganic compounds

LEARNING OBJECTIVES

■ Understand the causes of anthropogenic climate change, including the role of greenhouse gases in the greenhouse effect.

Carbon dioxide is one of the **greenhouse gases**; others include methane and water vapour. These gases have a very important role in the atmosphere and are necessary for life on Earth.

Greenhouse gases reduce heat loss from the surface of the Earth in a way that is similar to how glass panels reduce heat loss from a greenhouse. This is known as the **greenhouse effect** (see **fig A**). When radiation from the Sun reaches the Earth, some is reflected back into space by the atmosphere and by the surface of the Earth and some is absorbed by the atmosphere. The key wavelength is infrared. This is the radiation that we feel as warmth. Infrared radiation that reaches the Earth's surface has a relatively short wavelength. This is absorbed by the surface of the Earth and then light is radiated from the surface at a longer wavelength. Some of this radiation is absorbed and re-radiated back to the Earth's surface by greenhouse gas molecules in the atmosphere. This maintains the temperature at the surface of the Earth at a level which is suitable for life. Without greenhouse gases in our atmosphere, the surface of Earth would probably have an average surface temperature of −63 °C instead of 14 °C and look a lot more like the surface of Mars.

However, in theory, if the levels of greenhouse gases increase, the 'greenhouse' becomes more effective and the temperature of the atmosphere will rise.

EXAM HINT

Many students think only of the greenhouse effect as harmful. Don't forget that it is essential to life on Earth. It is the *enhanced* greenhouse effect caused by human activities such as overuse of fossil fuels that is harmful.

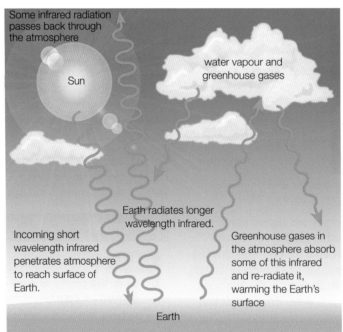

Some infrared radiation passes back through the atmosphere

Sun

water vapour and greenhouse gases

Earth radiates longer wavelength infrared.

Incoming short wavelength infrared penetrates atmosphere to reach surface of Earth.

Greenhouse gases in the atmosphere absorb some of this infrared and re-radiate it, warming the Earth's surface

Earth

▲ **fig A** The greenhouse effect.

GLOBAL WARMING, CLIMATE CHANGE AND WEATHER

There are a number of different terms which are used when we discuss the effects of people on the ecosystems we live in. It is important to understand the differences between them and to use the appropriate term.

- **Global warming** is a measurable increase in the temperature of the Earth's atmosphere or temperature at the surface of the Earth. Global warming and global cooling have happened many times in the history of the Earth, but the increase in the temperature of the Earth's surface that we are currently observing is unusually rapid and linked to human activities which increase the levels of greenhouse gases in the atmosphere.

- **Climate** is the average weather in a relatively large area (such as a country) over a long period of time. Measurements that are used include mean temperature, precipitation (the amount of rain, snow, sleet, hail, etc. that falls), wind, humidity and atmospheric pressure over a period of time. The World Meteorological Organization defines the period of time over which we measure the climate as 30 years, but we can make measurements over much longer periods of time.

- **Weather** is the conditions in the atmosphere at any particular time. So, for example, it might be sunny or windy or rainy when you go outside. Weather forecasts describe the expected weather on a particular day. Weather can be very variable from day to day, whereas the climate of an area is expected to remain relatively stable over time. There is a popular saying 'Climate describes what you expect, weather is what you get.' There have been an increasing number of what are known as extreme weather events around the world in recent years. In these events, the weather is much more extreme than normal and may include more and stronger cyclones, longer and more severe droughts, or more violent storms and flooding (see **fig B**). There is increasing evidence that this increase in severe weather events is linked to global warming and possible climate change.

▲ **fig B**   In December 2017, cyclone Ockhi hit Sri Lanka and India causing massive damage and the loss of several hundred lives. It then moved on as a major storm to other areas, including the Maldives.

- **Climate change** describes a large-scale change in global or regional weather patterns that occurs over a period of many years. So, for example, the average temperature of the Earth's surface has been 14 °C for around 11 000 years, but it increased by 0.89 °C between 1901 and 2012. This is a very rapid change in terms of the history of the Earth. Rainfall patterns are also changing over time. These shifts in weather patterns over time are what we call climate change, but because by definition climate change happens over long periods of time it is not easy to measure or draw rapid conclusions. There are natural variations in the weather, and a single year that is unusually hot, dry or wet does not count as climate change. But if the average temperature, or amount of rainfall, varies in the same way over many years, then climate change may be happening.

THE ROLE OF METHANE

Methane is a potent greenhouse gas. Over a period of 20 years, it has an effect on warming the atmosphere that is about 72 times

greater than carbon dioxide (CO_2). Fortunately, much less methane is produced than carbon dioxide. Its main sources are from the decay of organic material by some kinds of bacteria, particularly in wet conditions, and from the digestive systems of ruminant herbivores such as deer and cows. It naturally breaks down high in the atmosphere in a series of reactions that eventually form carbon dioxide and water molecules.

Methane levels have risen by about 150% since 1750 for several reasons. Rice paddy fields are waterlogged during much of the time the rice is growing. Bacteria in this waterlogged soil release methane as they grow. Levels of rice production have increased steadily to feed the ever-increasing world population, so more methane is produced. Further, the growth in human population has resulted in more animals that we depend on for food, including cattle. So the amount of methane released from their digestion increases too. Scientists have calculated that up to 60% of the methane in the atmosphere is now produced because of human activity in some way.

Table A shows the estimated contributions of the main greenhouse gases to the greenhouse effect. The variation depends on the amount of water vapour present and the concentrations of other greenhouse gases.

GREENHOUSE GAS	PERCENTAGE ESTIMATED CONTRIBUTION
water vapour	36–72
carbon dioxide	9–26
methane	4–9
ozone	3–7

table A   Contributions of greenhouse gases to the greenhouse effect

MANAGING METHANE EMISSIONS

DID YOU KNOW?

Managing methane, changing cows!

Cows burp a lot (release air from their stomachs via their mouths) and every time they burp, they release methane gas. It is estimated that a cow produces from 100 to 700 dm³ of methane per day. This varies depending on factors such as the breed of cow, the type of food eaten and whether the cow is giving milk. There are an estimated 1.2 billion cows in the world so a lot of methane is being produced. The Intergovernmental Panel on Climate Change (IPCC) estimates that 16% of the methane produced as a result of human activities comes from livestock (farm animals), and dairy cows produce the most.

A number of research teams are working to breed or engineer new strains of grass which cows can digest more easily, thus reducing methane emissions. And an Irish research team looked at whether changing the way we farm cows could reduce overall methane emissions. Older cows produce more milk, but less methane per litre. Therefore, the average methane emissions of the entire herd can be reduced by keeping cows alive, healthy and giving milk for longer.

If you add concentrates to the diet for cows, it reduces methane emissions per cow, because cows can digest the concentrates more easily. It also helps cows to produce milk for longer.

However, a balance needs to be found because generating the electricity needed in the manufacture of the concentrates produces carbon dioxide.

Current evidence suggests that with a combination of good husbandry (management of animals), careful breeding and possible genetic engineering of food plants, people may be able to drink milk, eat beef and reduce the production of methane at the same time.

If people can be persuaded to eat a wider variety of meat including more goat, sheep and chicken, and use more goat and sheep's milk, we can reduce the methane levels even more (see **fig C**).

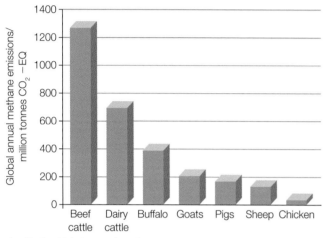

▲ **fig C**   Global annual methane emissions by species. Source: *Tackling climate change through livestock – A global assessment of emissions and mitigation opportunities.* 2013. Food and Agriculture Organization of the United Nations.

CHECKPOINT

1. The term *greenhouse effect* is widely used to suggest something negative. Explain why this is an inaccurate use of the term.

2. (a) State the difference between *weather* and *climate*.

    (b) Explain why a single extreme weather event such as a flood, a cyclone or very low temperatures cannot be used as evidence for global climate change.

> **SKILLS**   CRITICAL THINKING, PERSONAL AND SOCIAL RESPONSIBILITY

3. ▷ There is one very simple way of reducing the methane emissions from cattle. Briefly explain what it is and why you think it is not widely suggested.

SUBJECT VOCABULARY

greenhouse gases gases found in the atmosphere, including carbon dioxide, methane and water vapour, which are involved in the greenhouse effect

greenhouse effect the process by which gases in the Earth's atmosphere absorb and re-radiate the radiation from the Sun, which has been reflected from the Earth's surface, maintaining a temperature at the surface of the Earth that is warm enough for life to exist

global warming a measurable increase in the temperature of the Earth's atmosphere or temperature at the surface of the Earth

climate the average weather in a relatively large area (such as a country) over a long period of time

weather the conditions in the atmosphere at a particular time (for example, if it is sunny or windy or rainy when you go outside)

climate change a large-scale change in global or regional weather patterns that happens over a period of many years

LEARNING OBJECTIVES

■ Understand the different types of evidence for climate change and its causes, including records of carbon dioxide levels, temperature records, pollen in peat bogs and dendrochronology; recognise correlations and causal relationships.

The Meteorological Office in the UK has daily weather records since 1869, but written evidence from diaries and ships' logs go back for another 100 years. Their recent weather records, and records from other weather stations all over the world, suggest that the Earth's surface temperature is increasing.

All over the world the evidence is growing that:

- levels of carbon dioxide and other greenhouse gases in the atmosphere are increasing
- mean global temperatures are increasing (global warming)
- the global climate seems to be changing
- extreme weather events such as flooding, droughts and hurricanes are increasing.

TEMPERATURE RECORDS

In 1998, the Intergovernmental Panel on Climate Change (IPCC) collected a lot of data to produce a graph of temperature in the Northern Hemisphere, which they have updated regularly since then (see **fig A**).

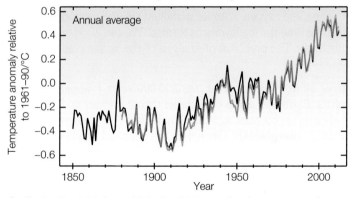

Observed globally averaged combined land and ocean surface temperature anomaly 1850–2012

▲ **fig A** The IPCC data published in 2012 uses data from a variety of sources to show how the temperature at the surface of the land and the oceans has changed since 1850.

We have data of measured temperatures only since the mid-1800s. Further temperatures are estimated from other data that can give an indication of the temperature but not an exact value. This has resulted in a famous graph known as the 'hockey stick graph' (see **fig B**). The fluctuating black line indicates the mean values. The other sources of data are called temperature proxies and the error lines shown in grey on the graph indicate how accurate we consider these values to be. Temperature proxies include tree rings, corals, ice cores and peat bog data.

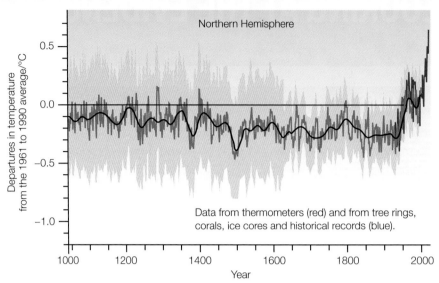

▲ **fig B** The IPCC 'hockey stick graph' originally published in 2001 uses data from a variety of sources to show how the temperature has changed over centuries. The term *hockey stick* comes from the shape of the graph. Note the grey error lines on each measurement (that is, the range in which the real value is thought to lie, as we cannot be very accurate).

Rising atmospheric carbon dioxide levels, global warming and climate change all threaten human life and the huge range of biodiversity on the planet. If we are to make sensible decisions and persuade governments, we need good scientific evidence to back up the theories of climate change. In this section, you will look at some of the evidence we use.

FROZEN ISOTOPES AND TEMPERATURE RECORDS

Antarctic and Greenland ice cores are often used as a source of temperature proxies. Scientists drill deep down into the ice and then analyse the air trapped in the different layers. This provides a record which goes back thousands of years. The oxygen isotopes in melted ice (the proportions of O^{18} to O^{16}) reflect the air temperature at the time the ice layer was formed. We can also use cores to measure atmospheric carbon dioxide levels. The precision of analysis of the air samples from the core ice is given as 0.2 ppm.

Fig C shows the results of the analysis of air from ice cores for over 300 000 years. It appears that about 140 000 years ago the surface of the Earth was about 6 °C cooler than it is today and the Earth was in an ice age. On the other hand, about 120 000 years ago the climate was 1–2 °C warmer than it is now. These warm periods are known as **interglacials**. Since then, we have had another ice age and some more warming.

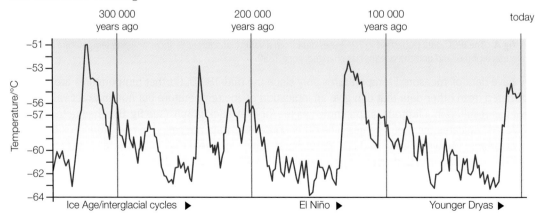

▲ **fig C** Data from ice cores are used to show how the temperature of the surface of the Earth has fluctuated over time.

DENDROCHRONOLOGY

Another temperature proxy is **dendrochronology** (more precisely, dendroclimatology). Dendrochronology is the dating of past events using tree ring growth. Trees increase in width as they get older by cell division of one particular layer in their trunks. When there is plenty of moisture and trees are growing quickly, for example in early summer in temperate regions, these new cells are large. The new cells that are produced in more difficult conditions are smaller. Eventually, growth stops for the year until the next spring. It is the contrast between the small cells at the end of one year and the large ones produced the next spring which give the appearance of rings. So by counting the rings, it is possible to find an approximate age for a tree or piece of wood. It is always approximate, because if conditions vary a lot during the year, a tree will produce more than one growth ring.

Dendroclimatology is the study of what dendrochronology can tell us about the climate in the past. However, growth in trees is dependent on many factors, including amount of sunshine, temperature, carbon dioxide levels and amount of rainfall, so evidence from dendrochronology may not always be accurate.

This is a relatively new science, but one which scientists hope will be very useful in the future as they refine their techniques and discover ways to determine which factors have affected the growth of trees in the past. One way to check the reliability of the data is by comparing results from different places (see **fig D**). If the rings are similar, then the climate was similar generally, not just in a small area. However, this does not explain what wider rings mean. Were they due to better climate, was the weather warmer, wetter or sunnier; was there more carbon dioxide, or was it a combination of factors?

Many people have questioned the validity of using tree-ring data to determine past temperature conditions and this has weakened some of the findings on climate change. Data from coral reefs can be used to confirm the evidence from trees, as the proportions of different isotopes taken up by the coral vary as the sea temperatures change and this gives another valuable proxy record of climate change. Scientists have found that there is considerable agreement among the different data sets.

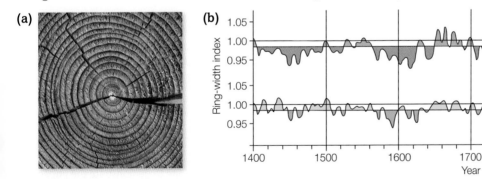

▲ **fig D** (a) Tree rings can tell us about the age of a tree, and much more. (b) Graphs of two trees of the same species but growing 200 km apart show very similar patterns, suggesting the changes affecting them happened over a large area.

PEAT BOG RECORDS

Another source of data for a temperature proxy comes from peat bogs. Peat bogs are made of partly decomposed plant material, mainly *Sphagnum* mosses. The peat is very acidic, cool and anaerobic, which prevents bacteria from decomposing organic material. As a result, peat preserves pollen grains, moss spores and even plant tissues. We can look back in time at the plants and mosses growing in and around that area from hundreds and even thousands of years ago by sampling cores of peat. As the types of plant that can grow in an area are affected by climate, the pollen/moss record can give a clear reflection of how the climate has changed with time. For example, the presence of cotton grass and some species of *Sphagnum* moss indicate cool wet conditions, whereas other species of *Sphagnum* and species of *Polytrichum* moss reflect a period of drier conditions in the bog. Pollen which has blown in from around the area adds to the picture. So pollen from horse chestnut trees indicates warmer conditions, whereas pollen from birch trees shows that it was cooler.

Peat growth rate depends on the prevailing conditions and varies widely. So, evidence from undisturbed peat bogs (and lake sediments) can give us a clear and unbroken record of the climate, and has resulted in a continuous record from about 10 000 years ago which gives clear evidence of periods of warming and cooling.

INCREASING DATA RELIABILITY

We can use both dendrochronology and peat bog dating to confirm radiocarbon dating. For example, we can date wood or peat bog samples of known age from radiocarbon measurements using the remains of plants and pollen grains. These give an indication of the climatic conditions at the time those plants were alive, which will link to historically recorded events such as a flood or extreme cold weather. We can compare the results to give a form of **calibration**. This gives scientists clear reference points which they can use to determine the accuracy of their estimations of age, making the data considerably more reliable.

In 2008, scientists looked at the hockey stick graph (see **fig B**, **page 66**) and recalculated the figures using more than 1200 temperature proxy records, going back 1300 years without using tree-ring data, and using two different statistical methods. They found that the hockey stick graph was valid whichever statistical method was used and whether tree-ring data was included or not.

EVIDENCE FOR INCREASING LEVELS OF CARBON DIOXIDE

Scientists have found evidence for the increasing levels of carbon dioxide in the atmosphere in many different ways. Some of the most famous evidence comes from what is known as the Keeling curve, a series of measurements which have been taken at regular intervals at the Mauna Loa observatory on Hawaii (see **fig E**). The air is sampled continuously at the top of four 7-metre tall towers and an hourly average of carbon dioxide concentration is taken (along with a number of other measurements). The air there is relatively free from local pollutants and scientists believe it is representative of the air in the Northern Hemisphere. Measurements started in 1958 and the monitoring methods and instruments used have remained very similar throughout that time. The records show that the level of atmospheric carbon dioxide has increased from 315.98 ppmv (parts per million by volume of dry air) in 1959 to 408.68 ppmv early in 2018. The annual fluctuations in the levels of carbon dioxide seem to be the result of seasonal differences in the fixation of carbon dioxide by plants, as in temperate regions plants lose their leaves in winter and absorb less carbon dioxide.

Ice core data also show clear changes in carbon dioxide concentration. **Fig E** shows data taken from the Law Dome ice cores. This is an area of particularly pure and undisturbed ice in the Antarctic. The shape of the curve is similar to the temperature curve, but it goes back much further in time.

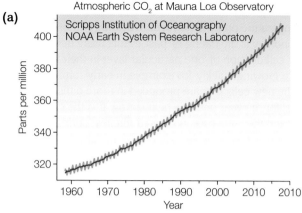

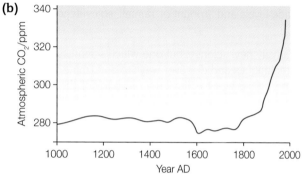

▲ **fig E**   Graph (a) shows levels of carbon dioxide in the atmosphere from the Mauna Loa data and graph (b) was produced using data from the Law Dome ice cores on the atmospheric carbon dioxide levels. They provide clear evidence of the rising trend in atmospheric carbon dioxide levels.

CHECKPOINT

1. Determine the overall percentage increase in atmospheric carbon dioxide from 1959 to 2006 based on the Mauna Loa data.
2. Why is the data from Mauna Loa regarded as reliable?
3. Explain what the data from the Law Dome ice cores can tell us that the Mauna Loa data cannot and comment on how reliable the data is.

   SKILLS   CREATIVITY

4. ▶ Some people use these data alone to demonstrate the impact of human activity on carbon dioxide levels in the atmosphere. Suggest why this is a misuse of the data.

   SKILLS   ANALYSIS

5. ▶ (a) Give a reason why the original 'hockey stick' graph produced by the IPCC in 1998 was challenged.

   (b) Explain how the 2008 update has helped to reassure many doubters.

SUBJECT VOCABULARY

interglacials the relatively warm periods between ice ages
dendrochronology the dating of past events using tree ring growth
calibration checking the measurement values given by one system of measurement against another of known accuracy

6 THE GLOBAL WARMING DEBATE: CORRELATION OR CAUSATION?

LEARNING OBJECTIVES

■ Understand the different types of evidence for climate change and its causes, including records of carbon dioxide levels, temperature records, pollen in peat bogs and dendrochronology; recognise correlations and causal relationships.

According to the Intergovernmental Panel on Climate Change (IPCC), atmospheric concentrations of carbon dioxide, methane and nitrogen oxides have increased to levels that have not been seen for at least 800 000 years. Carbon dioxide levels have increased by 40% since pre-industrial times. This is mainly as a result of burning fossil fuels and partly from changes of land use such as deforestation. The oceans have absorbed about 30% of the emitted **anthropogenic** (produced by people) carbon dioxide and this has caused increased ocean acidification. A lot of evidence from many studies now suggests a clear **correlation** between the rise in the levels of carbon dioxide and other greenhouse gases in the atmosphere and the increase in global temperatures. However, the correlation is so close (see **fig A**) that it is difficult to decide whether increases in greenhouse gases are *causing* the increase in temperatures or are the *result* of rising temperatures.

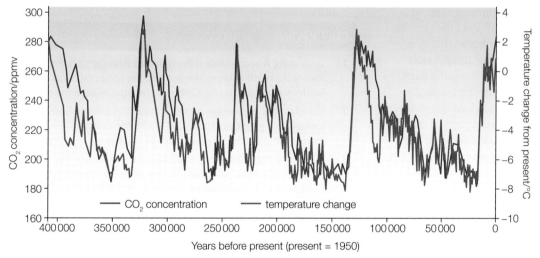

▲ **fig A**   Data from ice cores from Vostok (in the Antarctic) showing carbon dioxide content of the air trapped in the ice cores and the air temperature change relative to modern temperature based on measurements of the isotopic concentration of the gases.

IS IT CAUSAL?

To say that there is a **causal relationship** between rising carbon dioxide levels and rising temperatures (with the associated climate change) we need a mechanism to explain how one factor changes the other. There is a clear mechanism for greenhouse gases to raise the surface temperature of the Earth through re-radiation (see **Section 5C.4**). We understand how the natural greenhouse effect works. It seems logical to consider that humans are responsible for the increases in the levels of carbon dioxide in the atmosphere and an increase in global temperatures because the timing of both changes is so closely linked. In the last two centuries, we have burned increasing quantities of fossil fuels for energy and transport. More recently, we have burned fossil fuels in very large quantities to generate electricity. Every time we burn fossil fuels, carbon dioxide is produced and released into the atmosphere.

However, scientists must look at different theories. For example, some scientists have proposed a mechanism for global warming based on events at the Sun's surface. They suggest there is evidence that solar activity affects cloud formation and, therefore, surface temperature. Some data, such as that shown in **fig B**, seems to show a much closer correlation between solar activity and atmospheric temperature than that between carbon dioxide concentration and temperature. However, after looking at all the evidence, the IPCC reached the conclusion that sunspot and solar flare activities over the past 50 years would probably have produced cooling rather than warming. They decided that any influence these factors have on global climate change is small.

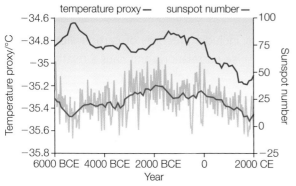

▲ **fig B** The relationship between solar activity and temperatures high in the atmosphere is not as simple as it may appear in this graph.

All of these theories about global warming and climate change are based on data that require detailed interpretation and the use of computer modelling to model a very complex system. It is almost impossible to prove a causal link, and we cannot conduct valid experiments on the atmosphere of our planet. But many studies on different aspects of global warming, using a wide variety of different computer models, suggest the same thing. They have looked at events ranging from polar ice melting to climate change in different regions of the world. All the studies suggest the increase in atmospheric carbon dioxide and other greenhouse gases is increasing surface temperature, and that human carbon dioxide emissions are responsible for at least some of the current warming (see **fig C**).

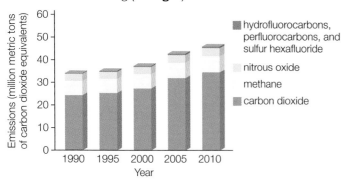

▲ **fig C** Global anthropogenic greenhouse gas emissions

In 2007, the IPCC looked at data and models of climate change presented by scientists from all around the world. They saw that anthropogenic carbon dioxide levels increased by 80% between 1970 and 2004, mainly due to the use of fossil fuels. The IPCC

finally decided that the balance of the evidence shows a 95% probability that human activities resulting in the build-up of greenhouse gases are at least partly responsible for the observed increase in global temperatures. In their 2013 report, they state that it is *extremely likely* that human influence has been the dominant cause of global warming since the mid-20th century.

The IPCC use language carefully designed to indicate how strongly the evidence backs up the hypothesis. They have decided that it is *very likely* that human activities have contributed to the rise in sea level in the second half of the 21st century, but only *likely* that these activities have also influenced the changes in wind and weather patterns that have been observed. At the moment, climate changes and environmental damage are occurring much faster than anyone imagined. It appears that some of the effects of human influences on global ecosystems through anthropogenic change cannot be reversed. The IPCC believe there is sufficient evidence now to state there is a causal link between anthropogenic production of greenhouse gases and global warming, but it will almost certainly become obvious that this is one of several causes. Global warming is likely to be multifactorial, but with anthropogenic greenhouse gases playing a very significant role.

CHECKPOINT

1. Look at **fig A** and explain if the different data support the theory that carbon dioxide levels and the temperature at the Earth's surface are linked. Comment on how reliable the data is.

2. Calculate the percentage increase in carbon dioxide from fossil fuel use using **fig C** and compare it with the overall percentage increase in greenhouse gases from all sources.

3. (a) Look at **fig B**. Research more about the evidence put forward for the 'solar flare' theory of global warming. Give the main strengths and weaknesses of this line of evidence.

    SKILLS   ANALYSIS

    ▶ (b) Compare the correlation between solar activity and atmospheric temperatures with the correlation between carbon dioxide levels and temperature as seen in **fig A** and any others you may find. Discuss the validity of the data and the strength of the correlations as a basis for the feasibility of any causal relationships.

SUBJECT VOCABULARY

anthropogenic produced by people
correlation a strong tendency for two sets of data to vary together
causal relationship one event happens as a direct result of another, with a clear mechanism by which one factor causes a given change

LEARNING OBJECTIVES

■ Understand that data can be extrapolated to make predictions and that these are used in models of future climate change.

■ Understand that models for climate change have limitations.

Scientists have developed huge computer models in order to investigate the historical relationship between carbon dioxide and climate change, especially relating to the Earth's temperature. These models have taken years to develop and scientists are still refining them as new research shows how different factors interact. There are many factors to consider: rates of photosynthesis across the world; rates of carbon dioxide production by natural causes; the exchange of carbon dioxide between the atmosphere and oceans; the effect of changing temperature on all of these. Scientists hope that by making models that fit the data from the past as accurately as possible, the models will provide reasonable tools for predicting the future.

CLIMATE CHANGE

In **Section 5C.4**, you saw that the term *weather* describes the state of the atmosphere at a particular time and place, with regard to features such as temperature, rainfall, humidity and level of wind. Climate is the mean weather pattern in an area over many years. Rising temperatures in the Earth's atmosphere affect weather and rainfall patterns and can also cause long-term changes in the climate. It is impossible to link any one weather event to global warming. However, statistical evidence suggests that there is an increase in the number of extreme weather events around the world and this does appear to be linked to the measured rise in global temperatures.

Rainfall patterns are complex, but they also seem to be changing. For example, there have been a number of years of lower than expected rainfall in Africa. In 2013, around 200 million people (about 25% of the population of Africa) experienced high water stress. It is predicted that if the low rainfall pattern continues, then by the year 2050 between 350 million and 600 million people in Africa will be short of water for their crops and for drinking. In contrast, in some areas rainfall has been both higher than usual and extremely heavy. This leads to flooding, which carries away vital topsoil. We have already seen a clear increase in very heavy rain leading to severe flooding in areas of China, Pakistan and India over several years. Low-lying communities and countries around the world are threatened by rising sea levels (see **fig A**).

▲ **fig A** St Mark's Square in Venice is frequently flooded in winter. Venice is one of many places that could disappear completely if global warming continues.

PREDICTING THE FUTURE

We can **extrapolate** the data on greenhouse gases and use them in models to make predictions about what will happen to temperature and other aspects of global climate in the future. We can use these extrapolations in other models to predict the long-term effects of increased temperature on the environment. These models can be very useful, as long as we have some evidence that their predictions are valid. So, for example, an international group of scientists collected monthly recorded rainfall measurements from around the world from 1925 to 1999 and compared what really happened to what various computer models predicted would happen as a result of global warming. They found that many of the changes they observed corresponded to those predicted if global warming was a factor affecting climate change (see **fig B**).

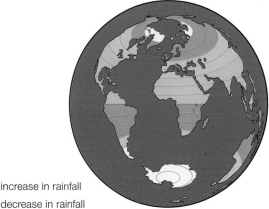

■ increase in rainfall
■ decrease in rainfall
■ disagreement between observed rainfall and climate models
□ insufficient data

▲ **fig B** This diagram, published in the journal *Nature*, shows how measured rainfall corresponded to climate change models in one piece of well-respected research.

Scientists are using models more and more as they become convinced of their reliability. For example, the IPCC is using some of the predictions shown in **fig C** to help plan international responses to the problems of rising carbon dioxide levels and global warming. But there are many limitations with even the best models. This is because it is impossible to tell the exact impact of carbon dioxide on global warming and also impossible to predict the impact of global warming on particular aspects of the world climate. In addition, extrapolations from past data cannot take into account unknown factors in the future, including how current trends in resource usage and technologies may change. So, assuming that the models are reasonably accurate and assuming that the predictions of temperature in the future based on carbon dioxide emissions are realistic, what effect might this have? You can see the results of one set on models in **fig C**.

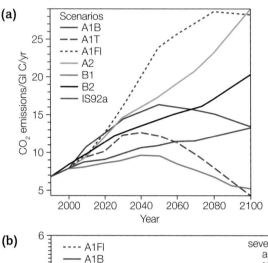

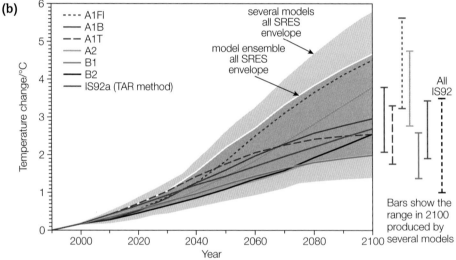

▲ **fig C** (a) The lines on this graph predict possible global carbon dioxide emissions in gigatons of carbon dioxide per year. (b) These figures are entered into other models that predict average global temperature compared with pre-industrial levels. As you can see there are large margins of error!

RISK OF FLOODING

In 2002, 500 billion tonnes of ice broke away from the Antarctic peninsula and eventually melted into the sea. Antarctic temperatures have increased by an average of 2.5 °C in the past 50 years. This is faster than anywhere else on the Earth. Many scientists believe that the fact that the ice is becoming thinner is a clear indication of global warming. The Antarctic ice contains around 70% of the world's fresh water.

The Arctic is changing too. The Arctic sea ice has been reducing by about 2.7% each decade since 1978 and many glaciers are also becoming smaller at a rate of about 50 m a year. The volume of water in the seas and oceans of the world will increase as the Antarctic ice melts, causing sea levels to rise. And as the water becomes warmer, its volume increases, resulting in an even bigger impact on sea levels. The implications for human life with the rise in sea levels are immense. There are around 100 million people who live less than 1 metre above current sea levels. You can see data on how various factors interact to change sea levels in **fig D**. This is the sort of information used to build the predictive models needed to help us plan for an uncertain future.

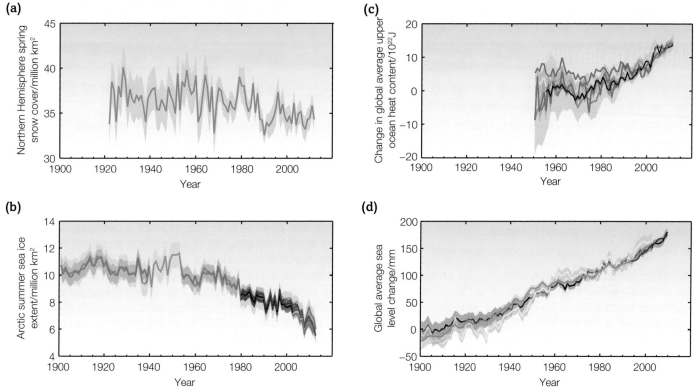

▲ **fig D** The reduction in snow cover (a), the melting of sea ice (b), the change in ocean temperature (c) and the rise in sea levels (d) all appear to be linked to anthropogenic climate change.

The risk of global warming to countries like the Republic of Maldives becomes very obvious with data such as those in **fig D**. These beautiful islands are the flattest country on Earth; the highest point is only 2.4 m above sea level. It is estimated that by 2100 around 77% of the islands' land mass will be under water because of rising sea levels. This will destroy the human, plant and animal populations that live there. Rising sea temperatures also threaten to kill the coral reefs that surround the islands. These reefs are a major source of global biodiversity, so the whole world will suffer from their loss.

CHECKPOINT

1. The models for global carbon dioxide stabilisation show a great deal of uncertainty and variety. Explain why the models are so uncertain.

2. (a) Using data from **fig D**, calculate the following:

    (i) the increase in average global heat content of the oceans per year between 1950 and 2010

    (ii) the percentage reduction in the extent of Arctic summer sea ice between 1950 and 2010

    (iii) Northern Hemisphere snow cover between 1930 and 2000.

    (b) What was the increase in average sea levels over the period 1950–2010?

    ▶ (c) Suggest how these data might be linked to rising carbon dioxide levels, to the decreasing snow cover in the Northern hemisphere and to each other.

> **EXAM HINT**
>
> Most students recall the effect of melting ice on sea levels. However, many forget that water expands as it warms up and this has a huge effect on sea levels as the global climate warms.

SKILLS ▶ INNOVATION, CREATIVITY

SUBJECT VOCABULARY

extrapolate apply already known trends to unknown situations to predict what will happen

8 THE BIOLOGICAL IMPACT OF CLIMATE CHANGE

LEARNING OBJECTIVES

■ Understand the effects of climate change (changing rainfall patterns and changes in seasonal cycles) on plants and animals (distribution of species, development and life cycles).

■ Understand the effect of temperature on the rate of enzyme activity and its impact on plants, animals and microorganisms, to include Q_{10}.

In the previous sections, you have looked at the evidence for anthropogenic climate change. This is happening because of:

- the increased levels of greenhouse gases added to the atmosphere by people around the world
- the way we destroy carbon sinks such as rainforests.

You have considered the effects of these changes on our climate, and on extreme weather events, and begun to think about how those changes might affect living organisms, including ourselves. Now you will look at some of these effects of climate change in more detail.

CHANGES IN TEMPERATURES AND SEASONS

Climate change already involves changes in temperature. In many parts of the world, this change will be an increase in temperature, but some places will get colder. Temperature has an effect on enzyme activity, which in turn affects the whole organism. In **Section 2B.2 (Book 1: IAS)**, you saw that a 10 °C increase in temperature will double the rate of an enzyme-controlled reaction. This effect of temperature on the rate of any reaction can be expressed as the **temperature coefficient (Q_{10})**.

You will remember that Q_{10} for any reaction between 0 °C and 40 °C is 2. This means that a 10 °C rise in temperature produces a doubling of the rate of reaction within the temperature range where most living things live. Similarly, a decrease in temperature will slow reactions down.

However, there is an **optimum temperature** for many enzyme-controlled reactions and if the temperature increases beyond that point the enzyme starts to **denature** and the reaction rate falls. As a result, increasing temperature could have different effects on processes, including the rate of growth and reproduction. If plants grow faster they will take up more carbon dioxide and may therefore reduce atmospheric carbon dioxide levels. In other places, temperature may exceed the optimum for some enzymes, and organisms there will die.

The majority of plant and animal species live in the tropics. Many of these species have very little tolerance for change because conditions in the tropics generally vary very little throughout the year. The temperature remains relatively stable, and so does day length. The main difference is in the amount of rain that falls in the wet seasons and the dry seasons. In contrast, temperate species are adapted to a range of temperatures of 40 °C or more through the year. Desert animals are adapted to huge temperature ranges within a single day. Tropical organisms, therefore, have no selection pressure to be able to adapt to changing conditions. A rise or fall in temperature of just a few degrees can mean the extinction of thousands of species of plants and animals. Experimental data suggest that a change of just two degrees could be fatal to many species. Insects, which are vital as pollinators of the many flowering plants, are particularly vulnerable. If they die, so do the plants, and then the animals which feed on the plants.

EXAM HINT

Remember that you will be expected to relate your studies at IAL back to what you learned in **Section 2B.2 (Book 1: IAS)**. Any time you see temperature having an effect, consider the activity of enzymes as a reason for the effect.

Many of the plants which provide humans with their staple diet (the food they normally eat) also have optimal temperatures and if the climate becomes warmer or colder, they will not grow as well. Climate change could cause major problems for people and other organisms due to temperature change combined with changes in rainfall, which bring drought to many important crop-growing regions and devastating floods to others.

In higher latitudes, seasonal cycles affect life cycles. Global warming appears to be affecting the beginning of the seasons. This affects both the life cycles and the distribution of species. Warmer temperatures mean that plants grow and flower earlier. Insects such as moths and butterflies become active earlier in this warmth, and the plant food they need for their caterpillars is available. Some birds can adapt to these changes. For example, the breeding cycle of the great tits in Wytham Woods near Oxford in the UK has moved forward, caused by the same temperature changes that mean the winter moth larvae that form the main food supply for the baby birds are also present on the leaves. The UK tits lay eggs about two weeks earlier now than they did 50 years ago. You can see the coordination between the birds and the insects in **fig A**. However, the great tit populations in the Netherlands are not doing so well. Their breeding time is becoming earlier every year but the caterpillars are emerging even earlier, so the birds are missing the peak population and raising fewer chicks.

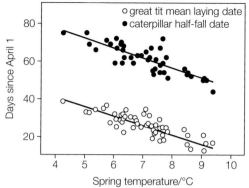

▲ **fig A** Mean temperatures are increasing and the incidence of egg laying in both the birds and the insects has changed in a very similar manner, so the life cycles remain coordinated. Not all population or species adapt to global warming so well.

Breeding earlier in the year may mean that some animal species can have more than one breeding cycle in each year, so those populations will increase. Changes in temperature could have dramatic effects on other organisms. For example, the embryos of some reptiles are sensitive to temperature during their development. Male crocodiles develop only if the eggs are incubated at 32–33 °C. If the eggs are cooler or warmer, females develop. If global warming means only female crocodiles develop then it could be the end of a species that has survived virtually unchanged for millions of years (see **fig B**).

▲ **fig B** Crocodiles have existed almost unchanged for millions of years, but global warming could signal the end of the species.

PRACTICAL SKILLS   CP12

It is possible to model the effect of increasing temperature on the development of living organisms in the laboratory. There are many different experimental procedures which you can use. You can investigate the effect of different temperatures on plants, by looking at the germination of seeds or the growth rate of young seedlings. You can investigate the effect of temperature on the growth of microorganisms, growing inoculated plates at different temperatures and counting the numbers and size of the bacterial colonies that grow at different temperatures. Finally, you can investigate the effect of temperature on animals, by investigating the hatching rate of brine shrimp eggs at different temperatures. You will need to plan the temperature differences you use and control the temperatures of your investigations very carefully.

If you use brine shrimps, make sure you take into account the ethical use of animals in practical investigations and treat them well. In **fig C**, you can see brine shrimps are tiny, simple animals which are not complex enough to suffer mental or physical stress when used in an investigation. They obviously cannot give consent to be used, so you must treat them with high standards of animal welfare to make sure they are not badly affected by being used in the laboratory.

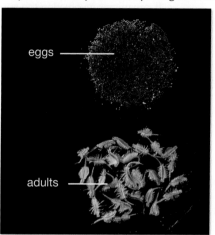

▲ **fig C** You can model the effect of global warming on tiny invertebrates like these brine shrimps.

> **!** Safety Note: After handling seeds, plants or microorganisms hands should be washed with soap and water. Disposable gloves should be worn where there is a risk of allergic reaction.

RAINFALL CHANGES

Global warming and its associated climate change seem to be altering the patterns of rainfall across the globe. Some countries, including Pakistan, are facing many difficulties. The monsoon rains are becoming heavier at times and in some places, and in recent years Pakistan has suffered a number of devastating floods. These floods destroy millions of homes, kill thousands of people and damage millions of acres of farmland. On the other hand, there are also increasing periods with no rain and so repeated droughts also affect the country, particularly around Balochistan and Sindh provinces. The lack of rain, combined with raised temperatures, has greatly reduced the ability of people to grow the crops they need to survive and it is also pushing many plants and animals towards local extinction.

Countries such as Jordan are also badly affected by changes in rainfall resulting from global warming. If nothing is done to reduce the global changes, scientists have estimated that the country will receive 30% less rainfall by 2100, and temperatures will be 4–5 °C higher than now. Reservoirs are already dangerously low and the winter rains have become erratic. This level of drought means even the well-adapted animals and plants in the country will struggle, and people will find it almost impossible to grow enough food to survive.

CHANGES IN SPECIES DISTRIBUTION

A change in climate could affect the area in which many different organisms can live. For example, alpine plants in many mountainous parts of Europe are becoming rarer. Most animals can move more easily than plants so they can often survive change more easily. Thus, as areas become warmer, some animals may be able to extend their ranges northwards while becoming extinct at the southern end. Others may be able to colonise a bigger area. In a study by Parmesan *et al.* in 1999 of 35 species of non-migratory European butterflies, the ranges of 63% of the species had shifted northwards by between 35 and 240 km in the past 100 years and only 3% (one species) had shifted south. The shift in butterfly populations paralleled a 0.8 °C warming over Europe during that time.

If organisms involved in the spread of disease are affected, patterns of world health could change as well. The World Health Organization (WHO) has warned that global warming could be responsible for a major increase in insect-borne diseases in areas such as Britain and Europe. The prediction is that by 2100, conditions could be ideal for disease-carrying organisms such as mosquitoes, ticks and rats in many new areas of the world. The WHO has predicted where malaria will have spread to by 2050 if global warming and climate change continue at the current rate (see **fig D**). The WHO is urging countries to make plans so that preventative measures can be implemented as the climate changes.

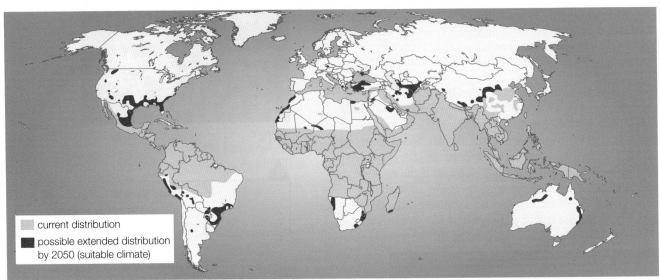

current distribution

possible extended distribution by 2050 (suitable climate)

▲ **fig D** The current distribution of the *Anopheles* mosquito (and, therefore, of malaria) could change if one of our models of continued global warming proves to be accurate.

Fig E summarises what the Intergovernmental Panel on Climate Change (IPCC) think could happen across the Earth as a result of global warming, with its associated climate change with different seasons and different patterns of rainfall. It is a frightening prospect for everyone.

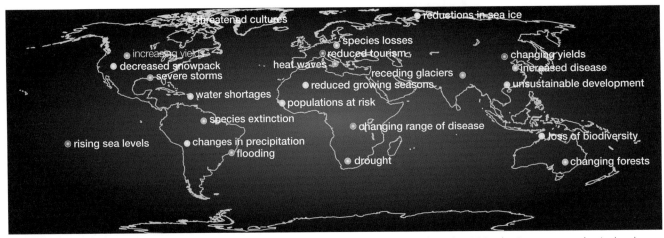

▲ **fig E** The impact of global warming threatens to be huge. This diagram, based on a report from the IPCC, takes into account what is already happening and predictions of what may happen in the future.

DID YOU KNOW?

UAE: vulnerable to climate change

The UAE is one of the countries in the world most vulnerable to climate change. This change will result in less rain, more droughts, higher temperatures, higher sea levels and more storms. All of these changes threaten the human population as well as the animals and plants which live in the UAE. Here are some examples of the problems which may be faced in the future.

- The UAE has around 1300 km of coastline where about 85% of the population live. Up to 6% of the coastline could permanently disappear as sea levels rise. Animals and plants will become extinct because of the salt water which will invade the land.

- Rising sea temperatures will bleach and kill the corals, which will have a huge impact on biodiversity in the seas around the country. There have already been several major bleaching events and the coral is struggling. **Fig F** shows the difference between healthy coral and coral that has been bleached and died.

▲ **fig F**   The difference between (a) living, healthy coral and (b) coral bleached as a result of rising sea temperatures is very easy to see.

- Some areas face flooding with loss of soil, habitat and the natural ecosystem, and people drowned and displaced.

- Other areas will have virtually no rainfall and almost permanent drought, which could wipe out any agriculture, threaten livestock and make human habitation impossible.

- Many animals and plants will shift their habitats to the northern and mountainous areas of the country where it will be cooler and perhaps there may be more water. But many species will not be able to move and are at risk of becoming extinct.

CHECKPOINT

1. Explain how a change in atmospheric carbon dioxide levels can have an impact on disease in both animals and humans.

2. Explain the effect of changes in temperature, rainfall patterns and seasonal timings on plants and animals. Give at least **one** clear example for each factor.

3. Create a table to summarise the main effects of global warming. Use the information in your table to develop a flowchart of events which illustrates the best current model of the events linked to global warming.

SUBJECT VOCABULARY

temperature coefficient (Q_{10}) a coefficient showing the effect of temperature on the rate of a reaction
optimum temperature the temperature at which an enzyme works most efficiently
denature to cause permanent changes in a protein by too high a temperature

LEARNING TIP

Remember that rising temperatures do not cause mutations. The alternative alleles have already arisen through random mutation. Rising temperatures cause environmental variation which results in selection pressure on the organisms. The organisms that survive will be those with the features best suited to the conditions.

▲ **fig A** Many animals are losing their habitat. (a) Polar bears are seriously threatened as the sea ice melts. But scientists have found that they are changing their habits and their range. (b) Some are meeting and breeding with grizzly bears to form fertile hybrids known as pizzly bears, which are less dependent on the ice than their polar bear parent.

In **Section 4C.4 (Book 1: IAS)**, you used the Hardy–Weinberg equilibrium to look at allele frequencies in populations, and you discovered that evolution (changes in frequency of alleles) can result from gene mutation and natural selection. You discovered the importance of selection pressure and found out that if the environment changes, a new selection pressure is exerted on a population of organisms. As a result of the change in conditions, certain alleles may become advantageous or disadvantageous. A new set of advantageous alleles is selected for and eventually a new species that is suited better to the new conditions may evolve.

You also learned in **Section 4C.4 (Book 1: IAS)** that when populations become isolated in some way, the normal gene flow between the individuals is reduced because they cannot meet and mate. This can lead to the formation of two different species as the populations respond to slightly different selection pressures. Through natural selection, the populations may become so different they can no longer breed successfully.

CLIMATE CHANGE AND SELECTION PRESSURES

The effects of anthropogenic global warming are being seen all over the world. In many places the climate is becoming warmer. In some places the climate is becoming colder or wetter or drier. These climatic changes are acting as selection pressures, driving changes in allele frequency and ultimately resulting in evolution through natural selection. The changes are also resulting in a much higher extinction rate because animals and plants cannot adapt quickly enough to the changing world around them.

Any population contains a variety of alleles. Some populations, for example the cheetahs of sub-Saharan Africa or the Northern elephant seals, have a very low genetic variability as a result of previous population bottlenecks (events that drastically reduce the size of a population). If their environment changes as a result of global warming, they have little chance of surviving. However, when populations have a lot of genetic variation, it is likely that within the population there will be combinations of alleles that will allow the species to adapt to the new conditions.

Some species are highly adapted to very specific habitats. For example, pandas are very vulnerable, with their diet of only bamboo. If climate change causes a loss of bamboo, pandas will not survive.

If an area becomes much wetter or drier, this will put selection pressure on the plants and animals that live there. For example, many desert plants only grow and flower every few years when rain falls. If the intervals between rains get longer, many seeds will die before they get the chance to grow and reproduce, so the selection pressure will be towards alleles that enable the seeds to survive longer and longer.

The organisms that are best suited to survive and adapt may well be those which are already surviving in extreme conditions, or conditions which have a lot of variation in factors such as temperature or water availability. Organisms living in very stable conditions such as the tropical rainforests or oceans of the world may have relatively little genetic variation because they have not needed to adapt to change. As the climate becomes hotter or drier or colder, many species will face extinction while others will adapt, evolve and survive.

CLIMATE CHANGE AND ISOLATION

One of the main drivers of speciation is isolation. **Allopatric speciation** is widely recognised as the main evolutionary process (see **Section 4C.5 (Book 1: IAS)**). It occurs when populations become physically or geographically isolated. Global warming, with its measurable effect on sea levels, will result in many populations becoming isolated as low-lying areas of countries become flooded and thousands of new islands are formed around the world. The plants and animals in these isolated populations will breed and selection pressure will drive the populations in different directions to adapt to the new conditions. For example, areas of the world such as Australia and Madagascar became separated from the mainland a very long time ago, and the organisms there have evolved in unique ways to cope with the selection pressures in their new environment. Islands such as the Galápagos archipelago have formed and the small number of organisms which have reached the islands have evolved to fill a wide range of niches. As rising sea levels drive the loss of low-lying areas and the creation of many new islands, the stranded animals and plants will evolve in different ways from the populations left on the mainland areas and speciation will occur again.

Sympatric speciation occurs when organisms become reproductively isolated by mechanical, behavioural or seasonal changes. Anthropogenic climate change is causing seasons to change around the world. Sometimes it is delay of rainy seasons in tropical regions, sometimes the early arrival of warmer temperatures in spring in temperate areas. Populations of the same species that live in different areas may experience differences in their seasonal changes, until eventually they can no longer reproduce.

Climate change will mean different plants will grow. This may be due to a natural spread of different species or it may be driven by human actions. If people find a new crop will grow as conditions change, they will definitely grow it. This can also lead to speciation, and we have a historical example which shows you exactly how these changes can occur.

Until the mid-1800s the tiny fruitfly *Rhagoletis pomonella* lived only on hawthorn bushes, laying its eggs on the fruits. The larvae respond to the smell of the hawthorns and return as adults to hawthorns to reproduce. However, around 150 years ago many huge apple orchards were planted along the Hudson River Valley in the United States. Genetically, apple trees are quite closely related to hawthorns. Some female *Rhagoletis pomonella* laid their eggs on the apples, either by mistake or because they could not find hawthorns. The larvae did not do particularly well, but some survived to adulthood. These flies responded to the smell of apples, not hawthorns, and a breeding group of apple-dwelling flies evolved.

Now there are two breeding groups of *Rhagoletis pomonella* in the Hudson River Valley. One feeds on hawthorns, the other on apples. The two populations show increasing reproductive isolation because they mate only with flies on the same food source. The apple-dwelling flies have adapted to life on apple trees so they now emerge from their pupae at a different time of the year (see **fig B**). Apples provide more food for the maggots and better protection against parasitic wasps. Scientists have analysed the frequency of a number of alleles in the flies and have found they are becoming increasingly different. It seems likely that two entirely different species will evolve which will no longer interbreed as their reproductive cycles will be completely out of synchronisation. This type of change and speciation will become increasingly common in response to the changes global warming and climate change will bring.

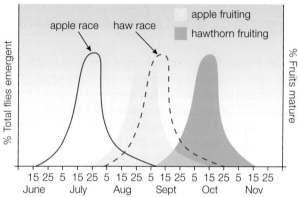

▲ **fig B**   The apple race of flies now emerges earlier than the hawthorn race, so there is less chance for flies to interbreed and a greater chance that this will lead to sympatric speciation.

EXAM HINT

Examples such as speciation in fruit flies are useful and can be used to illustrate examination responses. However, remember that in the examination you may be given other examples and you may be asked to describe, explain or deduce what is happening.

CHECKPOINT

SKILLS   CREATIVITY

1. ▶ (a) Suggest why the planting of orchards in the mid-1800s in the Hudson Valley triggered the beginnings of speciation in the fly *Rhagoletis pomonella*.

   ▶ (b) Explain how this mimics situations which are developing as a result of anthropogenic climate change.

2. Create a table to summarise the main ways in which global warming and climate change are likely to impact on changes in allele frequencies and speciation.

SUBJECT VOCABULARY

allopatric speciation an evolutionary process that occurs when populations become physically or geographically isolated
sympatric speciation an evolutionary process that occurs when organisms become reproductively isolated by mechanical, behavioural or seasonal changes

LEARNING OBJECTIVES

- Understand the way in which scientific conclusions about controversial issues such as what actions should be taken to reduce climate change, or to what degree humans are affecting climate change, can sometimes depend on who is reaching the conclusions.
- Understand how reforestation and the use of sustainable resources, including biofuels, are examples of the effective management of the conflict between human needs and conservation.

Global warming is happening and few people now dispute that. However, how much is caused by human activities and what actions, if any, should be taken to reduce it are much more controversial issues. The politicians and policy makers who make decisions that affect the problem and its solution are not always willing to accept issues such as this. The way scientific conclusions are interpreted and used can depend very much on who is making the decisions.

DID YOU KNOW?

Differing scientific conclusions

One problem is that scientists often disagree among themselves. Whatever the general agreement or scientific view in any area, there are almost always some scientists who do not agree. The great majority of scientists are convinced that atmospheric carbon dioxide levels, including human emissions, are contributing to global warming in the 21st century, but some others have a different view. There is also considerable debate among scientists as to the long-term effects of the increase in greenhouse gases on the climate in the future and about what, if anything, can be done. Different models predict different possible futures.

This is accepted as normal in the scientific community. Science works by debate, discussion and disagreement until a new model emerges from the results of investigations or the old model is confirmed. However, it can be hard for non-scientists to understand such disagreements. This leads to uncertainties and may cause non-scientists to select the theory which best suits their world view.

WHO DECIDES?

Politicians usually make the decisions about energy usage and carbon emissions and they are influenced by factors such as their political view, as well as the scientific evidence. They are also influenced by pressure groups and lobbyists who will be biased by their own interests. Environmental campaigners and scientists are desperate that politicians tackle carbon dioxide emissions to reduce the impact of global warming. On the other hand, it is in the interest of many industrialists, particularly in the field of electricity generation and the petrochemical industry, to promote alternative theories for global warming to avoid legislation that changes their industry. For example, when politicians have strong links to the petrochemical industry, it can be easier to obtain

research funding for projects looking into alternative theories of global warming than to obtain funding for research into anthropogenic greenhouse gas emissions.

Politicians also generally make decisions based on the short-term gains they need when they seek re-election. Consequently, policies that might be unpopular with voters may be ignored. For example, policies that affect individual car use by limiting petrol availability or raising the price of fuel, or those that raise the cost of electricity, may be abandoned even if they make environmental sense.

So, in the real world, the conclusions reached from scientific evidence depend very much on who is reaching those conclusions, who has funded the original research and on the financial and political pressures that the decision-makers experience.

PLANNING FOR THE FUTURE

The scientific evidence is growing steadily to show that the **sustainability** of biological resources is dependent on human beings changing their behaviour. It is easy for a country with plenty of food, readily available education and healthcare, and strong infrastructure to strongly criticise the cutting down of a rainforest. However, it is easy to understand that, for people who have very little, an activity which makes money to buy the food, healthcare and education they so badly need for themselves and their children is more important than keeping the rainforest.

Increasing carbon dioxide levels are a fact, and we need to find ways to stop using all the biological resources before it is too late. Environmentalists are proposing many possible solutions to reduce human emissions; for example by controlling the use of fossil fuels and making industrial processes and car engines cleaner and less polluting. People are becoming increasingly aware that we need to use **sustainable resources** to effectively manage the conflict between human needs and conservation. Sustainability demands a decent standard of living for everyone now, without compromising the needs of future generations or the ecosystems around us. This is not easy to achieve. There are too many human needs, self-interests and conflicts of interest to make this a simple problem to solve.

In most cases, the conclusions based on the science are not completely clear and people have differing priorities. For example, many people support the use of **biofuels** to replace fossil fuels. If you look at the carbon cycle, it makes sense to grow plants and then use products from those plants as fuel. This seems like an ideal, carbon-neutral solution to the problems of fossil fuels. The plants take carbon dioxide out of the air as they grow and it is released when they are burned as fuels, leading to no overall loss or gain of carbon dioxide in the atmosphere. Unfortunately it is not that simple. You need to convert a car to use biofuels, and the amount of land needed to grow the plants could threaten food production around the world. For example, it has been calculated that to replace 5% of US diesel consumption with biodiesel would take 60% of the land used to grown soy beans in that country. If poorer countries try to increase income by growing crops for fuel rather than food, it could mean millions of the world's poorest people will not have enough to eat (see **fig A**).

However, globally a huge amount of waste plant material is produced in agriculture and forestry. Scientists are investigating the possibilities of using this waste plant material (for example, straw or wood pulp from paper making) as the raw materials for biofuels. This could reduce the use of fossil fuels without taking up space which could be used for growing crops, and also use more of the plant material we already grow.

The use of renewable energy sources, such as wind or nuclear, can also reduce carbon emissions, but cause their own problems. Naturally, different interest groups have different opinions on how these should be developed.

▲ **fig A** Growing crops for biofuels in countries such as Bangladesh, where many people are short of food, may not be the ideal solution to replace fossil fuels around the world.

REFORESTATION

Reforestation is another way of trying to tackle the issues of global warming. Worldwide, deforestation (the destruction of natural forests and woodlands) has been happening at an alarming rate. This contributes to the increase in atmospheric carbon dioxide levels in several ways. Trees take up large amounts of carbon dioxide from the atmosphere in photosynthesis. They act as carbon sinks, storing carbon dioxide in their tissues for hundreds of years. Once trees are cut down (perhaps as building material or to produce farmland to grow beef cattle), they can no longer remove carbon dioxide from the air. The carbon sinks are lost, and because the trees are often burned, carbon dioxide

is released into the atmosphere in huge quantities. Replanting trees to replace those that have been cut down makes wood a sustainable resource and also removes carbon dioxide from the atmosphere into the tissues of new trees (see **fig B**).

▲ **fig B** Between 1950 and 1983, Costa Rica lost more than 46% of its natural rainforests and cloud forests due to deforestation and poor management. Since then, the country has protected its remaining forests and planted new ones. Now, more than 50% of the country is covered in forest and the proportion is growing.

Anthropogenic greenhouse gas production raises questions for both scientists and society. It is important that everyone works together to find the answers, and that conclusions are reached as fairly and transparently as possible.

CHECKPOINT

> **SKILLS**   PERSONAL AND SOCIAL RESPONSIBILITY

1. ▶ Explain why it is difficult for politicians to enact legislation that would substantially reduce the use of fossil fuels on both a national and a worldwide basis. Consider social and ethical considerations as well as practical ones.

2. Give **two** advantages and **one** disadvantage of using the following to help manage the conflict between human needs and conservation:

   (a) biofuels

   (b) reforestation

3. Different groups give very different impressions of the causes and possible solutions to the problem of global warming. Discuss how the views of major organisations can vary so much.

SUBJECT VOCABULARY

sustainability the production of a decent standard of living for everyone now, without compromising the needs of future generations or the ecosystems around us
sustainable resources resources which can be grown and used in a sustainable way
biofuels fuels produced directly or indirectly from biomass
reforestation the replanting of trees in an area where trees have been lost

THE TURTLES OF TORTUGUERO

SKILLS › INTERPRETATION, DECISION MAKING, ADAPTIVE LEARNING, CREATIVITY, PERSONAL AND SOCIAL RESPONSIBILITY

Hawksbill turtles are critically endangered in the UAE; there are perhaps 100 nesting females surviving. Scientists are working hard to conserve them. Other places in the world have the same problem because most species of turtles are struggling for survival. Success can help to inspire more success in others. For example, in the beautiful but tiny Costa Rican area of Tortuguero, a major effort by scientists, conservationists, the local population and the government has led to a large increase in the nesting population of rare green turtles.

MAGAZINE EXTRACT

For 50 years, the Sea Turtle Conservancy (STC) has conducted annual sea turtle nest monitoring studies on the 21-mile black sand beach of Tortuguero, Costa Rica, the nesting site of more endangered green turtles than anywhere else in the Western hemisphere. Since being initiated by Dr Archie Carr in the 1950s, this monitoring program has provided much information on the reproductive ecology and migratory habits of sea turtles. A recent peer-reviewed analysis showed an encouraging trend in green turtle nesting activity. Through this five-decade-long conservation initiative, STC has reversed the decline of green turtles in the Caribbean.

Research methods include turtle tagging, turtle track surveys, collection of biometric data, fibropapilloma examination, determination of nest survivorship and hatching success, collection of physical data, and collection of data on human impacts to the nesting beach and the turtles. Protection methods include a cooperative effort with Tortuguero National Park officials and law enforcement to reduce poaching of eggs and turtles. Training methods include training research assistants, recruited heavily from Latin American countries, and training Tortuguero National Park guards as well as local eco-tour guides in sea turtle biology and conservation. Public outreach methods include teaching Tortuguero school children, local adults and tourists about sea turtles and working with the international media to raise awareness about sea turtles and threats to their survival.

The overall goal of STC's sea turtle research and conservation work in Tortuguero is to conserve the area's nesting green and leatherback turtle populations so that these species fulfill their ecological roles. The strategies used to achieve this goal include the following: (1) monitoring and studying Tortuguero's nesting turtles; (2) working with the Costa Rican government, the community of Tortuguero and others to protect nesting turtles from poachers; (3) training young scientists, conservationists, and others to help ensure the continuation of sea turtle protection efforts in Tortuguero and elsewhere; and (4) educating the public about sea turtles and the threats to their survival.

But there are some things no-one can change. In the last few years, exceptionally high tides have eroded the beaches and reduced the opportunities for the turtles to nest. However hard the local people work, they cannot protect the turtles of Tortuguero and other Costa Rican nesting beaches from the far reaching impact of anthropogenic climate change.

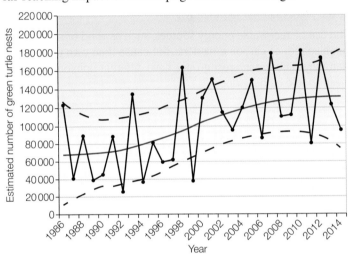

▲ **fig A** Trend in green turtle nests in Tortuguero, Costa Rica, between 1986 and 2014.

▲ **fig B** Tourists who visit the turtle beaches at Tortuguero learn about the turtles and how they are protected, and the money they pay is used to help fund the conservation efforts.

Source: Sea Turtle Conservancy, *http://www.conserveturtles.org/*

SCIENCE COMMUNICATION

The Sea Turtle Conservancy website provides information on sea turtles, the global threats they face and the work of the Sea Turtle Conservancy as well as updates on the latest efforts to save these amazing reptiles.

- Parts of the site are designed to inform the general public of what is going on, for example: https://conserveturtles.org/information-about-sea-turtles-an-introduction/
- Other parts are reports aimed at the scientific community, the Costa Rican and other governments and potential donors.

1 After reading the extract on the previous page, make a judgement whether it is aimed at the general public or the scientific community. Give reasons for your decision.

2 Find a page on the website that is aimed at the scientific community. Compare the writing styles of the page you found and the page listed above. Explain why they are so different.

3 '... A recent peer-reviewed analysis showed an encouraging trend in green turtle nesting activity ...' State what *peer review* means and explain why it is important for an organisation such as the Sea Turtle Conservancy to be able to quote a peer-reviewed analysis of its work.

BIOLOGY IN DETAIL

You already know about biodiversity, ecology, the effects of different factors on population size, ways of measuring population sizes, human influences on ecosystems, how sustainability of resources depends on effective management of the conflict between human needs and conservation, and the importance of the peer-review process in validating scientific evidence. Use this knowledge, as well as the Sea Turtle Conservancy website, to help you answer these questions.

4 Many different aspects of human behaviour have had an effect on sea turtle population numbers.
   (a) Discuss **three** ways in which human activities can have a negative effect on sea turtle numbers.
   (b) Discuss **two** ways human in which activities can have a positive effect on sea turtle numbers.

5 Look at the data in **fig A**.
   (a) What is the lowest number of turtle nests recorded in the period 1986–2013 and when was it recorded?
   (b) What is the highest number of turtle nests recorded and when was it recorded?
   (c) What is the percentage difference between the highest and the lowest figures?
   (d) Although the overall trend has been a steady increase in the number of turtle nests over the recorded period, there have been fluctuations from year to year. Suggest reasons for these variations in nest numbers.

ACTIVITY

Develop a news report, slideshow presentation or a public information poster on **one** of the following:

- the natural history of sea turtles
- threats to the future of sea turtles
- saving sea turtles from extinction in the UAE.

Use the Sea Turtle Conservancy website http://www.conserveturtles.org/ as your first source of information. You can also use other online resources, scientific magazines and journals, scientific papers and books. In each case, judge the reliability of your source before you use it and reference all the sources you use.

1 (a) Only approximately 10% of energy is transferred between trophic levels. Which of these is **not** a reason for energy loss between trophic levels? [1]

 A respiration

 B roots

 C undigested food

 D photosynthesis

(b) The table below shows the fresh biomass of green plants and consumers on an area of grassland.

Organism	Fresh biomass/g
green plants	2250.0
primary consumers	240.0
secondary consumers	38.0

 (i) Calculate the percentage loss in fresh biomass between the green plants and the primary consumers. Show your working. [2]

 (ii) State **two** ways in which biomass can be lost between the primary and secondary consumers. [2]

(c) Only a small percentage of the light energy that falls on the green plants is used in photosynthesis. Explain why blue and red light would be more useful to a plant than green light. [3]

(Total for Question 1 = 8 marks)

2 A group of students carried out an investigation to determine the energy flow in a rainforest food chain. They collected invertebrate animals from trees, identified them and placed them in the appropriate trophic level.

For each trophic level, the animals were counted, weighed and the energy content determined. Using their data and other sources of information the students produced the diagram below to show the energy flow along the food chain.

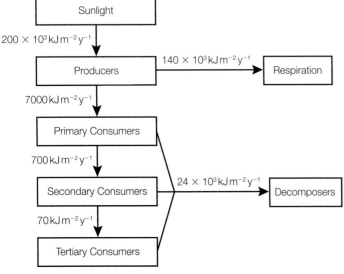

Energy flow along a food chain in a rainforest.

(a) What is meant by the term *net primary production (NPP)*. [1]

 A net primary production is growth due to photosynthesis

 B net primary production is gross primary production + respiration losses

 C net primary production is gross primary production − respiration losses

 D net primary production equals gross primary production

(b) Using the data in the diagram:

 (i) What is the percentage of light energy falling on a leaf that is converted to plant growth? [1]

 A 3.50%

 B 3.57%

 C 35%

 D 35.7%

 (ii) Suggest what happens to the light energy falling onto a leaf that is not converted into chemical energy. [3]

 (iii) Explain why only 10% of the energy locked up in the secondary consumers is transferred to the tertiary consumers. [3]

(Total for Question 2 = 8 marks)

3 Farmers clear plots of rainforest to use for crops. The trees are felled and burned. The soil is left covered in ash, which is rich in nutrients. However, the nutrients are soon used up by the growing crops. Within two or three years, the plot is abandoned and the farmer moves on to a new plot.

The abandoned plot is colonised by tree species and appears to recover to form secondary rainforest. This sequence of events is shown in the chart below, which also shows that the total biomass of the rainforest trees is made up of leaves, stems and branches, roots and leaf litter. These components of the biomass change as the rainforest is cleared, farmed and then abandoned.

Biomass component	Primary rainforest	Farming	Secondary rainforest	
			10 years	25 years
Leaves	50	10	50	50
Leaf litter	400	50	200	300
Branches	50	10	50	50
Roots	200	50	50	75

Components of biomass in million tonnes per hectare.

(a) (i) Using the data in the chart, compare and contrast the components of the biomass of the primary rainforest with those of the secondary rainforest after 25 years. [5]

(ii) In the past, plots of land were not reused for many years and had time to recover. However, as a result of deforestation plots are now reused after only 25 years. Using the information provided, justify why this form of farming is now considered to be unsustainable. [3]

(b) Suggest how the biodiversity of the surrounding rainforest could be changed by these farming practices. [5]

(Total for Question 3 = 13 marks)

4 Wheat seeds may be stored for several months after they are harvested. During storage, a proportion of the seeds will be wasted. Most of this waste is due to two causes:

- flour beetles (*Tribolium* sp.) eat live dormant seeds
- bacteria decompose dead seeds.

Flour beetles are eaten by insects called *Xylocoris* sp.

The diagram below shows the relative energy flow within a wheat store.

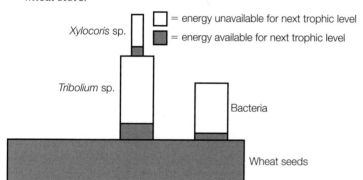

Relative energy flow within a wheat store.

(a) What is the best unit to use to record the energy content of the wheat seeds? [1]

A g

B g kJ^{-1}

C kJ g^{-1}

D kJ per seed

(b) To what trophic level do the *Tribolium* beetles belong? [1]

A producers

B primary consumers

C secondary consumers

D top carnivores

(c) (i) Explain why some of the energy that passes to *Tribolium* sp. from the wheat seeds is not available to be passed on to *Xylocoris* sp. [2]

(ii) Explain why all of the energy in the wheat seeds is available to be passed on to the next trophic level. [3]

(Total for Question 4 = 7 marks)

5 (a) Which of these contributors to the greenhouse effect is anthropogenic? [1]

A methane from cows

B methane from rice fields

C carbon dioxide from volcanic activity

D carbon dioxide from burning fossil fuels

(b) Which of the following statements provide evidence that global warming is occurring? [1]

1 earlier hatch times for winter moths

2 an increase in the carbon dioxide concentration in ice core samples

3 wider tree rings

A statement 1 only

B statements 1 and 2

C statements 1, 2 and 3

D statements 1 and 3

(c) Describe and explain **two** effects of global warming on biodiversity. [4]

(d) Suggest **two** possible ways to combat climate change and explain the effect each will have. [4]

(Total for Question 5 = 10 marks]

TOPIC 6 MICROBIOLOGY, IMMUNITY AND FORENSICS

6A MICROBIOLOGY

The 2014 International Longitude prize offers a reward of £10 million to the scientist or scientists from any country who develop an easy-to-use test for bacterial infections. The last time the prize was awarded, it was for the development of the technology that allows us to measure latitude and safely navigate the oceans of the world.

Why is a test for bacterial infections so important? Globally, bacteria cause many serious and even fatal diseases. The development of antibiotics has enabled people to survive many of these infections which were dreaded for many years. But in recent years, more and more bacteria have become resistant to commonly used antibiotics. At the same time, only a small number of new antibiotics have been developed. Fewer people are researching into antimicrobial drugs, partly because they generate less income for the companies that produce them than drugs for chronic conditions which are used for a lifetime. Yet new drugs, with new modes of actions, are badly needed. When we have these new drugs, it is important only to use them when they are really needed, to avoid antibiotic resistance in the future. Infections caused by viruses cannot be treated by antibiotics, so a successful test for bacterial infections would help us avoid using these precious drugs when they cannot help us.

In this chapter, you will be finding out about the organisms that cause infectious diseases, especially bacteria and viruses. You will discover more about how we can grow microorganisms in the laboratory, and how we can calculate the numbers we are working with. You will discover how bacteria cause disease by producing different types of toxins or invading body tissues, and how viruses take over the biochemistry of individual host cells.

You will learn how pathogens can be spread from one person to another, and how this awareness can help us avoid spreading diseases. The bactericidal and bacteriostatic actions of different antibiotics are explained in this chapter, and you will discover more about how antibiotic resistance develops. Finally, you will look at ways in which the problems of diseases caused by antibiotic-resistant organisms can be tackled and perhaps eventually overcome.

MATHS SKILLS FOR THIS CHAPTER

- **Recognise and make use of appropriate units in calculations** (*e.g. calculating serial dilutions*)
- **Estimate results** (*e.g. size of bacterial populations*)
- **Use ratios, fractions and percentages** (*e.g. calculating serial dilutions*)
- **Use exponential and logarithmic functions** (*e.g. growth of bacterial colonies*)
- **Use logarithms in relation to quantities that range over several orders of magnitude** (*e.g. growth of bacterial colonies*)
- **Translate information between graphical, numerical and algebraic forms** (*e.g. calculating the exponential growth rate constant*)
- **Plot two variables from experimental data** (*e.g. the growth of bacterial colonies*)
- **Determine the intercept of a graph** (*e.g. the growth of bacterial colonies*)

What will I study in this chapter?

- The ultrastructure of prokaryotic cells and viruses
- The difference between Gram-positive and Gram-negative bacterial cell walls and why each type reacts differently to antibiotics
- The evolutionary race between pathogens and the development of medicines to treat the diseases they cause
- The aseptic techniques used in culturing organisms
- The principles and techniques involved in culturing microorganisms, including the use of different media
- Different methods for measuring the growth of a bacterial culture, including cell counts, dilution plating, and mass and optical methods
- The different phases of a bacterial growth curve and the calculation of the exponential growth rate constant using the equation
  $$k = \frac{\log_{10}N_t - \log_{10}N_0}{0.301 \times t}$$
- Bacteria as pathogens, and the way that pathogenic effects can be produced by exotoxins, endotoxins and the invasion of the host tissue
- Viruses as pathogens and the way their different life cycles affect host cells
- The action of bactericidal and bacteriostatic antibiotics
- The development and spread of antibiotic resistance in bacteria, and the methods used and difficulties of controlling the spread of antibiotic resistance
- How *Mycobacterium tuberculosis* and human immunodeficiency virus (HIV) infect human cells and cause the symptoms of disease

What prior knowledge do I need?

Chapter 3A (Book 1: IAS)

The ultrastructure of prokaryotic cells

The importance of staining specimens in microscopy

What will I study later?

Chapter 6B

- The non-specific response of the body to infection, including inflammation, lysozyme action, interferon and phagocytosis
- The response of the body to infection, including the action of the macrophages, neutrophils and lymphocytes
- The development of the humoral and the cell-mediated immune responses
- The role of T and B memory cells in the secondary immune response
- How immunity can be natural or artificial, and active or passive
- How vaccination is used in the control of communicable diseases and the development of herd immunity, including the potential problems in a population where a proportion of people choose not to vaccinate their children
- The causes and ways of preventing hospital-acquired infections

LEARNING OBJECTIVES

■ Be able to compare the structure of bacteria and viruses (nucleic acid, capsid structure and envelope) with reference to Ebola virus, tobacco mosaic virus (TMV), human immunodeficiency virus (HIV) and lambda phage (λ phage).

Bacteria are important microorganisms. They play many important roles in the ecosystems around you, especially as decomposers. You will find out more about this in **Chapter 6C**. They keep your skin and your gut healthy, and they help animals such as cows, goats and sheep to digest the cellulose in the cell walls of the plants they eat.

However, some bacteria cause diseases in animals, including in human beings, and in plants. Microorganisms that cause disease are known as **pathogens**.

In **Chapter 3A.4 (Book 1: IAS)**, you learned about the structure of bacteria (see **fig A**). The main features of bacteria as typical prokaryotes are summarised below.

All bacteria have:

- a bacterial cell wall containing peptidoglycan
- a cell surface membrane similar to those in eukaryotic cells
- a nucleoid (a single, circular strand of DNA that is the genetic material of the bacterium)
- 70S ribosomes which are the site of protein synthesis.

EXAM HINT

Remember that these are all features of prokaryotes.

Some bacteria have:

- pili – thread-like projections from the surface of the cell wall
- flagella – long rapidly rotating whip-like structures which can move the bacteria about
- a capsule or slime layer – a thick slippery substance around the outside of the cell wall
- mesosomes – internal extensions of the membrane which fold into the cytoplasm and may be the site of cellular respiration
- plasmids – small circles of additional DNA that code for specific characteristics.

Some bacteria act as parasites, some act as mutualistic organisms and some are saprophytes (organisms that feed on dead and decaying organic material).

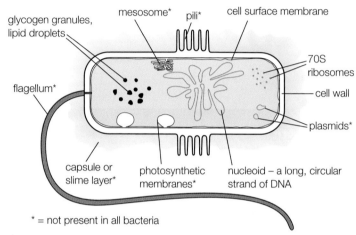

▲ **fig A** The structure of a typical bacterium

But bacteria are not the only microorganisms that act as pathogens. Viruses in the natural world *only* act as pathogens; there are no viruses which have a positive role. However, scientists now use viruses to produce genetically modified organisms, and they hope to use viruses to attack some of the bacteria that cause disease.

WHAT IS A VIRUS?

Viruses are the smallest of the microorganisms; they range in size from 0.02–0.3 μm across, which is about 50 times smaller than the average bacterium. Viruses are not cells. They are arrangements of genetic material and protein that invade living cells and take control of the cellular biochemistry to make more viruses. This reproduction, and the fact that they change and evolve in an adaptive way, means that they are classed as living organisms.

Most scientists who study viruses class them as obligate intracellular parasites. This means that they exist and reproduce as parasites only in the cells of other living organisms. Because viruses invade and take control of living cells to reproduce, they all cause some kind of damage and disease. They can resist drying and long periods of storage and, at the same time, maintain their ability to infect cells. There are very few drugs which have any effect on viruses. The drugs that do work are only effective in very specific situations (e.g. acyclovir can help prevent outbreaks of cold sores and shingles).

DID YOU KNOW?

Discovering viruses

In the late 19th century, people suspected the presence of viruses as disease-causing agents. Viruses were developed as a model to explain the way certain diseases were transmitted from one individual to another, but it was not until 1935 that Wendell Stanley identified the first virus.

The leaves of tobacco plants are susceptible to an unpleasant disease which destroys the plants, but no one could find the cause. Stanley pressed the juice from around 1300 kg of diseased tobacco leaves. He purified the juice and eventually produced pure, needle-like crystals. When he dissolved the crystals in water and painted the liquid onto tobacco leaves, the plants produced the symptoms of the disease (see **fig B(a)**). He called the particles **tobacco mosaic virus (TMV)**. It was obvious that the crystals were not living in the usual sense of the word; however, they retained the ability to cause disease. You cannot see viruses using a light microscope because they are usually smaller than half a wavelength of light. But with the development of the electron microscope we can see the TMV particles (see **fig B(b)**). They are rod-like structures with a protein coat formed around a core of RNA.

(a)

(b)

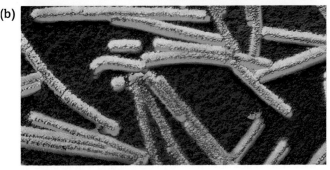

▲ **fig B** (a) The visible damage to the leaves that you can see is caused by (b) the very small rod-shaped particles of the tobacco mosaic virus.

THE STRUCTURE OF VIRUSES

Viruses are usually geometric shapes and have similar basic structures. However, their genetic material varies and the structure of their protein coat also varies considerably.

All viruses have the following characteristics.

- A protein coat or **capsid** which consists of simple repeating protein units known as **capsomeres**, arranged in different ways. The amount of genetic material needed to code for coat production is minimised by using these repeating units. These units also make assembling the protein coat in the host cell as simple as possible.

- Nucleic acids acting as genetic material. This may be DNA or RNA, and the nucleic acid is sometimes double stranded and sometimes single stranded. The viral genetic material is used differently in the host cell to make new viruses depending on which form it is in.

- Specific proteins (antigens) known as **virus attachment particles (VAPs)** which target proteins in the host cell surface membrane. This is how viruses attach to the cells which they infect. The VAPs respond to particular molecules on the host cell surface. Viruses are often quite specific in the tissue they attack.

Some viruses have the following characteristic.

- A lipid **envelope** which is produced from the host cell membrane and which covers the genetic material and protein coat. The presence of the envelope makes it easier for the viruses to pass from cell to cell, but it does make them vulnerable to substances such as ether which will dissolve the lipid membrane.

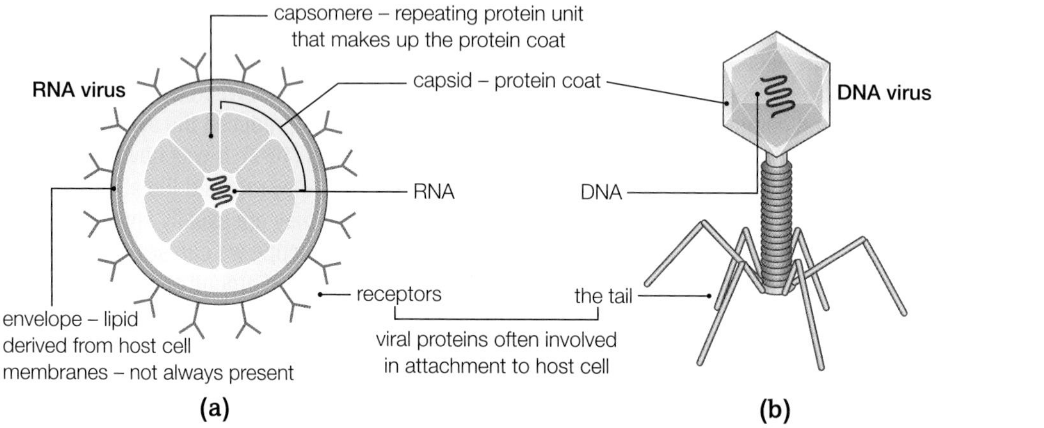

▲ **fig C**   This diagram shows you the general virus structure of (a) an RNA virus and (b) a DNA virus such as λ phage.

CLASSIFYING VIRUSES

Viruses are classified by their genome and their mode of replication (see **Section 6A.2**). As you saw above, viral genetic material can be DNA or RNA, and the nucleic acid is sometimes double stranded and sometimes single stranded. The form of the genetic material influences how the virus makes new viruses in the host cell.

- **DNA viruses** have DNA as their genetic material. The viral DNA acts directly as a template (model) for both new viral DNA and for the mRNAs needed to induce synthesis of viral proteins. Examples of DNA viruses include the smallpox virus, adenoviruses (which cause colds) and bacteriophages (viruses which infect bacteria; for example, the λ (lambda) phage in **fig C**).

- **RNA viruses** have RNA as their genetic material. They are much more likely to mutate than DNA viruses. RNA viruses do not produce DNA as part of their life cycle. The majority of RNA viruses contain a single strand of RNA. Examples of plant and animal diseases caused by RNA viruses include tobacco mosaic viruses, Ebola fever, polio, measles and influenza.

- **Retroviruses** are a special type of RNA virus. They have a protein capsid and a lipid envelope. The single strand of viral RNA controls the synthesis of a special enzyme called **reverse transcriptase**. This is responsible for making DNA molecules corresponding to the viral genome. This DNA is then incorporated into the host cell DNA and is used as a template for new viral proteins and ultimately a new viral RNA genome. **Human immunodeficiency virus (HIV)** is a retrovirus. Some forms of leukaemia are also caused by this type of virus.

CHECKPOINT

1. Explain how viruses are adapted to their particular way of life.

   SKILLS   CRITICAL THINKING

2. ▶ Suggest valid arguments for the case that:
    (a) viruses are living organisms
    (b) viruses are **not** living organisms.

3. Make a table to compare the structure of bacteria and viruses.

LEARNING OBJECTIVES

■ Understand what is meant by the terms *lytic* and *latency*.

People are constantly fighting against the viruses that cause disease. It is important to understand how viruses reproduce in the human body so that we can understand how viruses make us ill.

VIRUS LIFE CYCLES

Viruses reproduce only within the cells of their host. They attack host cells in a number of different ways. For example, bacteriophages inject their genome into the host bacterial cell, but the main part of the viral material remains outside the bacterium. The viral DNA forms a circle or **plasmid** within the bacterium. The viruses that infect animals enter the cells in several ways. Some types enter the cell by endocytosis (with or without the envelope) and the host cell then digests the capsid, releasing the viral genetic material. Most commonly, the viral envelope combines with the host cell surface, releasing the rest of the virus inside the cell membrane. Plant viruses usually enter the plant cell using a **vector** (often an insect, such as an aphid) to penetrate the cellulose cell wall.

Once a virus is in the host cell, there are two different routes of infection:
- the lysogenic pathway
- the lytic pathway.

LYSOGENIC PATHWAY

Many DNA viruses do not cause disease when they first get into the host cell. They insert their DNA into the host DNA so it is replicated every time the host cell divides. This inserted DNA is called a **provirus**. Messenger RNA is not produced from the viral DNA because one of the viral genes causes a repressor protein to be produced. This makes it impossible to translate the rest of the viral genetic material. The virus is said to be **latent** during this period of **lysogeny**, when the virus is part of the reproducing host cells.

Under certain conditions, for example when the host is damaged, viruses in this lysogenic state are activated. The amount of repressor protein decreases and the viruses enter the lytic pathway and become virulent (see **fig A**).

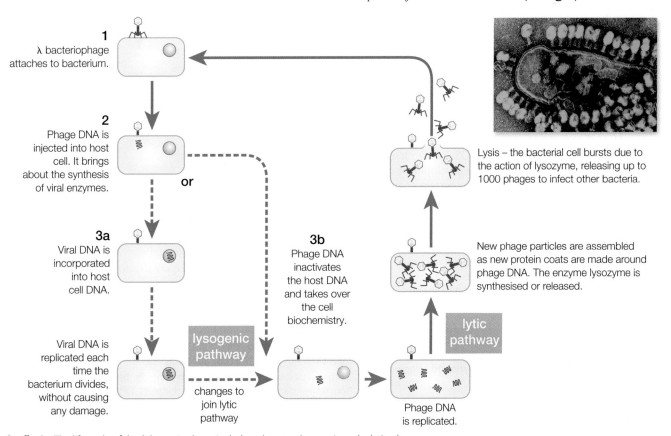

1
λ bacteriophage attaches to bacterium.

2
Phage DNA is injected into host cell. It brings about the synthesis of viral enzymes.

or

3a
Viral DNA is incorporated into host cell DNA.

Viral DNA is replicated each time the bacterium divides, without causing any damage.

lysogenic pathway

changes to join lytic pathway

3b
Phage DNA inactivates the host DNA and takes over the cell biochemistry.

Phage DNA is replicated.

lytic pathway

New phage particles are assembled as new protein coats are made around phage DNA. The enzyme lysozyme is synthesised or released.

Lysis – the bacterial cell bursts due to the action of lysozyme, releasing up to 1000 phages to infect other bacteria.

▲ **fig A** The life cycle of the λ bacteriophage includes a latent, a lysogenic and a lytic phase.

LYTIC PATHWAY

Sometimes, the viral genetic material is replicated independently of the host DNA immediately after it enters the host. Mature viruses are made and eventually the host cell bursts, releasing large numbers of new virus particles to invade other cells. The virus is now disease-causing and the process of replicating and killing cells is known as the lytic pathway.

A MORE COMPLEX LIFE CYCLE

Retroviruses (for example, the HIV virus that causes AIDS and the Rous sarcoma virus that causes cancer in chickens) have a rather different and complex life cycle. Their genetic material is viral RNA. This cannot be used as mRNA but is translated into DNA by the specific enzyme reverse transcriptase in the cytoplasm of the host cell. This viral DNA enters the nucleus of the host cell where it is inserted into the host DNA. Host transcriptase enzymes then make viral mRNA and new viral genome RNA. New viral material is synthesised and the new viral particles leave the cell by **exocytosis** (see **Section 2A.4 (Book 1: IAS)**). The host cell continues to function as a virus-making factory, and at the same time the new viruses continue to infect other cells. This process is summarised in **fig B**.

VIRUSES AND DISEASE

Viruses cause the symptoms of disease by the lysis of the host cells, by causing the host cells to release their own lysosomes (see **Section 3A.2 (Book 1: IAS)**) and digest themselves from the inside or by the production of toxins that inhibit cell metabolism.

Viral infections are often specific to particular tissues. For example, adenoviruses cause colds and they affect the tissues of the respiratory tract but do not damage the cells of the brain or the intestine. This specificity seems to be due to the presence or absence of cell markers on the surface of host cells. Each type of cell has its own recognition markers and different types of virus can only bind (attach) to particular markers. The presence or absence of these markers can even affect whether a group of living organisms is vulnerable to attack by viruses. For example, the angiosperms (flowering plants) are vulnerable to viral diseases, but the gymnosperms (conifers and their relatives) are not.

Viruses cause diseases like flu, measles and AIDS. Research also shows that, in some cases, they play a role in the development of cancers. Viral infection has been clearly linked to certain animal cancers, and in humans there seems to be a link in certain specific cases.

LEARNING TIP

The name of an enzyme often tells you what the enzyme does. Transcription is making RNA from a DNA template. So reverse transcription is making DNA from a RNA template; it needs an enzyme called reverse transcriptase.

CHECKPOINT

1. Describe the main differences between the lytic and lysogenic pathways of infection.

2. (a) State the main structural difference between retroviruses and other viruses.

   (b) Explain how this structural difference affects their life cycle.

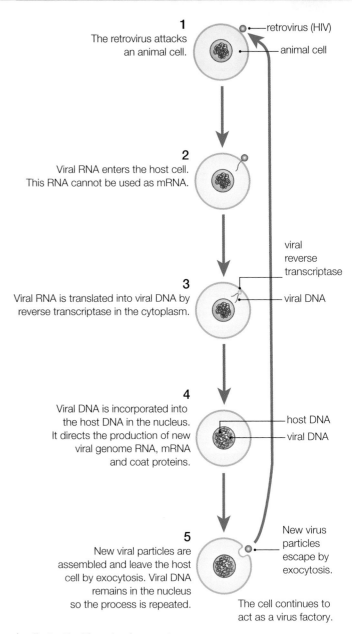

1 The retrovirus attacks an animal cell.
retrovirus (HIV)
animal cell

2 Viral RNA enters the host cell. This RNA cannot be used as mRNA.

3 Viral RNA is translated into viral DNA by reverse transcriptase in the cytoplasm.
viral reverse transcriptase
viral DNA

4 Viral DNA is incorporated into the host DNA in the nucleus. It directs the production of new viral genome RNA, mRNA and coat proteins.
host DNA
viral DNA

5 New viral particles are assembled and leave the host cell by exocytosis. Viral DNA remains in the nucleus so the process is repeated.
New virus particles escape by exocytosis.
The cell continues to act as a virus factory.

▲ **fig B**   The life cycle of a retrovirus.

SUBJECT VOCABULARY

plasmid small, circular piece of DNA that codes for a specific characteristic of the bacterial phenotype

vector living organisms or environmental factors which transmit pathogens from one host to another

provirus the DNA that is inserted into the host cell during the lysogenic pathway of reproduction in viruses

latent the state of the non-virulent virus within the host cell

lysogeny the period when a virus is part of the reproducing host cell, but does not affect the host adversely

retrovirus a type of RNA virus that controls the production of DNA corresponding to the viral RNA and inserts it into the host cell DNA

exocytosis the energy-requiring process by which a vesicle fuses with the cell surface membrane so the contents are released to the outside of the cell

LEARNING OBJECTIVES

■ Understand the different phases of a bacterial growth curve (lag phase, exponential phase, stationary phase and death phase) and be able to calculate exponential growth rate constants.

Bacteria can reproduce in two main ways. The most common method is by asexual reproduction, in which the bacterium divides in two (**binary fission**). Once a bacterium reaches a certain size, determined by the ratio of nuclear material to cytoplasm, the DNA is replicated and the old cell wall begins to break down around the middle of the cell. During this process the DNA appears to be associated with the cell surface membrane. If a mesosome is present it may hold the DNA in position.

Enzymes break open the circular piece of DNA, allowing the strands of DNA to unwind and be replicated. New cross-walls are also formed between the two new daughter cells, and again the mesosome appears to play a part in cells if it is present. New cell membrane and cell wall material extend inwards, forming a septum which eventually divides the cell into two new daughter cells, each containing a circular chromosome attached to the cell membrane. Plasmids often divide at the same time, so the daughter cells usually each contain both a copy of the original genome and any plasmids present in the parent cell (see **fig A**).

The time between cell divisions is known as the **generation time**. Bacteria can reproduce at very high speed. The small size and relative simplicity of the prokaryotic cells mean that they can undergo binary fission every 20 minutes when conditions are favourable, but it often takes longer.

In certain situations, some types of bacteria can reproduce using what appear to be different forms of sexual reproduction. These are very rare events, and none of them involves true sexual reproduction with the formation and transfer of gametes. However, they are methods by which genes can be transferred between bacteria, not necessarily of the same species. This process has important implications in our fight against disease.

THE GROWTH OF BACTERIAL COLONIES

You have seen that in ideal conditions bacteria can reproduce very rapidly. There are enormous implications of this rapid rate of reproduction in bacterial cells. If a single bacterium has unlimited space and nutrients, and if all its offspring continue to divide at the same rate, then after 48 hours there will be 2.2×10^{43} bacteria, weighing 4000 times the weight of the Earth! Clearly this sort of growth never occurs because, although the number of bacteria in a colony can increase exponentially at first, limited nutrients and an accumulation of waste products always slow down the rate of reproduction and growth.

USING A LOGARITHMIC SCALE

When you consider the growth of bacteria, the numbers involved are so enormous it rapidly becomes impossible to show what is happening. If we use a logarithmic scale, the data become much easier to manage. Log numbers are used to represent the bacterial population because the difference in numbers from the initial organism to the billions of descendants is too great to represent using standard numbers. So with the growth scale on the y-axis being logarithmic, and time on the x-axis kept on a normal scale, this would be a semi-logarithmic graph.

In a logarithmic scale, the numbers on the scale are actually logarithms (also known as powers) of a base number. So continuing with our example, the time scale stays the same (non-logarithmic) but the numbers of bacteria are represented on a $\log_{10}$ scale. This means that every number on the y-axis represents a power of 10. So 1 is actually 10^1 (10), 2 is 10^2 (100), 3 is 10^3 (1000) and so on. Look at **fig B(a)** showing a simple bacterial growth curve over 12 hours; now look at **fig B(b)** showing the logarithmic graph. You can see the difference this makes.

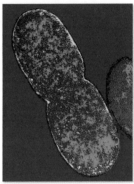

▲ **fig A**   Binary fission in *E. coli*. You can see how the cell wall and membrane begin to extend inwards to separate the two new cells.

EXAM HINT

Remember that variation is an essential component of natural selection and evolution. Therefore, exchange of genetic material is important for bacteria to evolve.

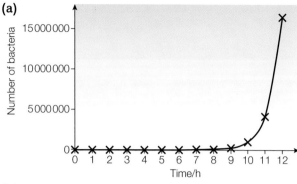

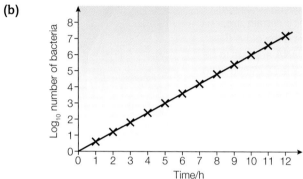

▲ **fig B** Graphs to show bacterial growth, assuming bacteria divide every 30 minutes; (a) using a non-log scale and (b) using a log scale.

EXAM HINT

When looking at a graph always read the title and axes labels first. This will tell you if a normal scale or a log scale is being used.

EXPONENTIAL GROWTH RATE CONSTANTS

You can calculate the number of bacteria in a population using the following formula:

$N_t = N_0 \times 2^{kt}$ where:

N_t = the number of organisms at time t

N_0 = the number of organisms at time 0 (the beginning of the experiment)

k = the exponential growth rate constant

t = the time the colony has been growing

Using this, you can calculate the number of bacteria in a colony at any point if you know the initial number of bacteria and the time the colony has been growing under constant conditions. However, to use the equation, you need to know the exponential growth rate constant. This constant equals the number of times that the population will double in one unit of time. You can calculate it using the following formula:

$$k = \frac{\log_{10} N_t - \log_{10} N_0}{\log_{10} 2 \times t}$$

LEARNING TIP

This formula can be derived from the equation for population growth. You do not need to be able to derive the formula yourself, but if you are confident using logarithms you might like to try!

WORKED EXAMPLE

Using the data from **fig B**, if the number of bacteria at the beginning is 1 and the colony has 1024 bacteria after growing for 5 hours, calculate k:

$$k = \frac{\log_{10}(1024) - \log_{10} 1}{\log_{10} 2 \times 5}$$

$$k = \frac{3.010}{0.301 \times 5}$$

$$k = 2$$

You can then substitute k into the equation to find out the population at any time.

For example, after 15 hours:

$$N_{15} = 1 \times 2^{15k} \text{ so } N_{15} = 2^{30} = 1\,073\,741\,824$$

This answer makes sense. If the bacteria divide every 30 minutes, the size of the population will double twice each hour. So in 15 hours, the initial population size of 1 will have doubled 30 times.

ANALYSING THE DATA

In **Section 6A.5**, you will learn how to measure the growth of bacterial colonies in a number of different ways. When the numbers of cells in a bacterial culture are measured over time, a pattern emerges (see **fig C**). This is called a growth curve; it has four stages:

- the **lag phase** when bacteria are adapting to their new environment and are not reproducing at their maximum rate
- the **log phase/exponential phase** when the rate of bacterial reproduction is close to or at its theoretical maximum, repeatedly doubling in a given time period
- the **stationary phase** when the total growth rate is zero as the number of new cells formed by binary fission is equal to the number of cells dying
- the **death phase/decline phase** when reproduction has almost stopped and the death rate of cells is increasing.

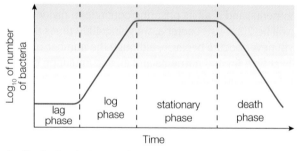

▲ **fig C** Graph showing the growth curve of a bacterial colony over time.

EXAM HINT

Remember this as a sigmoid (S-shaped) curve.

Exponential growth in a bacterial culture does not continue. Reasons the rate of growth slows down include the following.

- A reduction in the amount of nutrients available. At the start of the culture, there are more than enough nutrients for all the microorganisms. As the numbers multiply exponentially in the log phase of growth, the excess food is consumed. If no fresh nutrients are added, the level of nutrients available will become insufficient to support further growth and reproduction and so will limit the growth of the population of bacteria in the culture.

- An accumulation of waste products. At the beginning of the growth cycle of a bacterial culture, the waste products are minimal. As cell numbers rise, toxic material accumulates and becomes sufficient to inhibit further growth and even to poison and kill the culture. In particular, carbon dioxide, which is produced by the respiration of the bacterial cells, accumulates and the pH of the colony falls to a point at which the bacteria can no longer grow.

DID YOU KNOW?

From psychrophiles to hyperthermophiles…

For a long time, temperature has been recognised as one of the limiting factors on the growth of a colony of bacteria. Very little growth occurs at low temperatures. High temperatures are regularly used to destroy bacteria and make sure that culture media and medical instruments are sterile. However, scientists have now discovered bacteria living in the most extreme conditions in the depths of the ocean and deep in cores of ice. In the mineral-rich waters of the deep ocean are hydrothermal vents known as 'black smokers'. They are almost 3000 m below the surface of the ocean, the pressure is high and the temperature is around 350 °C. In these conditions, 'black smoker' bacteria grow vigorously; they stop reproducing if the temperature falls much below 100 °C.

Bacteria have also been found under almost 4000 m of ice, at temperatures and pressures which are usually not suitable for life and in which most enzymes are no longer active.

These bacteria challenge many of our accepted theories about the biochemistry of enzymes and cells, and illustrate clearly the ability of bacteria to take advantage of environments in which nothing else can survive.

CHECKPOINT

1. Explain why bacteria:

   (a) usually use asexual reproduction

   (b) occasionally use some form of sexual reproduction.

2. If a single bacterium starts dividing every 30 minutes, how long will it take for the colony to contain 16 777 216 bacteria? Show your workings to get a record of the bacterial numbers at each stage.

3. Use your answer to question **2** to:

   (a) explain the use of log scales in graphs to show the rate of bacterial growth

   (b) plot a graph of the data that you have generated

   SKILLS   ANALYSIS

   ▶ (c) sketch another graph to show what would have happened if the culture had been allowed to continue in a closed culture.

SUBJECT VOCABULARY

binary fission asexual reproduction in bacteria in which the bacteria split in half

generation time the time span between bacterial divisions

lag phase when bacteria are adapting to a new environment and are not reproducing at their maximum rate

log phase/exponential phase when the rate of bacterial reproduction is close to or at its theoretical maximum, repeatedly doubling in a given time period

stationary phase when the total growth rate is zero as the number of new cells formed by binary fission is equal to the numbers of cells dying due to factors including competition for nutrients, lack of essential nutrients, an accumulation of toxic waste products and possibly lack of, or competition for, oxygen

death phase/decline phase when reproduction has almost ceased and the death rate of cells is increasing so that the population number falls

LEARNING OBJECTIVES

■ Understand the principles and techniques involved in culturing microorganisms, using aseptic techniques.

Most microorganisms are so small that they are impossible to see with the naked eye. To investigate microorganisms (for example, to diagnose a disease or for scientific experiments) you need to culture them. This involves growing large numbers of the microorganisms so they can be measured in some way. They need to be provided with the right amount of nutrients and oxygen and also the ideal pH and temperature for growth. Bacteria and fungi are the most commonly cultured organisms (see **fig A**).

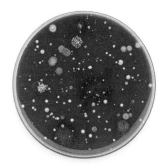

▲ **fig A** Different types of bacteria cultured on an agar plate.

When you grow a pure strain of a microorganism, the entry of any other microorganisms from the air or your skin into the culture will contaminate it. It is important to take great care when culturing microorganisms because:

- even if the microorganism you are planning to culture is completely harmless, there is always the risk of a mutant strain developing that may be pathogenic
- there is a risk of contamination of the culture by pathogenic microorganisms from the environment.

Health and safety precautions must always be followed very carefully when handling, culturing or disposing of microorganisms. All of the equipment must be **sterile** before you start the culture. It is particularly important that a culture you have grown does not leave the lab. You should dispose of all cultures safely by sealing them in plastic bags and sterilising them at 121 °C for 15 minutes under high pressure, before throwing them away. There are no ethical issues associated with the culturing of microorganisms from the perspective of the microorganisms themselves. However, you should always consider the danger of infecting other people with pathogens by accident.

ASEPTIC CULTURE TECHNIQUES

Culturing microorganisms involves a number of steps. First, you need to decide which microorganisms you want to culture, and obtain a culture of them. Then you need to provide the microorganisms with the right nutrients in order to grow. Most microorganisms require a good source of carbon and nitrogen as well as specific minerals. You can use a **nutrient medium** in the form of **nutrient broth**, where the nutrients are in liquid form for a **liquid culture** in a flask or test tube, or in a solid form, usually **nutrient agar**. Agar is a jelly extracted from seaweed. It is very useful because, although it solidifies as a jelly at 50 °C, it does not melt again until it is heated to 90 °C. Both solid and liquid media must be kept sterile until ready for use. You learned the basic principles of culturing bacteria in aseptic conditions on agar plates in **Section 4A.6 (Book 1: IAS)**.

Safety Note: Always conduct a risk assessment prior to handling and culturing microorganisms, as there can be both safety and ethical issues when you carry out an investigation using organisms which can potentially cause disease.

Some microorganisms will grow on pure agar but most need added nutrients. The majority of microorganisms grow on or in a medium enriched with good protein sources such as blood, yeast extract or meat extract (see **fig B**). Some need a very precise balance of nutrients. By producing a nutrient medium with very specific ingredients you provide a **selective medium**, a medium in or on which only a select group of microorganisms with those particular requirements will grow. Selective media are important in identifying particular mutant strains of microorganisms and antibiotic resistance. For example, YM media have a low pH, which encourages the growth of fungi and moulds but discourages bacteria, while MacConkey agar is designed to grow Gram-negative bacteria.

Selective media are also useful for identifying microorganisms that have been genetically modified. When scientists genetically modify a microorganism and add a new, useful gene (see **Chapter 8C**), they often also add a gene for the requirement for a particular nutrient or for antibiotic resistance. This acts as a marker gene to show which microorganisms have been genetically modified.

▲ **fig B** Cultures grown in broth are usually prepared in flasks or test tubes, and agar-based cultures are prepared in Petri dishes. The hot sterile liquid agar is poured onto plates which are immediately resealed and cooled so they solidify ready for use.

PRACTICAL SKILLS   CP13

The medium used for a liquid culture is often called a broth. Once you have prepared a suitable liquid medium for your culture, the next step is to introduce your microorganisms, for example, bacteria. Introducing the bacteria into your broth is called **inoculation** (see **fig C**). Many inoculations are done with an inoculating loop, removing bacteria from a solid medium surface such as an agar plate or immersing an inoculating loop into a known liquid culture. You can also use an inoculating broth, which involves making a suspension of the bacteria to be grown and mixing a known volume with the sterile nutrient broth in the flask.

Once you have inoculated your broth, you must put the sterile cotton wool stopper back in the flask again as quickly as possible to prevent contamination from the air.

Always make sure your flask is clearly labelled.

Incubate your flask at a suitable temperature (never higher than 20 °C in a school laboratory). This helps to make sure that no microorganisms which cause human diseases grow. They grow best at human body temperature of around 36–37 °C, so it is important to keep them cool. Shake your flask gently and regularly to make sure that the broth is aerated, allowing oxygen to reach the growing bacteria.

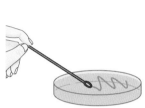
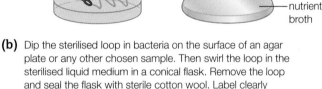

nutrient broth

(a) Sterilise the inoculating loop by holding it in the Bunsen burner until it glows red hot and then leave it to cool.

(b) Dip the sterilised loop in bacteria on the surface of an agar plate or any other chosen sample. Then swirl the loop in the sterilised liquid medium in a conical flask. Remove the loop and seal the flask with sterile cotton wool. Label clearly before incubating at a suitable temperature.

▲ **fig C**   Inoculating a flask with bacteria.

Safety Note: In school laboratories, always culture microorgansims at temperatures well below human body temperature to minimise the risk of growing potential human pathogens.

GROWING A PURE CULTURE

If you want a pure culture of just a single type of bacterium or fungus, you must isolate the desired microorganism. The most common way to isolate an organism is to use information either about its specific needs or about the requirements of possible contaminating organisms.

- Growing a culture under anaerobic conditions will ensure that only anaerobic bacteria will survive. Similarly, growing organisms with oxygen means that only aerobic organisms can survive. However, some bacteria will grow under both conditions. You may not be able to completely separate the microorganisms necessary for a pure culture, but you can reduce the variety considerably.

- The nutritional requirements of different microorganisms vary greatly. You can produce a medium that will favour the growth of the organism you want to culture and inhibit the growth of others. This allows you to identify the colony you want and then re-inoculate it to produce a single pure culture. You may need to control the range of nutrients available, or introduce selective growth inhibitors, antibiotics or antifungal chemicals that will reduce or prevent the growth of all microorganisms except the one you want.

- There are indicator media that cause certain types of bacteria to change colour. Colonies that do (or do not) change colour can be isolated and cultured (see **fig D**).

- We can only culture around 1% of the known species of bacteria. We do not know the right conditions to grow the remaining 99% of species in the laboratory.

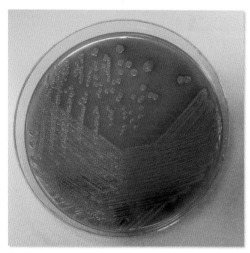

▲ **fig D** Indicator media such as this MacConkey agar are used to show up different types of bacteria. In this case, lactose-fermenting bacteria are red/pink and the non-fermenters form colourless colonies which are much harder to see. It is a simple step from here to isolate and further culture either type of bacteria.

It is particularly important to be able to isolate disease-causing organisms from those of the normal body flora, so that a disease can be diagnosed and appropriate treatment planned. The techniques described above make this possible.

CHECKPOINT

1. Make a flowchart to explain how to produce a culture of microorganisms:

   (a) in a flask of broth

   (b) on a plate of nutrient agar.

2. Describe **three** safety precautions that you should always take when culturing microorganisms and explain their importance.

SKILLS ▷ EXECUTIVE FUNCTION

3. ▷ Explain how different media may be used to produce a pure sample of microorganisms for culturing.

SUBJECT VOCABULARY

sterile a term used to describe something that is free from living microorganisms and their spores

nutrient medium a substance used for the culture of microorganisms, which can be in liquid form (nutrient broth) or in solid form (usually nutrient agar)

nutrient broth a liquid nutrient for culturing microorganisms, commonly used in flasks, test tubes or bottles

liquid culture growing microorganisms in a nutrient broth in a flask or test tube rather than on an agar plate

nutrient agar a jelly extracted from seaweed and used as a solid nutrient for culturing microorganisms, commonly used in Petri dishes

selective medium a growth medium for microorganisms containing a very specific mixture of nutrients, so only a particular type of microorganism will grow on it

inoculation the process by which microorganisms are transferred into a culture medium under sterile conditions

5 MEASURING THE GROWTH OF BACTERIAL CULTURES

LEARNING OBJECTIVES

■ Understand the different methods of measuring the growth of microorganisms, as illustrated by cell counts, dilution plating, mass and optical methods (turbidity).

You need to be able to measure the number of cells present in the culture at various time intervals in order to measure the changes in a bacterial culture and the growth in the population. Bacteria are so small that it is impossible to do this with the naked eye, and so you need to use a variety of different methods to count bacteria either directly or indirectly. It is important to choose the most suitable method of measuring for the investigation you are conducting.

CELL COUNTS

HAEMOCYTOMETER

You can count bacteria and single-celled fungi which are cultured in nutrient broth directly using a microscope and a **haemocytometer**. A haemocytometer is a specialised thick microscope slide with a rectangular chamber that holds a standard volume of liquid of $0.1\,mm^3$. The chamber is engraved (marked) with a grid of lines. It was originally designed for counting blood cells.

You dilute the sample of nutrient broth by half with an equal volume of trypan blue, a dye that stains dead cells blue so you can identify and count only the living cells. Then you can view and count the cells using a microscope.

Each corner of the haemocytometer grid has a square divided into 16 smaller squares. The number of cells in each of these four sets of 16 squares is usually counted (see **fig A**) and the mean is calculated. The haemocytometer is calibrated so that the number of bacterial or fungal cells in one set of 16 squares is equal to the number of cells $\times 10^4$ per cm^3 of broth. In this way, you can calculate the number of microorganisms in a standard volume of broth. You can construct a picture of the changing cell numbers by taking measurements at regular time intervals throughout the life of a bacterial colony.

OPTICAL METHODS (TURBIDITY)

An alternative way of measuring the number of cells in a culture is by **turbidimetry**, a specialised form of colorimetry. As the numbers of bacterial cells in a culture increase, it becomes increasingly **turbid** (cloudy). As a solution becomes more turbid, it absorbs more light, so less light can pass through it. A colorimeter measures how much light passes through a sample and thus shows how much light is absorbed. This indirectly indicates how many microorganisms are present. You can produce a calibration curve by growing a control culture and taking samples at regular time intervals. You can measure the turbidity of each sample and also count the cells using a haemocytometer for each sample. This gives you a relationship between the turbidity of the culture and the number of bacterial cells present. You can then use this calibration curve to measure the number of microorganisms simply using turbidimetry (see **fig B**). For example, you might want to investigate the effect of different conditions on the growth rate of the microorganism.

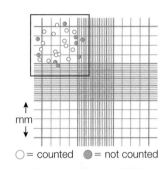

mm

○ = counted   ● = not counted

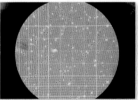

▲ **fig A** Using a haemocytometer. You need to decide for each line whether you will count cells which are touching the line or not. In this case, the cells that are on or touching the top and left lines are counted, but the ones on or touching the right or bottom lines are ignored. You do not count any dead cells, which show up as blue. There are none in this example.

EXAM HINT

Don't forget to describe how you deal with cells that are on the grid lines.

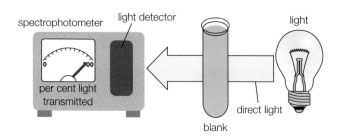
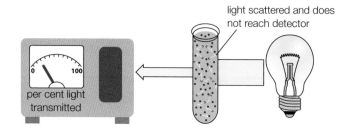

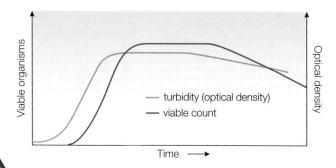

These test tubes show different levels of turbidity because they contain different levels of bacteria, from no bacteria in suspension in the tube on the left to a high level of bacteria in the tube on the right.

Safety Note: Always conduct a risk assessment prior to handling microorganisms.

▲ **fig B** Turbidimetry is a quick and easy way to measure the cell population of a culture once the initial calibration curve for an organism has been established.

DILUTION PLATING

Another way of counting the microorganisms in a culture is the technique of **dilution plating**, which is used to find the **total viable cell count**. This technique is based on the idea that each of the colonies on an agar plate has grown from a single, viable microorganism on the plate (see **fig C**). So if you have two bacterial colonies after culturing, you can presume that there were two initial living bacteria on the plate. For example, if you count 30 patches of fungal mycelia, you can say that 30 fungal cells were on the plate when it was inoculated. However, a solid mass of microbial growth is often present after culturing and it is not possible to identify the individual colonies. You can solve this problem by diluting the original culture in stages until you reach a point when you can count the colonies. You can calculate the total viable cell count for the original sample by multiplying the number of colonies by the dilution factor. Because two or more plates are often used to count individual colonies, you can obtain a mean, giving a reasonably accurate number of the cells in a particular sample. You can check the accuracy by using a haemocytometer to count the cells in the original culture.

EXAM HINT

Don't forget to revise your work from **Section 4A.6 (Book 1: IAS)**. The IAL examination will expect you to make links back to this work and explaining aseptic techniques is an obvious link.

EXAM HINT

As part of your study of this topic, you will carry out **CP13**: Investigate the rate of growth of microorganisms in a liquid culture, taking into account the safe and ethical use of organisms.

Make sure you understand this practical and know how to measure the numbers of bacteria in a liquid culture in several ways because your understanding of the experimental method may be assessed in your examination. You may also be asked how to make sure the method is repeatable or how you can be sure the results are valid.

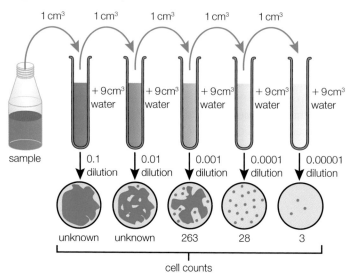

▲ **fig C** The technique of dilution plating.

AREA AND MASS OF FUNGI

When you culture fungi rather than bacteria, you can assess growth in a simple way by measuring the diameter of the individual areas of mycelium. You can use this method to compare growth rates in different conditions. For example, to find the optimum temperature for growth follow these steps.

- Inoculate identical Petri dishes containing identical growth medium with the same number of fungal spores.
- Culture the Petri dishes at different temperatures, with several identical dishes grown at each temperature.
- After a specific period of time, measure the diameter of each fungal colony and calculate the mean value for the diameter at each temperature.

The optimum temperature for growth is close to the temperature at which you obtained the largest mean diameter.

You can use the same technique for bacteria but the growth of the colony tends to be slower and therefore less easy to measure because the microorganisms themselves are so small.

One very effective way to discover the best array and concentration of nutrients or the optimum pH at which to grow fungi is to test the dry mass of the microorganism. The best method to do this is by using a liquid growth medium. You can remove samples of broth at regular intervals and separate the fungi from the liquid by centrifugation or filtering. Then you dry the material thoroughly until you record no more loss of mass, for example, in an oven overnight at around 100 °C. This gives you a measure of the dry mass of biological material in a certain volume of the culture medium, and an increase or decrease in the dry mass gives an indication of the increase or decrease in the mycelial mass. The conditions which produce the greatest dry mass of fungus are the optimum conditions for growth.

EXAM HINT

Be ready to evaluate experimental techniques. In this experiment, the dry mass grown may be so small that it is difficult to measure accurately and a small error can create a large standard deviation in your results.

CHECKPOINT

1. ▶ Look at **fig B**. Explain the difference between the curve for turbidity and the curve for viable cells.
2. Express each dilution in **fig C** as a ratio (e.g. 0.1 = 1 : 10).
3. Using **fig C**, work out the number of cells in the original sample using the fourth and fifth tubes.
4. Use your answers to question **3** and the third tube in **fig C** to work out the mean cell count in the original sample.

SKILLS ▶ CREATIVITY

SUBJECT VOCABULARY

haemocytometer a thick microscope slide with a rectangular indentation and engraved (marked) grid of lines that is used to count cells
turbidimetry a method of measuring the concentration of a substance by measuring the amount of light passing through it
turbid a term used to describe something that is opaque because of suspended matter
dilution plating a method used to obtain a culture plate with a countable number of bacterial colonies
total viable cell count a measure of the number of cells that are alive in a specific volume of a culture

Most infections are **communicable**. This means that the infection is capable of spreading from one person to another. Diseases can spread in many different ways. Sometimes part of the normal **bacterial flora** of the body will change and cause disease in response to a change in the body environment, but more often new microorganisms enter the body and cause problems.

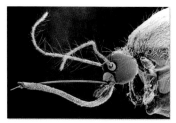

▲ **fig A** Insects such as mosquitoes carry many different diseases, including malaria, West Nile fever and dengue, from infected individuals to healthy people.

METHODS OF SPREAD

For any disease to be spread, the pathogen needs to enter the body of the new host. This can happen in a number of different ways. The body openings provide relatively easy access; examples are the eyes, nose, mouth, ears, anus and urinogenital openings. The alternative is for microorganisms to enter directly into the blood through the skin. This is a more difficult but more direct route. Pathogens are transmitted in a variety of ways.

- **VECTORS** A living organism that transmits infection from one host to another is known as a vector. Many insects are vectors of disease (see **fig A**). Examples include malaria, dengue and yellow fever.

- **INHALATION** When you cough, sneeze or talk, millions of droplets are expelled from your respiratory tract. If you have an infection of the respiratory system, those droplets all contain pathogens. Part of the water in these droplets evaporates, leaving very small droplets which are full of pathogens and small enough to remain suspended in the air for a long time. When these droplets are inhaled by another individual the pathogens enter into a new respiratory tract and another infection is established. Examples include flu, measles and tuberculosis.

- **INGESTION** Many of the pathogens that cause gut diseases are transmitted by faecally contaminated food or drink. As many of these organisms are destroyed when food is heated completely, the risk of infection is greatest from raw or undercooked food. Often only a small number of disease-causing organisms need to be absorbed to cause disease. Examples include most forms of diarrhoea, hepatitis A and *Salmonella* poisoning.

- **FOMITES** Fomites are inanimate objects that carry pathogens from one host to another. Hospital towels and bedding can be a risk, or even using someone else's cosmetics. Examples include *Staphylococcus* infections.

- **DIRECT CONTACT** Direct contact is often important in the spreading of skin diseases in small children. Many sexual diseases are also spread by direct contact of the genital organs and then the pathogens pass through mucous membranes. Examples include impetigo, HIV, gonorrhoea, syphilis and Ebola.

- **INOCULATION** A pathogen can be inoculated into the body directly through a break in the skin. This transmission might be through sexual contact if the skin is damaged, via an injury from contaminated medical instruments or shared needles in drug abuse. An infected animal may bite, lick or scratch you, or you may be exposed to pathogens through a wound from a contaminated knife or a sharp stone. Examples include hepatitis B, HIV, rabies and tetanus.

BARRIERS TO ENTRY

The body does not make it easy for pathogens to enter. Pathogens have to cross several natural barriers before infection can occur.

EPITHELIAL DEFENCES

Your skin is an impenetrable layer strengthened by keratin, a fibrous structural protein (see **Section 1A.5 (Book 1: IAS)** to remind yourself of what a fibrous protein is). It forms a physical barrier between the many pathogens in the environment and the delicate, blood-rich tissues beneath the skin. An oily substance produced by the skin, called **sebum**, contains chemicals that inhibit the growth of microorganisms and this forms a second layer of skin defences. Sebum does not harm the natural skin flora which are adapted for survival on your skin surface and also play a role in preventing disease. Your natural skin flora compete successfully for a position on the skin and also, in some cases, produce substances that inhibit the growth of other microorganisms. In fact, washing too often and using antibacterial soaps can reduce your resistance to disease by destroying the natural pH balance and surface flora of the skin (see **fig B**). The pharynx and the large intestine have protective coverings which are similar to skin.

▲ **fig B** Washing your hands after using the toilet and before preparing food is good, but too much washing can destroy the natural flora of microorganisms that live on your skin and prevent pathogens from invading.

The surfaces of internal tubes and ducts found in areas such as your nose, respiratory system, gut, urinary and reproductive tracts are more vulnerable to the invasion of pathogens than the skin on the outside of your body. This is because these internal surfaces are very thin and not strengthened with keratin. However, these epithelial layers also produce defensive secretions. Many produce mucus, a sticky substance that traps microorganisms. Mucus contains **lysozymes**, enzymes which are capable of destroying microbial cell walls. These enzymes are particularly effective against Gram-positive bacteria. They destroy the cross-linkages in the peptidoglycans in the bacterial cell wall. Lysozymes are also present in tears, the secretions which keep your eyes moist and protect them from the entry of pathogens. They are part of the non-specific defence of the body against disease (see **Chapter 6B**).

The mucus produced in the respiratory system is constantly moved towards the outside of the body, and the lining of the urinary system is constantly washed with urine. Cilia also move mucus in the reproductive tracts. For example, they move mucus along the fallopian tubes leading from the ovaries to the uterus. Phagocytic white blood cells which can engulf and digest pathogens are also often present on the epithelial surfaces. These mechanisms are so successful that the respiratory, reproductive and urinary systems usually have no bacteria in them except for the areas nearest to the outside environment. However, if microorganisms are present in sufficient numbers, they may cross these defences and reach the areas where the cells are vulnerable to infection. Examples here are the cold and flu viruses that can enter the body through the respiratory system.

EXAM HINT

Where mucus and cilia are present, it is the mucus that traps pathogens. The cilia simply move the mucus along the tubes.

When we cut ourselves, the tissues under the skin are exposed. This is when we become vulnerable to the entry of potentially disease-causing microorganisms directly into the blood. Such pathogens include herpes viruses, HIV and MRSA bacteria. In addition, some organisms, such as hookworms, ticks and biting insects, can penetrate the skin themselves. They then either cause disease directly or introduce disease-causing pathogens as in malaria and Lyme disease. The first response of the body to a break in the skin is to seal the wound through the mechanism of blood clotting. The clotting process involves fibrinogen, thrombin, blood platelets and red blood cells forming a solid clot, blocking the open wound and effectively preventing the entry of any further pathogens. You can remind yourself of this process in **Section 1B.2 (Book 1: IAS)**.

ENTERING THE GUT

The gut is also important in the natural defences of the body against disease. The saliva in your mouth has bactericidal properties. Some polypeptides produced in the salivary glands destroy bacteria and others slow down bacterial growth. Your stomach produces hydrochloric acid with a pH of approximately 2 and this effectively destroys the majority of microorganisms which you absorb through your mouth. The natural flora in the gut usually compete successfully for both nutrients and space with any microorganisms which manage to pass through the stomach. Like the skin flora, the gut flora produce anti-microbial compounds.

Another way in which the stomach helps protect against disease is vomiting. This is also a common symptom of illness or infection. Vomiting is when the contents of the stomach are forcibly removed from the body through the mouth. This complex process is partly coordinated through a number of different sites in the brain. Vomiting can be caused by many things, including absorbing toxins, bad smells, pregnancy and tumours. However, one of the most common causes is infection of the gut. If the stomach is infected with bacteria or viruses, vomiting effectively physically removes many of the microorganisms from the system. In most cases, the disease period is short and self-limiting. However, in very young children and weak elderly people, or when the disease lasts longer (as in cholera and dysentery), the loss of fluids and electrolytes, often combined with high fever, can result in severe dehydration and death.

DID YOU KNOW?

Ulcers, changing the model of a disease

For many years, people thought that gastric and duodenal ulcers were caused by stress increasing acid production in the stomach. These high levels of acid then damaged the lining of the stomach, causing an ulcer, which is a painful eroded area on the lining of the gut. Treatment involved drugs to reduce and/or neutralise the amount of stomach acid, allowing the ulcer to heal. However, ulcers often returned, presumably reflecting the continued stressful lifestyle of the patient.

In the 1980s, an Australian pathologist, Robin Warren, observed bacteria (*Helicobacter pylori*) in the stomach cells of about 50% of the stomach biopsies he saw, and noticed that inflammation was always visible as well. The scientifically accepted view then was that bacteria could not live in the stomach because of the acidic conditions. Barry Marshall was a young doctor who found these results very interesting and he and Warren started to work together. Marshall suspected the bacteria might be causing stomach ulcers and designed a series of tests to try to demonstrate this. These included infecting himself with the bacteria and observing the changes in his stomach. His work showed that ulcers were caused by bacteria, and that they could be treated with antibiotics that destroyed the bacteria so the ulcers did not return (see **fig C**).

Gastrologists who spent a lot of time treating chronic ulcer sufferers, and the drug companies who produced the best-selling drugs to relieve ulcer pain and block acid production, resisted this cheap and simple solution. It took 10 years for around 10% of doctors to be convinced by Marshall's work. Now, most doctors accept that *H. pylori* is the major cause of ulcers, not stress. Barry Marshall continued his research to investigate a link between *H. pylori* and certain stomach cancers, which is now an accepted causative relationship. This was the first time bacteria had been linked to cancer, and the World Health Organization now classifies *H. pylori* as a class 1 carcinogen. In 2005, Robin Warren and Barry Marshall received a Nobel Prize for their work.

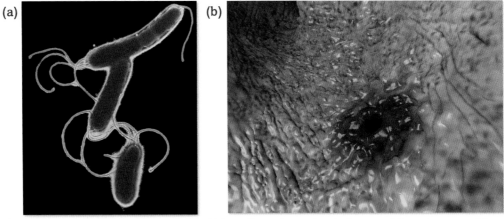

(a) (b)

▲ **fig C** (a) *Helicobacter pylori* are bacteria which can live and reproduce in the very acidic environment of the stomach. We now know they cause stomach and duodenal ulcers (b), which are very painful.

HOW DO BACTERIA CAUSE DISEASE?

Invading pathogens sometimes enter your body, in spite of all the barriers your body presents. You have seen in **Section 6A.2** how viruses cause the symptoms of disease when they cause the lysis of body cells and the release of virus particles, but how do bacteria make you feel ill?

Once pathogenic bacteria enter the body, they cause the signs and symptoms of disease in a number of ways. Some bacteria cause the symptoms of disease simply as a result of the way they invade and destroy the host tissues. However, most bacteria make people unwell because they produce toxins as a by-product of their metabolism. The toxins may prevent normal functioning of the whole host organism or just its immune system. Toxins are classified as **endotoxins** and **exotoxins**.

ENDOTOXINS

Endotoxins are **lipopolysaccharides** that are an integral part of the outer layer of the cell wall of Gram-negative bacteria. Scientists have discovered that the lipid part of the lipopolysaccharides acts as the toxin, and the polysaccharide stimulates an immune response. Bacterial endotoxins have their effect around the site of bacterial infection. The pathogenic effects often include symptoms such as fever, vomiting and diarrhoea. These can be caused, for example, by strains of *Salmonella* spp. and *E. coli*. The diseases caused by bacterial endotoxins are often not fatal themselves, but the symptoms they produce can lead indirectly to death (for example, by dehydration as a result of severe diarrhoea). Antibiotic treatments that destroy the bacterial cells by lysis of the cell wall can also lead to further endotoxin release, due to the lipopolysaccharide component of the cell wall.

EXOTOXINS

Exotoxins are usually soluble proteins that are produced and released into the host's body by bacteria as they metabolise and reproduce in the cells of their host. Exotoxins are produced by both Gram-positive and Gram-negative bacteria, and have greater effects than endotoxins as they often act at sites far from the infecting bacteria (see **table A**). There are many different types of exotoxin with specific effects. For example, some damage cell membranes causing cell breakdown or internal bleeding; some act as competitive inhibitors to neurotransmitters; others directly poison cells. Exotoxins rarely cause fevers, but they cause some of the most dangerous and fatal bacterial diseases. For example, *Clostridium botulinum* produces botulinum toxin, one of the most toxic natural substances known. It is estimated that 1 mg of the pure toxin could kill 1 million guinea pigs.

EXOTOXIN	LETHAL DOSE /mg	HOST ANIMAL	LETHAL TOXICITY COMPARED WITH:		
			STRYCHNINE	SNAKE VENOM	BACTERIAL ENDOTOXIN
botulinum toxin	0.8×10^{-8}	mouse	3×10^6	3×10^5	3×10^7
tetanus toxin	4×10^{-8}	mouse	1×10^6	1×10^5	1×10^7
diphtheria toxin	6×10^{-5}	guinea pig	2×10^3	2×10^2	2×10^4

table A Comparison of bacterial exotoxins with other toxins and bacterial endotoxins

HOST TISSUE INVASION

The third way bacteria act as pathogens is by invading host tissues and damaging the cells. The response of the host organism to the cell damage causes the symptoms of disease. Often this cell damage is also linked to the production of exotoxins or the presence of endotoxins in the bacterial cell walls. One of the most widespread and damaging bacterial infections of this type is tuberculosis (see **Section 6A.7**).

CHECKPOINT

1. Produce a table to summarise the main ways in which pathogens enter the body. For each way, explain how the natural barriers of the body help to prevent infection.

2. The enzyme lysozyme is almost universally present in animals. Explain its role in the non-specific defence against disease.

3. Explain the role of physical barriers, chemical barriers and biological defences in the way the skin and the gut help prevent the invasion of pathogenic microorganisms.

SKILLS CRITICAL THINKING

4. Bacterial endotoxins usually have a relatively local effect but exotoxins can affect many areas of the body. Explain this observation.

SUBJECT VOCABULARY

communicable diseases caused by pathogens which can be passed from one organism to another

bacterial flora the combination of different species of microorganisms found in or on a specific region of the body

sebum an oily substance produced by the skin which contains chemicals that inhibit the growth of microorganisms

lysozymes enzymes found in tears and other body secretions that are capable of destroying microbial cell walls

endotoxins lipopolysaccharides that are an integral part of the outer layer of the cell wall of Gram-negative bacteria and act as toxins to other cells

exotoxins soluble proteins produced and released into the body by bacteria as they metabolise and reproduce in the cells of their host; these proteins act as toxins in different ways

lipopolysaccharides chemicals made up of a combination of lipids and polysaccharides (complex carbohydrates)

6A 7 CASE STUDIES OF DISEASE: TUBERCULOSIS

LEARNING OBJECTIVES

■ Understand how *Mycobacterium tuberculosis* and human immunodeficiency virus (HIV) infect human cells, causing symptoms that may result in death.

It seems amazing that microscopic pathogens can cause animals as large as human beings to become seriously ill and even to die. In this section, you will look at a bacterial disease which infects, damages and kills millions of people each year.

TUBERCULOSIS (TB)

Tuberculosis (TB) is one of the most common human infections (see **fig A**). World Health Organization figures show that in 2016 there were around 10.4 million new cases of TB worldwide, and approximately 1.7 million people died of the disease. The numbers who died included 0.4 million people who also had HIV/AIDS (see **Section 6A.8**). Up to a third of the population of the world has been infected by TB, but many of these people have no symptoms. At least one in 10 of these symptomless people will develop active TB later and, without effective treatment, 50% of people with TB die of the disease.

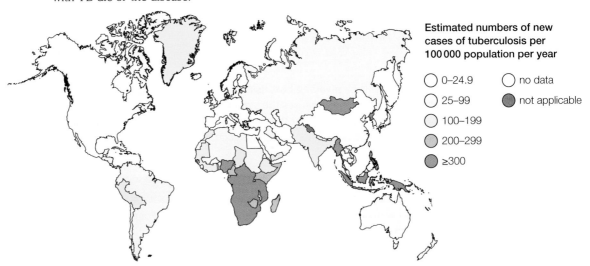

Estimated numbers of new cases of tuberculosis per 100 000 population per year

- ○ 0–24.9
- ○ 25–99
- ○ 100–199
- ○ 200–299
- ○ ≥300
- ○ no data
- ● not applicable

▲ **fig A**   TB affects people all over the world, but seven countries account for between 60% and 70% of all new cases. They are China, India, Indonesia, Philippines, Pakistan, Nigeria and South Africa (based on data from World Health Organization).

THE CAUSES AND SYMPTOMS OF TB

TB is commonly caused by the bacterium *Mycobacterium tuberculosis,* which is spread by droplet infection (see **Section 6A.6**). People living or working in crowded conditions have an increased probability of being infected because the disease is spread through people breathing, coughing and sneezing near to each other. People who are malnourished, ill or have problems with their immune systems are more vulnerable to the disease and are much more likely to develop active TB than healthy, well-fed individuals. People living with HIV/AIDS are particularly vulnerable to infection with *M. tuberculosis* as a result of their reduced immune response.

Globally, the other common source of infection is from the bacterium *Mycobacterium bovis,* which particularly affects cattle. People become infected by drinking infected milk or living and working in close contact with cattle.

TB often affects the respiratory system, damaging and destroying lung tissue. It also suppresses the immune system, making the body less able to fight the disease. The well-known symptoms of TB, which are weakness and coughing up blood, appear at the end of the disease process.

Only about 30% of the people exposed to TB will become infected. In the **primary infection**, the bacteria which have been inhaled into the lungs multiply slowly, often causing no obvious symptoms. If you have a healthy immune system, there will be a localised inflammatory response forming a mass of tissue called a **tubercule**, which contains dead bacteria and macrophages. After about eight weeks the immune system controls the bacteria, the inflammation disappears and the lung tissue heals. Primary TB infections often occur in childhood, and the majority (about 90%) heal without the infected individual ever realising they have had TB.

However, *M. tuberculosis* has an adaptation that allows it to avoid the immune system. This means some bacteria may survive the primary infection stage. The bacteria produce a thick waxy outer layer which protects them from the enzymes of the macrophages. Bacteria with an effective coating will remain deep in the tubercules in the lungs, dormant or growing slowly for years until the person is malnourished, weakened or their immune system does not work well. These bacteria then cause active tuberculosis. In this way, the most effective bacteria are selected and their DNA will be passed on. Once they become active again, they can grow and reproduce very rapidly, causing serious damage and disease.

The active phase is sometimes simply the result of new infection. If a really heavy load of bacteria enters the lungs, the infection has an effect before the immune system can respond properly. About 80% of people with active TB are suffering from a reactivation of an old, controlled infection rather than a new infection.

During active TB, the bacteria multiply rapidly in the lungs. They cause all the typical symptoms of TB, including fever, night sweats, loss of appetite and loss of weight. Patients often feel increasingly tired and lack energy. They cough because of the infection in the lungs and produce a liquid from their lungs known as sputum. As the infection becomes worse, lung tissue becomes damaged and blood may be coughed up in the sputum. Without treatment, the affected person will steadily lose weight, and the lungs will be destroyed as the alveoli break down to produce large, inefficient air spaces.

Mycobacteria also target T cells, reducing the production of antibodies and so disabling a critical part of the immune system. The bacteria are temperature sensitive and stop reproducing above 42 °C. A patient with active TB will often have a high temperature. However, human enzymes start to be denatured at around 40 °C, so the patient is at risk of dying before the bacteria are inhibited.

Eventually TB causes death, either because the individual cannot obtain enough oxygen from the air through their damaged lungs, or because their organs fail through lack of nutrition. And, since TB affects the immune system itself, sufferers often become vulnerable to opportunistic infections such as pneumonia, which may be the infection that actually kills them.

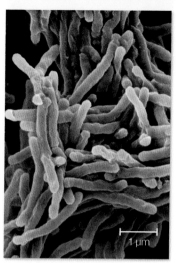

▲ **fig B** *Mycobacterium tuberculosis*: these bacteria kill around 250 000 children and almost 1.5 million adults every year.

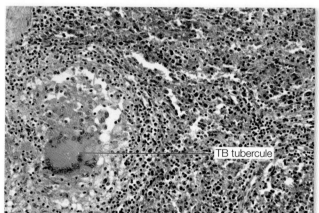

TB tubercule

▲ **fig C**   The bacteria in a tubercule are often destroyed by macrophages, but sometimes they will survive for months or even years before resulting in active TB.

EXAM HINT

Remember that you need to link your knowledge from **Section 2A.5–6 (Book 1: IAS)** to your IAL studies. You may be given a photomicrograph like this and be asked to describe, explain or suggest the effects on gas exchange.

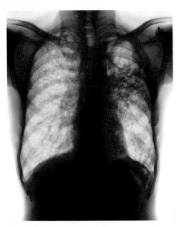

▲ **fig D** The shadowy area on the lung in this specially coloured X-ray is evidence of the damage caused by TB.

TREATMENT AND CONTROL

It is important to be able to diagnose TB as easily and cheaply as possible. A chest X-ray will show TB damage to the lungs as opaque areas and large, thick-walled cavities (see **fig D**). However, other diseases can give similar images. So chest X-rays can suggest that TB is present but not confirm it. New tests examining the DNA of the bacteria give relatively fast and reliable results that tell us if a person has TB and also if the bacteria are resistant to the antibiotic most commonly used to treat it.

The main treatment for tuberculosis is with antibiotics for many months. For the first two months, a mixture of different antibiotics is used. This is because the bacteria may be resistant to some of the commonly used drugs, but they will be affected by others in the mixture. This mixture destroys the rapidly reproducing bacteria and also attacks the bacteria that are hidden in the tissues of the body or within the immune system and are metabolising slowly. This is followed by treatment with two antibiotics for another four to seven months. Most patients will be completely free of TB after about nine months, but it can take a year to 18 months before all traces of active disease have disappeared.

The most effective ways of controlling TB involve improving living standards. People are less likely to pass on the disease if they are not living and working in crowded conditions. People who are well nourished and generally healthier have a lower probability of developing active TB after infection. Preventing and treating the disease in cattle, and pasteurising milk before it is drunk, help prevent the spread of *M. bovis*. In countries where TB is already rare, attempts are made to contact all the people in regular close contact with an infected individual so that they can be tested for TB and either immunised or treated as appropriate.

Immunisation has been very effective in the final reduction in cases of TB in countries such as the UK where the incidence is low. You will learn more about vaccination in **Chapter 6B**.

Until recently, particularly in developed countries, we seemed to be winning against *M. tuberculosis*. However, the number of cases of TB is increasing again. Reasons for this could include deteriorating social conditions in some areas, immigration and the movement of refugees between countries, more foreign travel, an increase in intravenous drug use and, most of all, increasing numbers of people infected with HIV/AIDS. One worrying development is the evolution of multidrug-resistant strains of *M. tuberculosis*, making treatment more difficult and more expensive and increasing the risk of people dying. You will learn more about antibiotic resistance in bacteria in **Chapter 6B**.

CHECKPOINT

1. Explain why *Mycobacterium tuberculosis* is such a successful pathogen.
2. Describe the symptoms of TB and explain how these can lead to death.
3. ▶ Look at **fig A**. Suggest reasons why some countries have such a problem with tuberculosis.

SKILLS ▶ ANALYSIS

SUBJECT VOCABULARY

primary infection the initial stage of tuberculosis, when *M. tuberculosis* has been inhaled into the lungs, invaded the cells of the lungs and multiplied slowly, often causing no obvious symptoms of disease
tubercule a localised mass of tissue containing dead bacteria and macrophages formed as a result of a healthy immune response to an infection by *M. tuberculosis*

LEARNING OBJECTIVES

■ Understand how *Mycobacterium tuberculosis* and human immunodeficiency virus (HIV) infect human cells, causing symptoms that may result in death.

HIV/AIDS

Acquired immunodeficiency syndrome (AIDS) is a relatively new disease: the first cases were identified in the 1980s. In 2016, 36.7 million people in the world were living with HIV/AIDS, and around 1 million people died of HIV-related illnesses. We know the infective agent is the **human immunodeficiency virus (HIV)** and we know how it is transmitted. Unfortunately, we have yet to stop the spread of the disease despite much funding for research.

HIV/AIDS starts with infection by HIV, a virus which attacks the cells of the immune system that normally protect us against disease. After time, the infection can lead to a disease called acquired immunodeficiency syndrome or AIDS. This disease has many symptoms, including fevers, persistent diarrhoea and weight loss. Secondary infections are common and include TB, a rare type of pneumonia and Kaposi's sarcoma. Death is frequently the result of secondary infection.

Infection by HIV does not immediately lead to AIDS. People infected by the virus, but showing no symptoms, are **HIV positive**. This means blood tests show the presence of HIV antibodies. **Table A** and **fig A** show the results from the Joint United Nations Programme on HIV/AIDS in 2016. The UN programme found that 90% of infected people live in the developing world. In sub-Saharan Africa nearly one in every 25 people is infected with HIV and AIDS was the single largest cause of death. However, trends are as important as absolute figures. The numbers of people infected with HIV and dying of AIDS in Africa and around the world are falling. New HIV infections among adults have decreased by 11% since 2010, and new infections in children have decreased by 47%, from 300 000 in 2010 to 160 000 in 2016. There is a long way to go: since the epidemic started, more than 70 million people have been infected and around 35 million have died, but the future looks encouraging.

> **EXAM HINT**
>
> Be clear about the terms.
>
> HIV is a virus or pathogen. AIDS is a medical condition caused by infection with HIV.
>
> AIDS comprises a wide range of symptoms and conditions resulting from secondary infection by a number of opportunistic pathogens.

Prevalence of HIV/AIDS in 2016 (% of population)

East Mediterranean
0.1 (<0.1–0.1)

West Pacific
0.1 (<0.1–0.2)

South-east Asia
0.3 (<0.2–0.4)

Europe
0.4 (<0.4–0.4)

The Americas
0.5 (<0.4–0.6)

Africa
4.1 (<3.4–4.8)

▲ **fig A** World map showing world distribution of HIV/AIDS in 2016 (based on data from the World Health Organization).

DATA FROM 2016	TOTAL / MILLIONS	ADULTS / MILLIONS	CHILDREN UNDER 15 YEARS / MILLIONS
NUMBER OF PEOPLE LIVING WITH HIV IN 2016	36.7 (30.8–42.9)	34.5 (28.8–40.2) of which 17.8 (15.4–20.3) are women	2.1 (1.7–2.6)
PEOPLE NEWLY INFECTED WITH HIV IN 2016	1.8 (1.6–2.1)	1.7 (1.4–1.9)	0.16 (0.1–0.2)
AIDS DEATHS IN 2016	1.0 (0.8–1.2)	0.88	0.12

table A The size of the AIDS problem worldwide in 2016. The numbers in brackets define the boundaries within which the number lies, based on the best available information.

HOW IS HIV TRANSMITTED?

HIV is very fragile and cannot survive in the air. It must be contained in human body fluids. The source of infection may be someone who is HIV positive (who may not be aware of their infective status) or someone with active AIDS. The virus can be transmitted from person to person in three main ways:

- direct exchange of bodily fluids; this is the most common way it is passed from one person to another

- inoculation through infected blood; people can also be infected by infected blood products (if these are not treated)

- from a mother to her fetus in the early stages of pregnancy, during birth or through breastfeeding.

HOW DOES HIV CAUSE AIDS?

Most of the symptoms of AIDS result from the effect of HIV on the immune system. You will learn more about the way the immune system works in **Chapter 6B**. HIV attaches to receptors on T helper cells and is then able to infect these cells (see **fig B**).

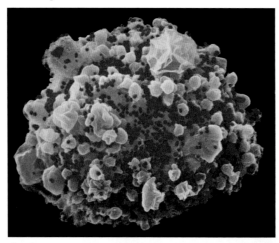

▲ **fig B** A scanning electron micrograph showing a large white blood cell of the immune system (a T helper cell) being attacked by HIV.

HIV is a retrovirus. This means that once it enters the T helper cell, it controls the host DNA and replicates (see **6A.1 fig C**, **6A.2 fig B** and **fig C** below). The host T helper cell is destroyed when the new viruses leave it. At the same time, other cells of the immune system called T killer cells recognise and destroy some of the heavily infected T helper cells. These two processes cause a large decrease in the number of T helper cells. As a result, the immune system cannot fight other pathogens. This means that individuals infected with HIV are vulnerable to secondary infections such as pneumonia and TB. It is often these secondary infections that kill patients with HIV/AIDS.

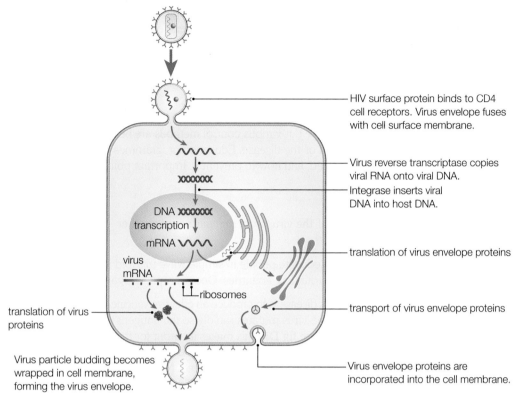

HIV surface protein binds to CD4 cell receptors. Virus envelope fuses with cell surface membrane.

Virus reverse transcriptase copies viral RNA onto viral DNA.

Integrase inserts viral DNA into host DNA.

translation of virus envelope proteins

transport of virus envelope proteins

Virus envelope proteins are incorporated into the cell membrane.

translation of virus proteins

Virus particle budding becomes wrapped in cell membrane, forming the virus envelope.

DNA XXXXXXX
transcription ↓
mRNA VVVV

virus mRNA

ribosomes

▲ **fig C** Scientists look at what happens when HIV controls a cell. They want to use the different stages as a target for new drugs or vaccines against HIV/AIDS.

THE COURSE OF AN HIV/AIDS INFECTION

The speed at which the disease progresses from HIV infection to full-blown (advanced) AIDS depends on many factors. A person who is fit, well-nourished and healthy will probably remain well longer than someone who is malnourished and unhealthy. Some people have a better genetic resistance to infection than others. Their immune system responds more effectively to HIV and so the disease takes longer to progress. Rapid medical treatment can also increase life expectancy by many years. There are four main stages in the progress of HIV/AIDS.

- **STAGE 1 ACUTE HIV SYNDROME** In the first few weeks after infection, some people feel unwell. Symptoms include fever, headaches, tiredness and swollen glands, but all of these are similar to much less serious viral and bacterial diseases. Some people infected with HIV have no symptoms. Between 3 and 12 weeks after infection, HIV antibodies appear in the blood, making the person test as HIV positive. This will happen even if they have not felt ill.

- **STAGE 2 THE ASYMPTOMATIC OR CHRONIC STAGE** Once the infection is established all symptoms disappear. In fit, young people with access to effective **anti-retroviral drugs** this stage can last many years. However, around the world many HIV-positive people have little food or medicine. For these people, the symptomless stage of the disease will be relatively short. During the asymptomatic stage, the virus replicates, infecting the T helper cells, but it is kept under control by the T killer cells. The individual is very infectious but they may not even know they have the disease. As this stage progresses, secondary infections develop because the immune system is unable to deal with the situation.

- **STAGE 3 SYMPTOMATIC DISEASE** Eventually the number of viruses attacking the immune system becomes so great that the whole immune system starts to fail. The normal T helper cell count falls from 500 to 200 per mm^3 of blood. Patients begin to suffer HIV-related symptoms, including weight loss, fatigue, diarrhoea, night sweats and low-grade infections such as thrush. This rapidly progresses to the final stage.

- **STAGE 4 ADVANCED AIDS** As T helper cell numbers fall, severe symptoms begin to appear such as major weight loss, dementia as brain cells become infected, cancers (e.g. Kaposi's sarcoma) and serious infections such as TB and cryptococcal meningitis. The worldwide increase in HIV/AIDS is one of the main causes of the worldwide increase in cases of TB. The final stage of advanced AIDS is always death.

TREATING AIDS

At the moment, AIDS is an incurable disease but various control methods are being investigated. The simplest method is to limit the spread of the disease. Education programmes help people to understand the ways in which HIV is spread, and how to prevent it. Important points include:

- celibacy
- only having one sexual partner
- using condoms to prevent the spread of the virus from one partner to another
- using clean needles if injecting drugs.

Some countries have made a big effort to educate people in simple ways to avoid infection and infection levels there are lower than in surrounding countries. One example is Uganda.

The usual approach to controlling infectious disease is to produce an effective vaccine (see **Chapter 6B**), but this is proving very difficult for HIV. The virus mutates rapidly so the antigens on the viral coat keep changing in the years after infection (see **fig D**). This makes it harder for the immune system to recognise the virus and destroy it. The rate of change only slows down as the T cell count starts to fall to very low levels so selection pressure on the virus is reduced. Natural selection also favours mutations that enable the virus to replicate particularly fast. This allows the virus to infect many cells very quickly. As a result, by the time a vaccine has gone through all its development and safety tests the virus has changed and the vaccine is not effective. In addition, AIDS is found mainly in humans. There are, therefore, not many animals that scientists can use as a model for their research. Rhesus macaques and chimpanzees are some of the only other animals that seem to be susceptible to the virus, and so work on animals to develop vaccines is very limited.

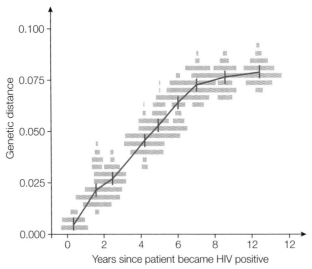

▲ **fig D** This graph shows you how quickly the genome of an HIV population changes in one patient. You can see how much it changes until eventually it slows down when the immune system collapses.

Drug therapy is also very important. Combined drug therapies which are anti-retroviral and stimulate the immune system mean that AIDS is becoming a treatable, long-term disease. The drugs are quite expensive and so they are mostly used in the developed world. The major pharmaceutical companies have supplied large quantities of drugs to many developing countries for no cost or at a minimal charge. Unfortunately, without comprehensive medical infrastructure and where there is great political instability, the drugs are often not available to those who need them.

None of the current drugs provides a cure, but they can delay the onset of advanced AIDS for many years. Recent UK figures show that if someone is diagnosed HIV positive at around 20 years old and given immediate anti-retroviral therapy, their life expectancy will be similar to that of a healthy person of the same age. The sooner treatment begins, the longer the patient can expect to live.

DID YOU KNOW?

HIV evolution vindicates doctors

In 2007, six medical workers were sentenced to death in Libya, accused of deliberately infecting 426 children with HIV-infected blood. In 2008, biologists produced genetic profiles (see **Chapter 6C**) from the HIV viruses infecting the children. By pinpointing the number of mutations which had taken place in the HIV viruses in the different children, they showed that the infection originated in a single case several years before the medical workers had arrived at the hospital. In addition, the viruses were closely related to West African strains, indicating the original virus was probably introduced accidentally by a West African worker or patient.

CHECKPOINT

1. Describe the symptoms of HIV infection and explain how these may result in death.

2. Compare HIV/AIDS and TB infections and comment on similarities and differences between them.

3. ▶ Explain how the mechanisms that allow HIV and *Mycobacterium tuberculosis* to evade the human immune system support the idea of an evolutionary race between pathogens and their hosts.

> **SKILLS** ▷ REASONING

SUBJECT VOCABULARY

acquired immunodeficiency syndrome (AIDS) the disease which results from the destruction of the T helper cells as a result of infection with HIV

human immunodeficiency virus (HIV) a retrovirus that causes AIDS

HIV positive someone who has antibodies to HIV in their blood, indicating that they have been infected with the virus and so are at risk of passing it on to other people

anti-retroviral drugs drugs which are effective against retroviruses such as HIV

EBOLA: A DEADLY VIRUS

SKILLS ▶ CRITICAL THINKING, ANALYSIS, INTERPRETATION, CREATIVITY, INITIATIVE, SELF-DIRECTION, ASSERTIVE COMMUNICATION

In 1976, a new and deadly disease appeared in Africa. Ebola virus (see **fig A**) causes ebola fever, a disease which has a 50–90% death rate mainly because of the extensive damage to body tissues. It is both contagious (spread by contact) and infectious (spread by droplets in the air). Symptoms can appear from two to 21 days after the initial infection. Reporting on this viral disease varies greatly, as you can see from the three extracts below.

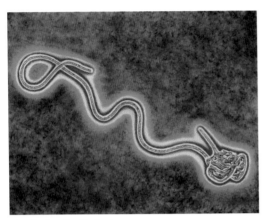

▲ **fig A** The Ebola virus

POPULAR NEWSPAPER

The deadly Ebola virus arrived in Nigeria on an international flight … where will it travel next?

- **The Ebola virus lays dormant in victims for up to three weeks.**
- **Up to 90% of the people infected with Ebola virus will die of it.**
- **Health organisations warn that anyone travelling from an affected country could carry the virus with them, and spread the disease around the world in hours.**
- **Nigeria shows that Ebola CAN be contained.**

On 20th July 2014 Patrick Sawyer, a Liberian-American lawyer, flew into Nigeria from Liberia carrying the Ebola virus. He rapidly became seriously ill and died 4 days later in hospital.

Nigeria's Ministry of Health immediately put an effective quarantine system in place. Anyone who had had contact with Patrick Sawyer, or the people who had nursed him, was quarantined for 21 days. As a result, only 19 people in total were infected and good nursing meant only 7 people died, only 37% of those infected.

INTERNATIONAL HEALTH ORGANISATION NEWS FEED

Surveillance and epidemiology

The resurgence of Ebola infections and the high level of deaths in Guinea raised concerns among scientists about unrecognised ways of transmitting the virus within the community. International organisations such as the World Health Organization (WHO) carried out ongoing monitoring of the Ebola virus disease (EVD) outbreak in Sierra Leone, Liberia, and Guinea in 2014. Between the 21st and 23rd July 2014, Guinea had 12 new cases of Ebola and 5 deaths after weeks of low viral activity. The results of the continued monitoring led to a renewed emphasis on contact tracing: infected individuals were prevented from returning into the community carrying the Ebola virus with them.

SCHOOL TEXTBOOK

What is Ebola?

Ebola virus disease is caused by infection with the Ebola virus. It is a serious, often fatal disease which we still cannot cure. It is mainly spread through contact with the bodily fluids of an infected person. Patients are given fluids and salts to keep their bodies hydrated and everything is done to maintain life until the immune system overcomes the virus.

The disease mainly appears in West Africa, although infected people have travelled and carried the virus around the world.

What are the symptoms?

Symptoms of infection by Ebola virus include a high fever, bad headaches, aching joints and muscles, muscle weakness, diarrhoea, vomiting and a very sore throat. The symptoms usually start suddenly, 2–21 days after you have become infected.

If you feel unwell on returning from a country affected by Ebola, ring the emergency services immediately. Explain the risk so isolation treatment can be prepared to minimise the spread of any potential infection and prevent the disease from spreading in other countries.

SCIENCE COMMUNICATION

Look at the different levels of information given in these pieces of writing, aimed at different audiences. The writing is for (i) a popular newspaper, (ii) the website of an international health organisation and (iii) a school textbook in a country not usually affected by Ebola.

1 Which article seems to be the most scientific and which is the least scientific? Explain your answer.

2 (a) Discuss the different objectives of the **three** pieces of writing.

(b) Describe the differences between the **three** pieces of writing and explain how the differences relate to the purpose of each article.

INTERPRETATION NOTE

Remember that a newspaper has to persuade people to buy copies or pay an online subscription. The other sources might be aimed at those with a genuine interest in researching detailed facts about the topics.

BIOLOGY IN DETAIL

Now let's examine the biology given in each piece of writing.

3 Look at the newspaper article and summarise the knowledge about Ebola at the end of the extract. How accurate is that information biologically?

4 The piece of writing for an international health organisation gives you little information about the Ebola virus or how it is spread. Visit the World Health Organization website and discover how Ebola is covered there.

(a) Summarise what the WHO website tells you about the life cycle of the Ebola virus.

(b) Explain why contact tracing is so important in preventing the spread of Ebola.

5 Summarise what the school textbook tells you about the Ebola virus and how it is spread. Suggest **three** other facts it would be useful for students to know about the disease.

ACTIVITY

Research is important in preventing the spread of viruses like Ebola. Discover as much as you can about the Ebola virus. Focus on how it is spread and the way it infects and takes control of the cells of the body.

• Think carefully about which stage of the virus life cycle you would target to try and prevent the spread of the disease.

• Imagine you have to bid for international funding to carry out your research. Create a poster presentation summarising the problem you have identified with Ebola. Use it to explain what you want to research and why you should receive the funding.

You can look at online resources, including encyclopaedias, scientific magazines and journals, scientific papers, books, etc. For each of them, judge the reliability of your source before you use it. In an academic presentation like this it is vital to reference all your sources.

THINKING BIGGER TIPS

Use your knowledge to be very clear about the different types of virus and the ways they reproduce in cells.

Refer to the information in your textbook and look on reliable websites to help you. Think very carefully about where and when a virus might be vulnerable to attack during its reproductive cycle.

1 (a) Which row correctly describes the components found in prokaryotic and eukaryotic cells? [1]

	Prokaryotic	Eukaryotic
A	mitochondria, chloroplasts and ribosomes	mitochondria, chloroplasts and ribosomes
B	ribosomes, plasmids, flagellum	nucleus, ribosomes, mitochondria
C	mitochondria, chloroplasts and ribosomes	nucleus, ribosomes, mitochondria
D	nucleus, plasmids, ribosomes	flagellum, chloroplasts, nucleus

(b) The figure below shows the apparatus used to make counts of bacterial cells.

   (i) What is the name of this apparatus? [1]

     A grid   **B** cell counter
     C haemometer   **D** haemocytometer

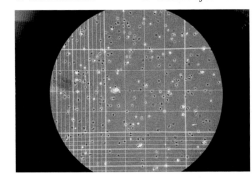

   (ii) Describe how this apparatus is used. [4]

(c) The figure below shows a diagram of part of the same apparatus.

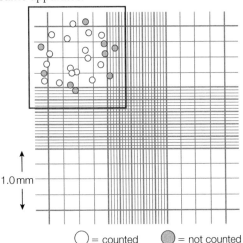

○ = counted   ● = not counted

   (i) Suggest why some of the cells have been stained. [2]

   (ii) State the number of viable cells per mm². [1]

   (iii) During an investigation a student would make this count in at least three separate squares. Explain why. [2]

(Total for Question 1 = 11 marks)

2 (a) Describe **three** ways in which HIV can be transmitted. [3]

(b) The table below shows the percentage of people with new HIV infections in four different parts of the world.

	Percentage of people with new HIV infections			
Year	**Western Europe**	**Eastern Europe**	**Far East**	**sub-Saharan Africa**
1980	0.0	0.0	0.0	0.0
1990	0.1	0.2	0.1	2.0
2000	0.4	0.5	0.3	8.5
2010	0.3	0.9	0.3	15.2

   (i) Describe the trends shown by the data in the table. [5]

   (ii) Suggest why the trend in Western Europe is different from the trend in other parts of the world. [4]

(c) Discuss the advantages and disadvantages of treating HIV/AIDS patients with antibiotics. [4]

(Total for Question 2 = 16 marks)

3 An investigation was carried out to study the effect of pH on the growth of *Escherichia coli* (*E. coli*) and *Lactobacillus bulgaricus* (*L. bulgaricus*).

- A set of five agar plates was prepared.
- Each plate contained agar at a different pH (5, 6, 7, 8 and 9).
- 0.1 cm³ of *E. coli* suspension was spread out over each of the agar plates.
- The agar plates were incubated for two days.
- After two days, the number of colonies on each plate was counted.
- The whole procedure was repeated with *L. bulgaricus*.

The graph below shows the results of this investigation.

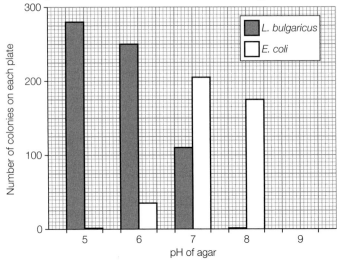

(a) State **two** precautions that should have been taken during this investigation to ensure that the results were valid. For each of the precautions you gave, explain how the results would have been affected if the precaution had not been taken. [4]

(b) (i) Which row of this table correctly states the conclusion that you can draw from this data? [1]

	Optimum pH for *E. coli*	Optimum pH for *L. bulgaricus*
A	7	5
B	uncertain	uncertain
C	uncertain	5
D	7	uncertain

(ii) Describe the effect of pH on the growth of *E. coli* and *L. bulgaricus*. [5]

(iii) Explain the effect of pH on the growth of both types of bacteria. [3]

(Total for Question 3 = 13 marks)

4 (a) During the production of yoghurt, the number of cells of the bacterium *Lactobacillus bulgaricus* changes. The graph below shows the changes in the $\log_{10}$ numbers of this bacterium and the changes in the pH during incubation.

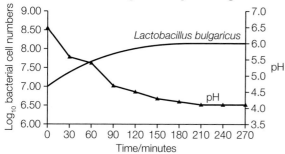

(i) Describe the relationship between the number of cells of *Lactobacillus bulgaricus* and the changes in the pH in the yoghurt. [3]

(ii) Calculate the number of generations of *Lactobacillus bulgaricus* produced during the first 120 minutes, using the formula below.

$$n = \frac{\log_{10} N_1 - \log_{10} N_0}{\log_{10} 2}$$

Where n is the number of generations

N_0 = number of bacteria at 0 hours

N_1 = number of bacteria at 120 minutes

$\log_{10} 2 = 0.301$

Show your working. [2]

(b) The following graph shows the number of cases of food poisoning caused by *Salmonella* and *Staphylococcus* in Japan between 1985 and 1998.

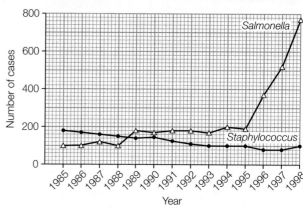

(i) Compare the number of cases of food poisoning caused by *Salmonella* with the number of cases of food poisoning caused by *Staphylococcus* for the years between 1985 and 1998. [3]

(ii) Food poisoning can be caused by the endotoxins of *Salmonella* and the exotoxins of *Staphylococcus*. State **two** differences between endotoxins and exotoxins. [2]

(Total for Question 4 = 10 marks)

5 (a) What type of pathogen is *Mycobacterium*? [1]

    A bacterium   **B** virus

    C fungus   **D** protist

(b) What is the disease caused by organisms in the genus *Mycobacterium*? [1]

    A malaria   **B** pneumonia

    C tuberculosis   **D** influenza

(c) Explain how *Mycobacterium* is transmitted and how it causes this disease. [3]

(d) The graph below shows the number of new cases of TB and the number of deaths reported each year in Scotland, between 1980 and 2004.

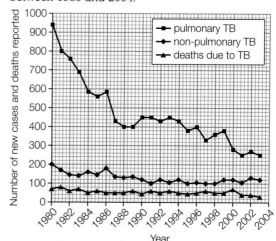

(i) Describe how the number of new cases of pulmonary TB changed between 1980 and 2004. [2]

(ii) State **two** reasons why the number of cases of TB has started to rise in recent years. [2]

(Total for Question 5 = 9 marks)

TOPIC 6 MICROBIOLOGY, IMMUNITY AND FORENSICS

6B IMMUNITY

In the 1980s, a new disease began to be contracted: the disease we now know as HIV/AIDS. It spread rapidly across the globe and still affects millions of lives. The human immunodeficiency virus causes an autoimmune deficiency syndrome. Although we can now keep the disease at bay with a mixture of drugs, including anti-retrovirals, we still have no effective vaccine or cure for the disease. Why is it so devastating? HIV attacks the immune system itself. It hides itself within key components of the immune response. This has two major effects. It reduces and damages the normal immune response of the body to other invading pathogens. Because it is hidden within the immune cells, it cannot be detected and destroyed by the immune system and so it remains safe inside the body to multiply and spread.

In this section, you will discover how the body responds to invading pathogens. You will consider the non-specific responses to infection, including the macrophages, neutrophils and lymphocytes. You will also look at the complexities of the specific immune system and the elegant way in which the humoral and cell-mediated immune responses work together. You will investigate the roles of the T and B memory cells in the immune response and use this insight to help you understand how we can induce artificial immunity to protect ourselves against many potentially deadly diseases.

MATHS SKILLS FOR THIS CHAPTER

- Recognise and make use of appropriate units in calculations (*e.g. calculating reduction rates in deaths due to communicable diseases over a stated time period*)

- Translate information between graphical, numerical and algebraic forms (*e.g. considering mortality rates during pandemic disease*)

What will I study in this chapter?

- The non-specific response of the body to infection, **including** inflammation, lysozyme action, interferon and phagocytosis
- The response of the body to infection, including **the action of the macrophages, neutrophils and lymphocytes**
- The development of the humoral and the cell-mediated immune responses
- The role of T and B memory cells in the secondary immune response
- How immunity can be natural or artificial, and active or passive
- How vaccination is used in the control of communicable diseases and the development of herd immunity, including the potential problems in a population where a proportion of people choose not to vaccinate their children
- How microorganisms develop resistance to antibiotics and the factors behind the spread of antibiotic-resistant hospital-acquired infections

What prior knowledge do I need?

Chapter 1B (Book 1: IAS)

The main types of cell in the blood

Chapter 2A (Book 1: IAS)

The structure of the cell membrane and cell signalling

Chapter 2B (Book 1: IAS)

The structure and replication of DNA and RNA

The control of protein synthesis

Chapter 3A (Book 1: IAS)

The ultrastructure of eukaryotic cells

The structure of prokaryotic cells

The importance of staining specimens in microscopy

What will I study later?

Chapter 6C

- The role of microorganisms in the process of decay

Chapter 8C

- The way genetic modification can be used to help produce vaccines against diseases that are difficult for the immune system to identify
- The way genetic modification can be used to help prevent non-communicable diseases
- How DNA sequencing can be used to identify specific pathogens

LEARNING OBJECTIVES

■ Understand the non-specific responses of the body to infection, including inflammation, lysozyme action, interferon and phagocytosis.

The body has many barriers, but millions of bacteria and viruses manage to enter it every day. Yet you do not constantly suffer from disease. This is because, in most cases, your body recognises that it has been invaded and destroys or inactivates the pathogens. This is the result of a number of different response systems in your body. Many of these responses depend on a process of cell recognition.

CELL RECOGNITION

The ability of the body to distinguish between its own cells ('self') and foreign cells or organisms ('non-self') is key to how the immune system works. The cells of different organisms have differing genetically determined protein molecules on their surface membrane that seem to be essential to cell recognition. Such proteins include glycoproteins, which are protein molecules with a carbohydrate component (see **Section 1A.3 (Book 1: IAS)**). These chains of sugar molecules are often different from cell to cell and are important in cell recognition in several ways.

The ability of the body to identify 'self' is vital when tissues and organs are forming in embryonic development. Similar sugar recognition sites may bind to each other, holding cells together. The ability to identify pathogens and any other foreign cells that enter the body as 'non-self' is equally important. Non-self glycoproteins act as **antigens** and are recognised by white blood cells (leucocytes) during the specific immune responses. An antigen is any substance that stimulates an immune response in the body. Antigens are often chemicals on the surface of a cell such as the proteins, glycoproteins or carbohydrates described here. They can also be toxins made by bacteria or whole microorganisms such as bacteria or viruses.

Pathogens that enter your body tissues will be met by a number of different defences. Some of these responses are non-specific and they are initiated by any pathogen. Others are specific to particular pathogens (see **Section 6B.2**). Many of the non-specific and specific responses of the body depend on different types of **leucocyte** (white blood cell).

THE LEUCOCYTES

Leucocytes are much larger than erythrocytes, but they can squeeze through tiny blood vessels because they can change their shape. They are responsible for many of the body's non-specific responses to pathogens. There are around 4000–11 000 leucocytes per mm^3 of blood and there are several different types (see **fig A**). They are formed in the bone marrow, although some mature in the thymus gland. Their main function is to defend the body against infection. They all contain a nucleus and have colourless cytoplasm, but some types contain granules that can be stained. Some are involved in the non-specific responses of the body to infection. Others are involved in the specific responses of the immune system.

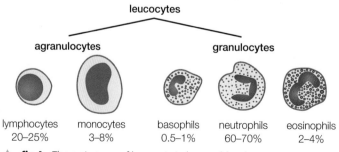

leucocytes

agranulocytes   granulocytes

| lymphocytes | monocytes | basophils | neutrophils | eosinophils |
| 20–25% | 3–8% | 0.5–1% | 60–70% | 2–4% |

▲ **fig A** The main types of leucocyte in human blood.

GRANULOCYTES

Granulocytes are leucocytes with granules in the cytoplasm. The granules absorb stain and you can see them under the microscope. These cells have lobed nuclei (with round projections). They are involved in the non-specific responses to infection. Granulocytes include the following cell types.

- **Neutrophils** are part of the non-specific immune system. They engulf and digest pathogens by phagocytosis. They have multi-lobed nuclei. Up to 70% of all leucocytes are neutrophils.
- **Eosinophils** are part of the non-specific immune system. They are stained red by eosin stain. They are important in the non-specific immune response of the body against parasites, in allergic reactions and inflammation, and in developing immunity to disease.
- **Basophils** are part of the non-specific immune system. They have a two-lobed nucleus. They produce histamines involved in inflammation and allergic reactions.

AGRANULOCYTES

Agranulocytes are leucocytes that do not have granules in their cytoplasm. Their nuclei are round and do not have lobes. They are involved in the specific immune response to infection. Agranulocytes include the following cell types.

- **Monocytes** are part of the specific immune system. They are the largest of the leucocytes. They can pass from the blood into the tissues to form cells called **macrophages**. Macrophages also play an important part in the specific immune system. They engulf pathogens by phagocytosis.
- **Lymphocytes** are small leucocytes with very large nuclei that are vitally important in the specific immune response of the body.

NON-SPECIFIC RESPONSES

Non-specific responses to infection are initiated by body cells breaking down and releasing chemicals, and by pathogens that have been labelled by the specific immune system.

INFLAMMATION

Inflammation is a common, non-specific way in which our bodies respond to infection (see **fig B**). It generally occurs in the case of an infection which is relatively localised, such as when you cut yourself and bacteria get into the wound. The inflammatory response involves a number of stages. Special cells called **mast cells** are found in the connective tissue below the skin and around blood vessels. When this tissue is damaged, these mast cells and basophils (a granular leucocyte) release chemicals known as **histamines**. Histamines cause the blood vessels in the area, particularly the arterioles, to dilate. This leads to local heat and redness. The locally raised temperature reduces the reproduction rate of any pathogens in the area.

The histamines also make the cells forming the walls of the capillaries separate slightly thus making the capillaries permeable. As a result, plasma (the fluid part of the blood) containing leucocytes (mainly neutrophils) and **antibodies** is forced out of the capillaries. This causes swelling (becoming larger than normal) and often pain. The antibodies disable the pathogens and the macrophages and neutrophils destroy them by phagocytosis. A

fairly common symptom of a more widespread infection is a rash. This is a form of inflammation or tissue damage that particularly affects the skin, causing red spots or patches.

▲ **fig B** The skin around this finger is inflamed as a result of a bacterial infection.

FEVERS

A **fever** is a common early non-specific response to infection. Normal body temperature is maintained by the hypothalamus and follows a regular circadian (roughly 24-hour) rhythm. It is lowest early in the morning and highest at about 10 pm. When a pathogen infects the body the hypothalamus resets to a higher body temperature, so we have a fever. A higher temperature helps the body combat infection in two ways.

- Many pathogens reproduce most quickly at 37 °C or below. A higher temperature will reduce the ability of many pathogens to reproduce effectively. Therefore, they cause less damage.
- The specific response of the immune system functions better at higher temperatures and so will be more successful at combating the infection if the temperature is raised.

In a bacterial infection, the temperature rises steadily and remains relatively high until treatment is successful or the body successfully controls the infection. In viral infections, the temperature tends to 'spike'. That is, it increases rapidly every time viruses break out of the cells and then decreases towards normal again. There are many designs of clinical thermometer to help monitor temperature (see **fig C**). Fevers are often beneficial, but if the temperature gets too high they can be damaging and even fatal. If your body temperature rises above 40 °C, the denaturation of some enzymes occurs and you may suffer permanent tissue damage. If the temperature is not lowered relatively quickly, death may result. Sweating is often associated with fever because the body sweats in response to the high temperature as a cooling mechanism (see **Chapter 7C**). If the fluid and electrolytes which are lost in the sweat are not replaced, dangerous dehydration and even death can result.

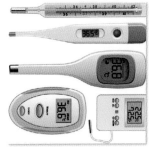

▲ **fig C** Taking the body temperature of a patient is a useful way of tracking the course of many infections. There are many different types of clinical thermometer to help do this.

PHAGOCYTOSIS

Phagocytosis is another non-specific response involving leucocytes. It is often associated with inflammation. **Phagocyte** is a general term used to describe white blood cells that engulf and digest pathogens (and any other foreign material in the blood and tissues). The process of digestion uses lysozymes (enzymes from the lysosomes, see **Section 3A.2 (Book 1: IAS)** and **Section 6A.7**). There are two main types of phagocyte.

- Neutrophils are granulocytes and make up 70% of the leucocytes in the blood. Each neutrophil can only ingest a few pathogens before it dies. These cells cannot renew their lysosomes and once the enzymes are used up, the cell cannot destroy any more pathogens **(see fig D)**.

- Macrophages are derived from monocytes, which are agranulocytes. They constitute about 4% of the leucocytes in the blood. Monocytes migrate (move) to the infected tissues and become macrophages, so there are large numbers of macrophages in the tissues. Macrophages have an enormous capacity to ingest pathogens because, unlike neutrophils, they can renew their lysosomes so they last much longer. They accumulate at the site of an infection to attack the invading pathogens. You will learn more about the actions of the macrophages in **Section 6B.2**.

When phagocytes engulf a pathogen, it is enclosed in a vesicle (membrane 'bag') called a **phagosome**. The phagosome then fuses (combines) with a lysosome. The lysozymes break down the pathogen. The phagocytes can sometimes be seen as pus, which is a thick yellowish liquid containing dead cells (mainly neutrophils). The pus may come out of a wound or spot, or it may be reabsorbed by the body.

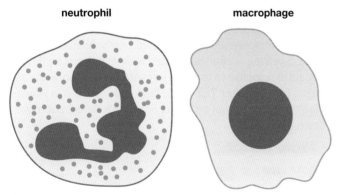

neutrophil   macrophage

▲ **fig D** Neutrophils and macrophages are important in the non-specific response of the body to invasion by pathogens. Macrophages also play a vital role in the specific immune response.

After a phagocyte has engulfed a pathogen, it produces chemicals called **cytokines** in the surrounding tissues. Cytokines are signalling molecules that stimulate other phagocytes to move to the site of an infection. They also raise the body temperature and stimulate the specific immune response.

INTERFERONS: A NON-SPECIFIC RESPONSE TO VIRUSES

All of the non-specific responses to infection mentioned above are particularly effective against bacteria. However, there is one body response that is effective *only* against viruses. Cells invaded by viruses begin to produce a group of chemicals called **interferons**. Interferons are proteins that inhibit viral replication within the cells. An interferon diffuses from the cell where it is made into the surrounding cells. It binds to receptors in the surface membranes of uninfected cells. This stimulates a pathway that makes these cells resistant to infection by viruses. In this way, interferons prevent the infection of more cells when the viruses break out of the first infected cell, so the viruses are unable to reproduce further.

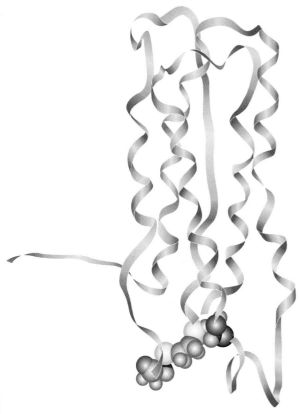

▲ **fig E**   This is a modified version of interferon which can be produced fairly cheaply and has been widely used to treat hepatitis C.

CHECKPOINT

1. The enzyme lysozyme is almost universally present in animals. Explain its role in the non-specific response to pathogens.

2. Describe inflammation and explain how it protects the body against disease.

3. Inflammation and phagocytosis are defences against pathogens that have invaded the body. Explain why they are referred to as part of the non-specific defences of the body.

SKILLS   CRITICAL THINKING

4. If someone is ill with a fever, people often try to decrease their temperature. Explain:

    (a) why this is not necessarily a good idea

    (b) why it can be very important to lower the temperature.

SUBJECT VOCABULARY

antigens glycoproteins, proteins or carbohydrates on the surface of cells, toxins produced by bacterial and fungal pathogens, and some whole viruses and bacteria that are recognised by white blood cells during the specific immune responses to infection; they stimulate the production of an antibody

leucocytes white blood cells which are larger than erythrocytes and can squeeze through tiny blood vessels as they can change their shape; there are around 4000–11 000 leucocytes per mm^3 of blood and there are several different types which carry out different functions in the body

granulocytes leucocytes with granules that absorb stain in the cytoplasm of the cells; this makes them visible under the microscope; they have lobed nuclei and are involved in the non-specific responses to infection

neutrophils the most common type of leucocyte; they engulf and digest pathogens by phagocytosis

eosinophils leucocytes important in the non-specific immune response against parasites, in allergic reactions and inflammation, and in developing immunity to disease

basophils leucocytes with a two-lobed nucleus; they produce histamines involved in inflammation and allergic reactions

agranulocytes leucocytes with round nuclei but without granules in their cytoplasm; they are involved in the specific immune response to infection

monocytes the largest of the leucocytes, they can pass from the blood into the tissues to form macrophages

macrophages cells that engulf pathogens by phagocytosis as part of the specific immune system

lymphocytes small leucocytes with very large nuclei that are vitally important in the specific immune response of the body; they make up the main cellular components of the immune system; they are made in the white bone marrow of the long bones

inflammation a common, non-specific response to infection involving the release of histamines from mast cells and basophils; this causes the blood vessels to dilate producing local heat, redness and swelling

mast cells cells found in the connective tissue below the skin and around blood vessels; they release histamines when the tissue is damaged

histamines chemicals released by the tissues in response to an allergic reaction

antibodies glycoproteins that are each produced in response to a specific antigen

fever a raised body temperature, often in response to infection

phagocytosis the process by which a cell engulfs another cell and encloses it in a vesicle to digest it

phagocyte cell which engulfs and digests other cells or pathogens

phagosome the vesicle in which a pathogen is enclosed in a phagocyte

cytokines molecules which signal between cells; they have several roles in the immune system, including stimulating other phagocytes to move to the infection site

interferons chemicals produced by cells in very small amounts when invaded by viruses; interferons act to prevent the viruses invading other cells

LEARNING OBJECTIVES

■ Understand the roles of antigens and antibodies in the body's immune response, including the involvement of plasma cells, macrophages and antigen-presenting cells.

■ Understand the differences between the roles of B cells (B memory and B effector cells) and T cells (T helper, T killer and T memory cells) in the host's immune response.

The **immune response** is the specific response of the body to invasion by pathogens. It enables the body to recognise anything that is 'non-self' and remove it as efficiently as possible. Each organism carries a unique set of markers (antigens) on its cell surface membrane(s). Some of these antigens are common to every member of a particular species, others are specific to a particular individual. When two individuals are closely related, they will have more antigens in common than individuals who are not related. Only genetically identical twins and clones have totally matching antigens.

The immune system of the body has four key characteristics.

- It can distinguish 'self' from 'non-self'.

- It is specific; in other words, it responds to specific foreign cells.

- It is diverse; that is, it can recognise an estimated 10 million different antigens.

- It has immunological memory; in other words, once you have encountered and responded to a pathogen, you can respond rapidly if you meet it again.

Lymphocytes and macrophages are the two main types of white blood cell involved in the specific immune system. Lymphocytes are made in the white bone marrow of the long bones. They move around the body in the blood and lymph, and they are involved in recognising and responding to foreign antigens. They leave the bloodstream and move freely through the tissues.

DIFFERENT KINDS OF LYMPHOCYTE

There are two main types of lymphocyte: B cells and T cells.

B CELLS

B cells are produced in the bone marrow. Once mature, they are found in the lymph glands and free in the body. B cells have globular receptor proteins attached (membrane-bound) on their cell surface membrane that are identical to the antibodies they will later produce. All antibodies are known as **immunoglobulins**. Around 100 million B cells are formed as an embryo grows, each with a different membrane-bound antibody. Each of these then divides to form a clone of cells, giving the baby an immune system with the potential to recognise and deal with an enormous range of pathogens.

When a B cell binds to an antigen, the following types of B cell are produced.

- **B effector cells** divide to form the plasma cell clones.

- **Plasma cells** produce antibodies to particular antigens at a rate of around 2000 antibodies per second.

- **B memory cells** provide the immunological memory to a specific antigen, allowing the body to respond very rapidly if you encounter a pathogen carrying the same antigen again.

T CELLS

T cells are produced in the bone marrow but mature and become active in the thymus gland. The surface of each T cell displays thousands of identical T cell receptors. T cell receptors bind to antigens on infected body cells and then several further types of T cells are produced and play different roles in the immune response.

- **T killer cells** produce chemicals to destroy infected body cells.

- **T helper cells** activate the plasma cells to produce antibodies against the antigens on a particular pathogen and also secrete **opsonins** to 'label' the pathogen for phagocytosis by other white blood cells.

- **T memory cells** are very long-lived cells that make up part of the immunological memory. When they meet a pathogen for the second time, they divide rapidly. This forms a large clone of T killer cells which then quickly destroy the pathogen.

The working of many of these cells depends on special proteins known as **major histocompatibility complex (MHC) proteins**. These proteins display antigens on the cell surface membrane, as you will learn later. The immune response to infection is extremely complex so we will look at it one stage at a time.

THE HUMORAL RESPONSE

The humoral response of the immune system reacts to:

- antigens found outside the body cells, including antigens on pathogens such as bacteria and fungi
- antigen-presenting body cells.

The humoral response results in the production of antibodies, which are not attached to cells but are carried around the body in the blood and tissue fluid. B cells produce the antibodies, but T cells are first involved in activating the B cells. The humoral response consists of two main stages: the T helper activation stage and the effector stage.

T HELPER ACTIVATION STAGE

When a pathogen enters the body, chemicals are produced that attract the phagocytes, including macrophages and neutrophils. Neutrophils engulf bacteria and destroy them in about 10 minutes. Macrophages take longer but do more: they prepare the way for the specific immune system (see **fig A**). The macrophage separates the antigens from the digested pathogen and combines them with the major histocompatibility complex. The complexes move to the surface of the macrophage cell outer membrane. A macrophage with these antigen–MHC protein complexes displayed on its cell surface is known as an **antigen-presenting cell (APC)**.

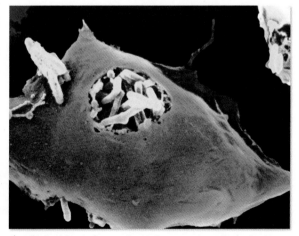

▲ **fig A** Macrophages play an important part in the immune system by engulfing bacteria (the small yellow rod shapes in this artificially coloured electron micrograph).

The next step involves T cells, which have receptors on the outer membrane that bind to the specific antigen of the antigen–MHC complex on the APC. The binding of the T cell with the APC triggers the T cell to reproduce and form a **clone** of cells. This process is summarised in **fig B**.

Most of these cloned cells become active T helper cells, which are used in the rest of the immune system. The rest of the cloned cells form inactive T memory cells, which remain in the body; they rapidly become active if the same antigen is encountered again.

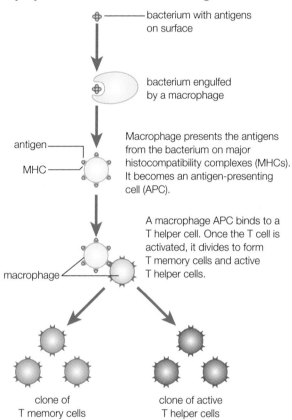

bacterium with antigens on surface

bacterium engulfed by a macrophage

antigen

MHC

Macrophage presents the antigens from the bacterium on major histocompatibility complexes (MHCs). It becomes an antigen-presenting cell (APC).

A macrophage APC binds to a T helper cell. Once the T cell is activated, it divides to form T memory cells and active T helper cells.

macrophage

clone of T memory cells

clone of active T helper cells

▲ **fig B** T helper activation stage of the humoral response. The production of the antigen-presenting cell is an important step in the immune system of the body.

EFFECTOR STAGE

Some of the millions of different B cells have immunoglobulins on their surfaces. These immunoglobulins are specific for the antigen presented by the pathogen and will bind to it. The B cell then engulfs the whole pathogen by endocytosis. The vesicle formed combines with a lysosome and enzymes break down the pathogen to leave fragments of processed antigen. These fragments become attached to MHC proteins within the cell, and the antigen–MHC complex is then transported to the cell surface membrane where the antigen is displayed. This is very similar to the macrophage process you saw in **fig B**.

A T helper cell from the active clone produced in the T helper activation stage recognises the specific antigen displayed on the MHC complex on the B cell and binds to it. This triggers the release of cytokines from the T helper cell. These cytokines stimulate the B cells to divide and form clones of identical cells.

This is known as **clonal selection** (see **fig C**). New clones of B effector and B memory cells are produced. The B effector cells differentiate to form **plasma cell clones**.

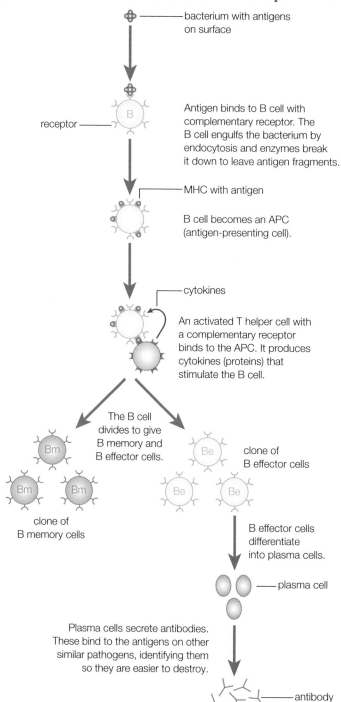

bacterium with antigens on surface

receptor

Antigen binds to B cell with complementary receptor. The B cell engulfs the bacterium by endocytosis and enzymes break it down to leave antigen fragments.

MHC with antigen

B cell becomes an APC (antigen-presenting cell).

cytokines

An activated T helper cell with a complementary receptor binds to the APC. It produces cytokines (proteins) that stimulate the B cell.

The B cell divides to give B memory and B effector cells.

Bm

Bm   Bm

clone of B memory cells

Be

clone of B effector cells

Be   Be

B effector cells differentiate into plasma cells.

plasma cell

Plasma cells secrete antibodies. These bind to the antigens on other similar pathogens, identifying them so they are easier to destroy.

antibody

▲ **fig C** Clonal selection. This process results in millions of antibody molecules, which bind to pathogens so they can be destroyed.

The plasma cell clones produce large amounts of antibodies that are identical to the immunoglobulin of the original parent B cell. An antibody is a special glycoprotein that is released into the circulation. It binds to the specific antigen on the particular pathogen that triggered the immune system. Antibodies destroy pathogens in one of several ways.

- **Agglutination** is when antibodies bind to the antigens on pathogens so the microorganisms agglutinate or stick together. This helps to prevent them from spreading through the body and also makes it easier for phagocytes to engulf them.

- **Opsonisation** is when an antibody acts as an opsonin, a chemical which makes an antigen or pathogen more easily recognised by phagocytes.
- **Neutralisation** is when antibodies neutralise the effects of bacterial toxins by binding to them.

The ability of most pathogens to invade the host cells is dramatically reduced when they are combined with antibodies.

Plasma cells live for only a few days, but as they can produce up to 2000 antibody molecules per second this is long enough to be effective. They have extensive endoplasmic reticulum and many ribosomes, which are adaptations for producing large quantities of protein antibodies. The antibodies remain in the blood for varying lengths of time. The memory cells may stay in the blood for years or even for a lifetime.

THE CELL-MEDIATED RESPONSE

Sometimes, the pathogen gets inside the host cells and the humoral response is not very effective against it. This is particularly true in viral infections. The cell-mediated response is important in this case. The cell-mediated response involves T killer cells, which respond to specific antigens.

T lymphocytes respond to cells that have been changed in some way. Examples include cells infected by a virus, antigen-presenting cells, cells changed by mutation to form cancer cells, and cells of a transplanted organ. After a body cell has been infected with a bacterium or virus, the pathogen is digested and the surface antigens become bound to an MHC in a process similar to that seen in macrophages. Consequently, the body cell effectively becomes an APC (see **fig D**). However, it is important to remember that it is still infected by the pathogen.

T killer cells present in the blood have a wide range of complementary receptor proteins on the surface of their cell surface membrane. T killer cells bind to the matching antigen–MHC complex on the surface of the infected body cell. The T cells are then exposed to cytokines from an active T helper cell and experience a rapid series of cell divisions. These cell divisions produce a clone of identical active T killer cells which can all bind to infected body cells. The T killer cells release enzymes that make pores (small holes) form in the membrane of the infected cells. This allows the free entry of water and ions, so the cells swell and rupture. Any pathogens that are released undamaged are labelled with antibodies produced by the plasma cells, and then destroyed. T killer memory cells are also produced.

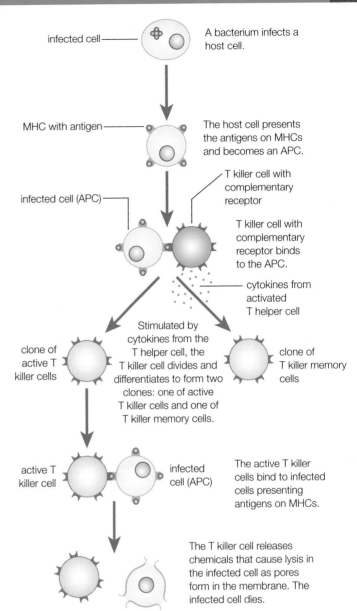

infected cell — A bacterium infects a host cell.

MHC with antigen — The host cell presents the antigens on MHCs and becomes an APC.

infected cell (APC) —

T killer cell with complementary receptor

T killer cell with complementary receptor binds to the APC.

cytokines from activated T helper cell

clone of active T killer cells

Stimulated by cytokines from the T helper cell, the T killer cell divides and differentiates to form two clones: one of active T killer cells and one of T killer memory cells.

clone of T killer memory cells

active T killer cell — infected cell (APC) — The active T killer cells bind to infected cells presenting antigens on MHCs.

The T killer cell releases chemicals that cause lysis in the infected cell as pores form in the membrane. The infected cell dies.

▲ **fig D** T helper activation stage of the cell-mediated response. The production of the antigen-presenting cell is an important step in the immune system of the body.

The response of your immune system to the invasion of pathogens is very complex. It works extremely well most of the time. The different responses are all integrated in the body. A summary of those interactions is shown in **fig E**.

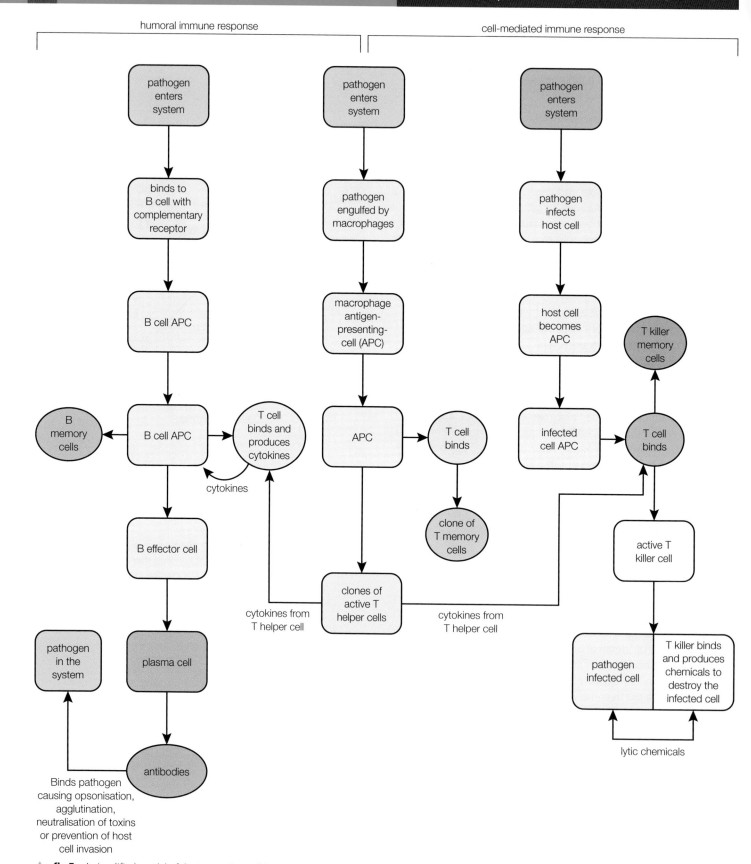

▲ **fig E**   A simplified model of the interactions of the immune system.

PRIMARY AND SECONDARY IMMUNE RESPONSE

If our immune system is so effective, why do we become ill? The primary immune response involves the production of antibodies by the plasma cells (produced from the B effector cells) and the activation of T killer cells. It is extremely effective. The problem is that it can take days or even weeks for the primary immune response to become completely active against a particular pathogen. This is why we develop the symptoms of disease; we feel ill when pathogens are reproducing freely inside our bodies, before the immune system has become fully operational against the pathogen concerned.

However, we have a secondary immune response which is quicker, greater and longer lasting. When the B cell APC divides, it also produces B memory cells. Unlike the plasma cells, B memory cells are very long-lived. They can remain in your body for years, and they are important in producing a rapid response to a second invasion by the same pathogen. When you have had a disease once, you usually do not catch it again. This is not because you never come into contact with the disease-causing pathogen again. Instead, when you do encounter it, the B memory cells recognise the antigens on the surface and help you produce the antibodies against it so rapidly that the pathogen is destroyed before the symptoms of the disease develop.

EXAM HINT

Remember that the secondary immune response does not prevent the pathogen entering your body. It enables your body to respond to the presence of the pathogen much more quickly. This means the pathogen does not have time to reproduce sufficiently to cause the symptoms of the disease.

In addition, at the same time as a clone of active T killer cells are formed, some cloned T cells become T memory cells. These persist in the blood so that the body can produce a rapid response if the same pathogen invades again. These memory cells release a large number of active T killer cells to engulf and destroy infected cells. At present, no one is completely sure exactly how memory cells provide immunological memory, but it is very effective.

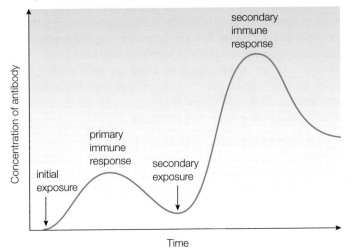

▲ **fig F** The primary and secondary immune responses: the primary response takes days or weeks, the secondary response takes hours or days, but the gap between exposures may be many years.

CHECKPOINT

SKILLS ▶ INTERPRETATION

1. Produce a large diagram to explain in full how the immune system works.
2. Explain why your immune system does not attack the other cells of your body.
3. Explain the role of antibodies in the immune system.
4. Distinguish carefully between the roles of the B cells and the T cells in the immune response of the body.
5. Discuss the difference between the primary and secondary immune responses and explain the importance of the secondary immune response.

SUBJECT VOCABULARY

immune response the specific response of the body to invasion by pathogens

lymphocytes small leucocytes with very large nuclei that are vitally important in the specific immune response of the body; they make up the main cellular components of the immune system; they are made in the white bone marrow of the long bones

B cells lymphocytes that are made and mature in the bone marrow; once they are mature, they are found in the lymph glands and free in the body

immunoglobulins antibodies

B effector cells lymphocytes that are made and mature in the bone marrow which divide to form the plasma cell clones

plasma cells cells that produce antibodies to particular antigens at a rate of around 2000 antibodies per second

B memory cells lymphocytes that are made and mature in the bone marrow and that provide the immunological memory to a specific antigen; they allow the body to respond very rapidly to the same pathogen carrying the same antigen a second time

T cells lymphocytes made in the bone marrow that mature and become active in the thymus gland

T killer cells lymphocytes that mature in the thymus gland and produce chemicals that destroy pathogens

T helper cells lymphocytes that mature in the thymus gland and are involved in the process that produces antibodies against the antigens on a particular pathogen

opsonins chemicals which bind to pathogens and label them so they are more easily recognised by phagocytes

T memory cells very long-lived cells which constitute part of the immunological memory

major histocompatibility complex (MHC) proteins proteins that display antigens on the cell surface membrane

antigen-presenting cell (APC) a cell displaying an antigen/MHC protein complex

clone a group of genetically identical cells which are all produced from one cell

clonal selection the selection of the cells that carry the right antibody for a specific antigen

plasma cell clones clones of identical plasma cells that all produce the same antibody

agglutination the grouping together of cells caused when antibodies bind to the antigens on pathogens

opsonisation a process that makes a pathogen more easily recognised, engulfed and digested by phagocytes

neutralisation the action of antibodies in neutralising the effects of bacterial toxins on cells by binding to them

LEARNING OBJECTIVES

■ Understand how individuals may develop immunity (natural, artificial, active and passive).

Once you have been exposed to a pathogen and your immune system has succeeded in fighting it, you will probably not become ill as a result of infection by that microorganism again. This is because your immune system has a 'memory', based on B cells, T cells and immunoglobulins (antibodies) (see **Section 6B.2**). You can gain this important immunity to diseases in a number of ways.

DIFFERENT TYPES OF IMMUNITY

As you learned in **Section 6B.2**, when your body comes into contact with a foreign antigen, your immune system is activated. It makes the antibodies needed against the antigens on the pathogen, and the pathogen is destroyed. This is known as **natural active immunity**, because your body actively makes the antibodies.

When a mammal such as a camel, a goat or a human is pregnant, pre-formed antibodies are passed from the mother to the developing fetus through the placenta. After birth, the baby receives additional protection from antibodies present in the mother's milk (see **fig A**). The milk made in the first few days after giving birth (colostrum) is particularly rich in antibodies from the mother. This gives the baby temporary immunity to many different diseases, until its own immune system becomes active. This is **natural passive immunity**. It is passive because your body does not make the antibodies. It tends to be quite short-lived because the antibodies are not replaced, but it is very important for the survival of new-born mammals.

▲ **fig A** Young mammals are protected from many different diseases by antibodies (immunoglobulins) that they receive across the placenta before birth, and through their mother's milk after birth.

EXAM HINT

Remember that natural immunity is when immunity is achieved in the normal course of life rather than by medical intervention.

INDUCING IMMUNITY

People spend a great deal of money on medicines to treat infections. Another way of controlling infectious diseases is to prevent them happening by using **immunisation**. Immunisation is the process of protecting people from infection by giving them passive or active artificial immunity. You develop immunity to a disease-causing pathogen by exposing your immune system to the specific antigens in a safe way that does not put you at risk of developing the disease. By exposing your body to a safe form of a pathogen, your body produces antibodies and memory cells that are ready to protect you if you are exposed to the live disease. **Vaccination** is the procedure by which you immunise people to produce immunity.

The aim of immunisation is to protect individuals against diseases that might kill or harm them. It also has a wider role in society. Immunisation can eradicate, eliminate or control diseases that cause large numbers of deaths, disabilities or illnesses within a population and which therefore create serious problems for the structure of that society.

Artificial passive immunity is given when antibodies formed in one individual are extracted and injected into another individual. The antibodies may be from a person who is already immune to a disease, or from a completely different species (e.g. horse) whose antibodies are effective. This does not confer long-term immunity, because the antibodies are gradually broken down and not replaced. However, it is extremely valuable if someone has been exposed to a rapidly acting antigen such as tetanus. Tetanus (lockjaw) results from a toxin produced by the bacterium *Clostridium tetani*. The toxin affects striated muscles, causing them to go into spasm (tetanus). This makes swallowing and breathing impossible and so causes death. If someone may have been exposed to tetanus, for example from a bad cut while tending crops or working with horses, they are injected with antibodies against tetanus. This prevents the development of the disease and death, but will not give long-term immunity.

Artificial active immunity is the basis of most immunisation programmes. Here, small amounts of a safe form of the pathogen (known as the vaccine) are used to produce immunity in a person. The pathogen is made non-infective without reducing its ability to act as an antigen. This can be done in a number of ways.

- If it is a toxin that causes the symptoms, a detoxified form (with one or more chemical groups changed) will be injected.

- Sometimes inactivated viruses or dead bacteria are used as vaccines.

- Sometimes **attenuated pathogens** (viable but modified so they cannot produce disease) are used.
- Modern vaccines are increasingly produced using fragments of the outer coats of viruses and bacteria; even DNA segments can be used.

With artificial active immunity, your immune system will produce antibodies against the antigen, and appropriate memory cells will be formed without you becoming ill. If you subsequently come into contact with the active antigen, it will be destroyed rapidly without you experiencing the symptoms of the disease it causes (see **fig B**).

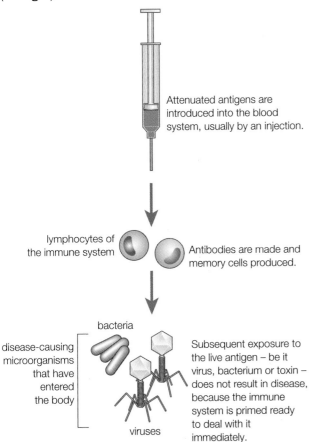

Attenuated antigens are introduced into the blood system, usually by an injection.

lymphocytes of the immune system

Antibodies are made and memory cells produced.

disease-causing microorganisms that have entered the body

bacteria

viruses

Subsequent exposure to the live antigen – be it virus, bacterium or toxin – does not result in disease, because the immune system is primed ready to deal with it immediately.

▲ **fig B** The process of vaccination stimulates the immune system so that an individual is immune to a disease before they ever become infected by a live pathogen.

EXAM HINT

Remember that active immunity is when the immune system is activated and makes its own antibodies. Passive immunity is when the body receives antibodies from another source.

ERADICATION OF DISEASE

For centuries, smallpox was greatly feared. It killed millions of people and disfigured many more around the world. Now it has disappeared. Smallpox is the only disease that has been completely eradicated by a vaccination programme. However, polio is close to being eradicated as well. Smallpox is no longer found either in people, animals or anywhere in the environment. The virus only exists isolated in two top-security labs.

Smallpox was very recognisable and had no non-human hosts, a long incubation period and a visible scar as evidence of immunisation (see **fig C**). These features all helped to make eradication possible.

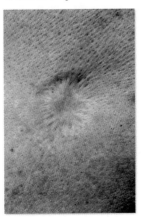

▲ **fig C** Smallpox vaccination leaves a distinctive scar. Because smallpox was eradicated worldwide in 1980, immunisation against it became unnecessary. Scars like these will not exist in the future.

ELIMINATION AND CONTROL OF DISEASE

There are many diseases for which the pathogen survives in soil, water or animal hosts. For such diseases, there is always a very large reservoir of potential infection so eradication is not realistically possible. The aims of immunisation are, therefore, elimination and control.

Elimination is the situation in which the disease disappears in defined areas but the pathogen remains in animals, the environment or in mild infections which are not recognised. In this case, immunisation must continue even when no clinical cases are observed.

For example, the vaccination programme against polio has not eradicated the disease, even in countries where the number of people who are vaccinated is high. A live vaccine is used, so very occasionally an infant or an unvaccinated carer develops the disease. In spite of this, polio remains endemic only in Afghanistan, Nigeria and Pakistan. Scientists and doctors are working hard to introduce universal vaccination programmes in these countries. For example, the UAE is working closely with the WHO to eradicate polio, especially focusing on Pakistan and Afghanistan. This 'Emirates Polio Campaign' has delivered in total over 86 million doses of polio vaccine to areas where they are desperately needed. If the threat of polio is removed, it will be because the global community is working together like this.

A disease that is controlled still occurs, but not frequently enough to be a significant health problem. For some serious infectious diseases such as malaria, the rapid evolution of the pathogen makes it very difficult to develop a vaccine because the surface antigens are always changing, but some progress is being made.

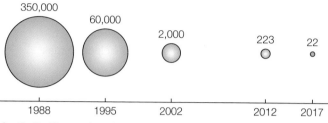

The decline of polio since 1988

350,000	60,000	2,000	223	22
1988	1995	2002	2012	2017

▲ **fig D** The number of polio cases around the world has fallen by over 99% in the last 30 years. The disease is still endemic in Afghanistan, Nigeria and Pakistan.

HERD IMMUNITY

For disease to be eradicated, eliminated or controlled, **herd immunity** is important. Herd immunity occurs when a significant proportion of the population is vaccinated against a disease. This makes it very difficult for the disease to continue affecting a population, because the pathogen cannot survive without hosts to infect, and very few people are vulnerable to it. Herd immunity can effectively stop the spread of a disease through a community. It is important that everyone who can be vaccinated is vaccinated. This protects those individuals against the disease, and also protects people who cannot be or have not been vaccinated. These include very young babies, very old people, people with compromised immune systems and people who are very ill with other diseases. Because everyone else is vaccinated, the people who cannot be vaccinated will probably never encounter the pathogen.

The percentage of the population that needs to be vaccinated to give herd immunity varies from one disease to another. It depends on factors such as how the disease is spread and how infectious it is. To give herd immunity against whooping cough (*pertussis*), 92–94% of the population need to be vaccinated. For measles, it is 83–94%. For polio only 80–86% of the population need to be vaccinated against the disease to protect everyone.

Some diseases are so serious that in many countries, including the Republic of Maldives, the UAE and the UK, all young children are offered vaccines against them. This gives herd immunity to the whole population. These diseases include diphtheria, tetanus, polio, whooping cough, TB and measles. Thousands of people were killed and disabled by these diseases every year before the arrival of vaccines and better living conditions for all.

In other cases, only vulnerable people are vaccinated as a general rule, with mass vaccination to provide herd immunity only introduced when there is a serious outbreak of disease (e.g. certain strains of flu).

CHECKPOINT

1. Make a table to compare natural active immunity, natural passive immunity, artificial passive immunity and artificial active immunity, giving an example of each.

2. What is herd immunity and why is it so important?

SKILLS CREATIVITY, REASONING

3. ▶ Some parents decide not to have their child vaccinated.
   (a) Explain how this puts their child at risk.
   (b) Explain how this puts other children at risk.

SUBJECT VOCABULARY

natural active immunity when the body produces its own antibodies to an antigen encountered naturally

natural passive immunity when antibodies made by the mother are passed to the baby via the placenta or mother's milk

immunisation the process of protecting people from infection by giving them passive or active artificial immunity

vaccination the introduction of harmless forms of organisms or antigens by injection or mouth to produce artificial immunity

artificial passive immunity when antibodies are extracted from one individual and injected into another (e.g. one form of the tetanus vaccine)

artificial active immunity when the body produces its own antibodies to an antigen acquired through vaccination

attenuated pathogens viable pathogens that have been modified so that they do not cause disease but still cause an immune response that results in the production of antibodies and immunity

herd immunity when a high proportion of a population is immune to a pathogen, usually through vaccination, thus lowering the risk of infection to all, including those not vaccinated

4 ANTIBIOTICS: TREATING BACTERIAL DISEASE

The idea that diseases can be caused by microorganisms is key to the way we look at health and medical treatments. As you learned in **Section 6B.3**, we can prevent some diseases by using immunisation to mimic the natural immunity we develop when we are exposed to diseases. However, we cannot be vaccinated against all of the diseases we will meet in our lifetimes. Our alternative to immunisation is to develop medicines to cure infectious diseases. To do that, we must destroy the pathogens that cause the disease. For thousands of years people used simple remedies to try to overcome disease. From the middle of the 19th century in the developed world, researchers focused more on known and recognised pathogens. Not surprisingly, the battle against disease became more successful. You learned about the way people have used plant-based remedies against both communicable and non-communicable diseases, and the way we develop drugs in the modern world, in **Sections 4A.6** and **4A.7 (Book 1: IAS)**. Here, you will discover more about the medicines we can use directly against bacterial pathogens and the ways in which those pathogens change to stay ahead in the evolutionary race between bacteria and humans.

DRUGS AGAINST MICROORGANISMS

The big advancement in the treatment of bacterial infections was the discovery and manufacture of the first **antibiotic** in the first half of the 20th century. Antibiotics are medicines that either destroy microorganisms or prevent them from reproducing. They are used to treat patients suffering from bacterial infections, and have had a dramatic impact on human survival and life expectancy. In the 21st century, it is hard for us to imagine how many people died of communicable diseases just over 100 years ago. In 1900, 53% of all deaths in the USA were from infectious diseases. By the year 2010, infectious diseases caused only 3% of US deaths. This is in spite of the emergence of HIV/AIDS and the linked rise in tuberculosis. The reason for this is partly the result of immunisation and partly because antibiotics mean we can cure many diseases which previously killed us.

All modern antimicrobial drugs work against microorganisms by the principle of **selective toxicity**. They interfere with the metabolism or function of the pathogen, with minimal damage to the cells of the human host. The most commonly used and best known antimicrobials are the antibiotics. They can also be used, but to a limited degree, in the treatment of some fungal infections.

The first antibiotic to be widely and successfully used in the treatment of bacterial diseases was **penicillin**. Penicillin does not work against all pathogens, so scientists searched for more antibiotics and continue searching today. Many other antibiotics

have been discovered, but globally penicillin and the many antibiotics derived from it are still commonly used to combat bacterial infections.

ANTIBIOTIC ACTION

Antibiotics are effective because they disrupt the biochemistry of the bacterial cells. Different classes of antibiotic interrupt different processes. The main classes of antibiotic drugs and the way they affect microorganisms are summarised in **table A**.

ANTIBIOTIC CLASS	ANTIMICROBIAL ACTION	EXAMPLES
BACTERIOSTATIC	Antimetabolites interrupt metabolic pathways; they block nucleic acid synthesis or the synthesis of vital nutrients such as folic acid in bacteria.	sulfonamides
	Protein synthesis inhibitors interrupt or prevent transcription and/or translation of microbial genes, so protein production is affected.	tetracyclines, chloramphenicol
BACTERICIDAL	Cell wall agents prevent formation of cross-linking in cell walls, so bacteria are killed by lysis (bursting).	beta-lactams (e.g. penicillins)
	Cell membrane agents damage the cell membrane, so metabolites can move out or water moves in, killing the bacteria.	some penicillins, cephalosporins
	DNA gyrase inhibitors stop bacterial DNA winding into rings, so it no longer fits within the bacterium.	quinolone

table A Some methods of antibiotic action

When you take an antibiotic, it may have one of two different effects. It may be **bacteriostatic**, which means that the antibiotic used completely inhibits the growth of the microorganism. This level of treatment is usually sufficient for the majority of everyday infections. The antibiotic combined with the actions of our immune systems will ensure that the pathogen will be completely destroyed. **Tetracycline** is an example of a bacteriostatic antibiotic. It is used to treat acne, urinary tract infections, respiratory tract infections and *Chlamydia*.

Sometimes a drug, or the given dose of a drug, is **bactericidal**. This means it will destroy almost all of the pathogens present. This type of treatment is often used in severe and dangerous infections. It is also used for treating infections in which the patient's immune system is suppressed. This is the case, for example, in transplant patients (who take immunosuppressant drugs to protect the transplanted organ) and in certain diseases such as TB and HIV/AIDS. Penicillins (drugs related to the original penicillin) are examples of bactericidal antibiotics which are used to treat skin infections, chest infections and urinary tract infections.

In fact, the terms *bacteriostatic* and *bactericidal* are misleading. How they work depends on the dose given. Bacteriostatic antibiotics often kill a lot of the bacteria, and will usually kill them all at high enough doses. Bactericidal antibiotics do not kill all of the bacteria; they kill 99% within a given time frame. Lower doses of bactericidal antibiotics are often bacteriostatic. So, the definitions are flexible.

A broad-spectrum antibiotic destroys a wide range of harmful bacteria, pathogens, and neutral and good bacteria alike. A narrow-spectrum antibiotic targets one or two specific pathogens.

Factors in the effectiveness of any antimicrobial drug include:

- the concentration of the drug in the area of the body infected; this will be affected by how easily the drug can reach the tissue and how quickly it is excreted
- the local pH
- whether either the pathogen or the host tissue destroy the antibiotic
- the susceptibility of the pathogen to the particular antibiotic used.

If the standard dose of a drug (what a doctor normally prescribes) successfully destroys the pathogen and cures the disease, the pathogen is said to be sensitive to that antibiotic. If the disease is cured only by using a dose of antibiotic that is much higher than the standard dose, then the pathogen is regarded as only moderately sensitive to that antibiotic. However, there are increasing numbers of cases in which a particular microorganism is not affected at all by an antibiotic, sometimes even one that may have been effective in the past. In these cases the microorganism has become resistant to the antibiotic. This microorganism is **antibiotic resistant** (see **Section 6B.5**).

LEARNING TIP

Always conduct a risk assessment prior to handling microorganisms.

PRACTICAL SKILLS   CP14

You can investigate the effect of different antibiotics on bacteria using standard microbiological techniques which you discovered in **Section 4A.6 (Book 1: IAS)** and **Chapter 6A**. You inoculate an agar plate with a known bacterial culture. You place filter paper discs containing different antibiotics, or different concentrations of the same antibiotic, on the agar and you cover but do not seal the plates, so oxygen is not excluded (see **fig A**). You grow a control culture of microorganisms with known sensitivity to the antibiotic at the same time under the same conditions. You measure the effectiveness of the drugs by the level of inhibition of bacterial growth.

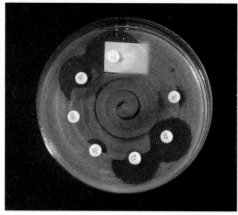

▲ **fig A** You can investigate the sensitivity of bacteria to a particular antibiotic by using discs soaked in known concentrations of different antibiotics, or in different concentrations of the same antibiotic.

 Safety Note: The paper discs should only be handled with sterile forceps, not with fingers.

EXAM HINT

You need to know the techniques involved and understand the necessity of aseptic techniques. You may be asked to explain why these are required in the exam.

CHECKPOINT

1. Using the information under the heading 'Drugs against microorganisms' (page 133), calculate the percentage reduction in deaths due to communicable diseases in the US between 1900 and 2010.

2. (a) Give two examples of how antibiotics affect bacteria.

   SKILLS ▶ **CRITICAL THINKING**

   (b) Some antibiotics are bactericidal and others are bacteriostatic. Explain the difference.

   SKILLS ▶ **INTEGRITY, COMMUNICATION, TEAMWORK**

3. Investigate the discovery and development of penicillin. Write a paragraph on each of the following people to explain their role in this: Alexander Fleming, Ernst Chain, Howard Florey, Ronald Hare, Cecil Paine, Norman Heatley, Mary Hunt.

SUBJECT VOCABULARY

antibiotic a drug that either destroys microorganisms or prevents them from growing and reproducing

selective toxicity a substance that is toxic against some types of cells or organisms but not others

penicillin the first antibiotic discovered; it is bactericidal and affects the formation of bacterial cell walls

bacteriostatic inhibits the growth of bacteria

tetracycline a bacteriostatic antibiotic that inhibits protein synthesis

bactericidal kills bacteria

antibiotic resistant a microorganism that is not affected by an antibiotic (even one that may have been effective in the past)

LEARNING OBJECTIVES

■ Understand how the theory of an evolutionary race between pathogens and their hosts is supported by evasion mechanisms shown by pathogens.

■ Know how an understanding of the contributory causes of hospital-acquired infections has led to codes of practice regarding antibiotic prescription and hospital practice that relate to infection prevention and control.

Antibiotic drugs provide an essential approach to controlling bacterial diseases, but they are not the whole answer to the problem. New problems are now occurring that result directly from the over-use of these valuable drugs.

EXAM HINT

Many candidates mistakenly refer to drug resistance in bacteria as immunity. This is incorrect because bacteria do not have an immune system.

CREATING DRUG-RESISTANT BACTERIA

There is a constant evolutionary race between pathogens and their hosts. We keep developing new medicines, and bacteria keep evolving resistance to these drugs. An antibiotic is only effective if the microorganism has a binding site for the drug and a metabolic process or biochemical pathway with which the antibiotic interferes. However, a mutation can occur during bacterial reproduction. Some mutations may help the microorganism resist the effects of the antibiotic. For example, a mutation may make the cell wall impermeable to the drug. As a result of natural selection, advantageous mutations become more common and the bacterial population becomes increasingly resistant to the drug. Mutations can also result in new biochemical pathways or can switch on or create a gene for the production of an antibiotic-destroying enzyme (see **Chapter 2C (Book 1: IAS)**). With these mutations, pathogens will be more competitive in the evolutionary race.

Extensive use of antibiotics accelerates this process. Using different antibiotics to tackle increasing resistance increases the selection pressure for the evolution of bacteria that are resistant to all of them. This evolutionary race is creating what are known as 'superbugs'. An example is **methicillin-resistant *Staphylococcus aureus* (MRSA)**. In the absence of an effective antibiotic, these resistant bacteria are quite capable of causing death. To prevent this continuing, we must reduce the selection pressure for antibiotic resistance. We can do this in a number of ways:

• by using antibiotics only when they are strictly necessary

• by making sure people understand that they must complete each course of antibiotics

• by using as few different antibiotics as possible, keeping some in reserve for use only if everything else fails.

To reduce the development of antibiotic resistance we need to vary the antibiotics we use and introduce new antibiotics when possible. The problem here is that we are developing and approving use for fewer and fewer new antibiotics. This means bacteria get repeated exposure to existing antibiotics, which increases the probability of them developing resistance (see **fig A**).

EXAM HINT

Remember that the basic concepts learned at IAS can be tested in the context of IAL topics. For example, you may be asked to explain the role of variation and natural selection (see **Chapter 4C (Book 1: IAS)**) in the context of developing drug resistance.

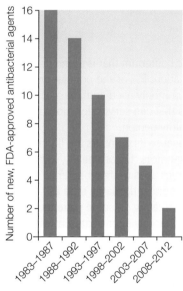

▲ **fig A** The number of new antibacterial drugs available for treatment is falling steadily.

HOSPITAL-ACQUIRED INFECTIONS

'Superbugs' are commonly found in hospitals and care homes, where people are ill or have had surgery and where antibiotic use is at its highest. MRSA and *Clostridium difficile* are causing particular problems. Patients who become infected with these **hospital-acquired infections (HAIs)** need to stay in hospital much longer. Based on recent UK figures, these extended hospital stays, combined with the treatment needed to overcome the infection, cost on average an extra £4000–10 000 per patient.

Some people die as a result of HAIs. A 2012 study of hospitals in the UK found that around 6% of patients in English hospitals have HAIs, including pneumonia and norovirus (causes diarrhoea and sickness) as well as MRSA and antibiotic-resistant *C. difficile*. This is down from 8.2% in 2006, but it represents many thousands of patients.

MRSA

About one-third of people have the bacterium *S. aureus* on their skin or in their nasal passages, without it causing problems. If the bacterium enters the body it can cause boils (a painful infected swelling under the skin) or infections throughout the body such as septicaemia. Many *Staphylococcus* infections have been treated very effectively with methicillin (a penicillin-related antibiotic). As a result of a mutation, some of the bacteria can now produce a penicillinase enzyme that breaks down methicillin. In hospitals and care homes, patients who are weak with other infections often develop opportunistic *S. aureus* infections and have been treated with methicillin, and with other antibiotics when methicillin has not worked. The result is that in hospitals and care homes, *S. aureus* is now winning the evolutionary race because almost all the bacteria produce penicillinases. The infections can be treated, but only with high doses of a very small number of antibiotics, which are used in small amounts to prevent the same thing happening again. Control measures reduced the incidence of MRSA from 1.8% of patients infected in 2006 to less than 0.1% by 2012.

CLOSTRIDIUM DIFFICILE

Clostridium difficile is an anaerobic bacterium that is found in small numbers in the large intestine of about 5% of the population. It is not affected by many of the commonly used antibiotics and produces extremely tough spores that can survive for months outside the human body. In a healthy person, *C. difficile* causes no problems, because its numbers are limited by competition with the normal gut flora. Unfortunately, some common broad-spectrum antibiotics used to treat bacterial infections destroy the normal gut flora at the same time as destroying the pathogens they are treating. In some patients, the numbers of *C. difficile* can increase rapidly after treatment with these antibiotics. The bacteria produce two different toxins that damage the lining of the intestines. This causes severe diarrhoea that can lead to bleeding from the gut and even death.

INFECTION PREVENTION AND CONTROL

Codes of practice have been established for doctors, nurses and all healthcare workers to try to prevent the spread of healthcare-associated infections, and to control them as effectively as possible when they do occur. All around the world these codes of practice include the following.

- **CONTROLLING THE USE OF ANTIBIOTICS** The careful use of antibiotics reduces the probability of resistant bacteria evolving. Antibiotics should only be used when absolutely necessary and every course of antibiotics should be completed. If a patient takes only part of a course, the

immune system is unable to cope with the numbers of bacteria remaining and does not destroy them all. Those bacteria that have some resistance will escape and can then infect other people. The use of different antibiotics just encourages a faster evolution of multiple resistance. It is important to educate the general public, who still tend to demand antibiotics for minor infections and often stop taking the medicine as soon as they feel better.

- **HOSPITAL PRACTICE** Good hygiene within hospitals and care homes can have a major impact on both preventing and controlling HAIs. Examples of good practice include the following.

  - **Hand-washing and hygiene practices** Doctors, nurses and other healthcare professionals wash their hands or use alcohol-based gels between patients to destroy as many pathogens as possible. For example, MRSA can easily be spread by personal contact, but it is destroyed by alcohol and washing. However, the spores of *C. difficile* are not destroyed by the alcohol gels and need a chlorine-based disinfectant to destroy them. Another aspect of good hygiene includes instructions about the clothing doctors and other staff should wear. For example, in some countries long ties which might hang down and carry bacteria from one patient to another are now banned. Wristwatches and even long-sleeved shirts are also banned in case they carry bacteria from one patient to another. However, there is very little scientific evidence that this makes any difference. In some countries, all healthcare workers wear scrubs (sterile surgical outfits), in others they do not. There is no clear difference in the levels of HAIs between the different practices. The spread of disease can also be controlled by thoroughly cleaning hospital wards, toilets and equipment. This is particularly important with diseases spread through faecal traces, such as *C. difficile*. Regular, thorough cleaning is needed as well as the routine daily cleaning that occurs.

  - **Isolation of patients** Patients affected by an HAI need to be isolated from other patients as quickly as possible. It is possible to minimise the spread of disease by nursing infected individuals in separate rooms with high levels of hygiene and infection control. This can be unpleasant for patients, as they can feel isolated from other people, but it can be very effective in preventing the spread of HAIs.

  - **Prevention of infection coming into the hospital** When patients are tested as they come into hospital, people who are carrying MRSA (around 3% of the general public) or other infections can be treated immediately and isolated until the bacteria have been destroyed. This reduces the chance of the person becoming clinically ill during treatment or of spreading the bacteria to others. Another problem is that people visiting patients may bring infections into the hospital. This is particularly the case with MRSA and many viral infections, such as norovirus (causes sickness and diarrhoea). Hospitals advise people not to visit if they are unwell but people often ignore the advice because they want to see seriously ill relatives. People visiting patients should follow good hygiene procedures and keep their hands clean and washed. Most hospitals provide alcohol gels for people to use as they come in and out of the hospital, but the biggest problem is persuading people to use them.

- **Monitoring levels of hospital-acquired infections** In some countries laws oblige hospitals to measure and report levels of MRSA and *C. difficile* infections. These results are published and are available to the general public as well to the government. This can focus attention on the problem and lead to increased efforts to reduce and overcome the infections. For example, government figures showing numbers of deaths involving (but not caused by) MRSA in England and Wales indicated that they had dropped from 40.2 per million population in 2005 to 6.3 per million in 2013.

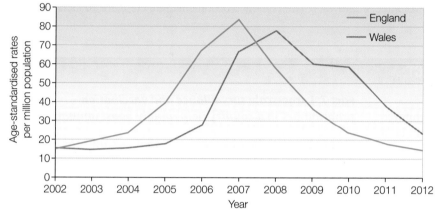

▲ **fig B**   Rates of *S. aureus* and MRSA per million population from 2002 to 2012 in England and Wales.

As a result of doctors and hospitals following strict codes of practice about both antibiotic use and hospital practice, there has been a marked decrease in the incidence of MRSA and *C. difficile* in recent years (see **fig B**). Continuing to give a high priority to good practice of infection control in hospitals should continue the decrease in levels of HAIs and related deaths.

CHECKPOINT

1.   Explain how bacteria develop resistance to common antibiotics.

2.   (a) Find out the current code of practice on the use of antibiotics in your country.

     (b) Explain how this can help to reduce the development of further antibiotic-resistant strains of bacteria.

3.   (a) Describe how the use of broad-spectrum antibiotics is implicated in the spread of *C. difficile*.

     (b) Explain how minimal use of narrow-spectrum antibiotics can help to avoid the problems of HAIs.

     (c) Discuss how the data in **fig A** has serious implications for the development of HAIs.

     SKILLS   ANALYSIS

4.   Explain how an understanding of the causes and methods of spread of HAIs relates to the codes of practice on hospital hygiene.

     SKILLS   CREATIVITY, REASONING

SUBJECT VOCABULARY

methicillin-resistant *Staphylococcus aureus* (MRSA) a strain of *S. aureus* that is resistant to several antibiotics, including methicillin

Clostridium difficile a type of bacterium that often exists in the intestines and causes no problems unless it becomes dominant as a result of antibiotic treatment that has removed or damaged the normal gut flora

hospital-acquired infections (HAIs) infections that are acquired by patients while they are in hospitals or care facilities; these infections may be the result of poor hygiene or the result of antibiotic treatment; the pathogens may be antibiotic resistant

THE BATTLE TO ELIMINATE POLIO

Polio is a viral disease once feared all over the world. Now, thanks to efforts led by the Global Polio Eradication Initiative, its incidence has been reduced by 99%. In the last 20 years, 20 million volunteers have vaccinated nearly 3 billion children. Here are 10 facts on polio eradication.

WORLD HEALTH ORGANIZATION WEBSITE EXTRACT

Fact 1: Polio still paralyses children.
Polio still exists in a few countries, mainly affecting children under 5. One in 200 infections leads to irreversible paralysis (usually in the legs) and 5–10% of those infected die when their breathing muscles stop working.

Fact 2: Polio is 99% eradicated globally.
In 1988 when the Global Polio Eradication Initiative was formed, polio paralysed more than 350 000 people every year. In 2016, the figure was 42.

Fact 3: There are just three countries where polio has never been stopped.
These are Afghanistan, Nigeria and Pakistan.

Fact 4: Polio is unusual because it can be completely eradicated.
This is because the virus cannot survive for long outside the human body and doesn't infect other species.

Fact 5: Cheap and effective vaccines are available to prevent polio.
They can be given by mouth rather than injection so don't need medically trained staff to deliver them.

Fact 6: The global effort to eradicate polio is the largest public-private partnership for public health.
It involves the WHO, Rotary International, the US Center for Disease Control and Prevention, UNICEF and the Bill and Melinda Gates Foundation.

Fact 7: Large-scale surveillance and vaccination programmes in a country help rapidly boost immunity.

Fact 8: Every child must be vaccinated to eradicate polio.
This includes children living in the most difficult and remote areas of the world.

Fact 9: Volunteers delivering polio vaccines can give other vaccines and vitamin A drops.

Fact 10: We CAN eradicate polio.
80% of the world now lives in certified polio-free regions.

▲ **fig A** Polio vaccines are saving children all over the world from this dreadful disease.

From WHO website: http://www.who.int/features/factfiles/polio/en/

SCIENCE COMMUNICATION

This list is based on one produced by the WHO.

1 (a) Who do you think is the intended audience for this list?

(b) Suggest **two** advantages of a list as a way of giving information.

2 Visit the original website and find the list.

(a) Give **two** ways in which the list on the website is better than the list given here.

(b) Look for information about polio on the website of one of the other organisations which make up the Global Polio Eradication Initiative. Compare the way the information is presented with the list on this page and write a paragraph explaining the strengths and weakness of both.

BIOLOGY IN DETAIL

Now you are going to think about the science in this resource.

3 (a) Summarise what this page tells you about polio.

(b) Suggest why polio is being eliminated by a vaccination programme not through the use of a medicine.

4 Explain how a polio vaccination works to protect a child against the disease. (Do some research to find out what type of vaccine is used.)

5 Investigate the eradication of smallpox globally and write an article to compare that campaign with the current efforts to eradicate polio. Highlight similarities and differences between the two diseases and the two campaigns.

ACTIVITY

- Work in a group. Plan a campaign to help the Global Polio Eradication Initiative finally eradicate polio.

- **Either:** Investigate the reasons why a particular country still has polio. Think carefully and develop an education pack which can be used in schools and with groups of mothers to explain why it is so important to get children vaccinated against polio, how it works and the benefits a polio-free world would bring.

- **Or:** Plan a fundraising campaign in your school or community to support the work of the Global Polio Eradication Initiative. You need to raise awareness of the problem, and excite people with how near we are to eliminating polio globally. You can then encourage them to give money or time to fundraising efforts, or maybe to train as volunteers to help deliver the vaccines where they are needed.

1 (a) One method of combating the influenza virus is by vaccination. How many of the following statements are true? [1]

   1 The virus cannot mutate.

   2 The virus is not affected by antibiotics.

   3 Vaccination must happen every year.

   4 The virus has antigens.

   A 1   **B** 2

   C 3   **D** 4

(b) Which of these lists correctly states three ways in which the influenza virus can be transmitted between humans? [1]

   A coughing, unprotected sexual intercourse, kissing

   B sharing a cup, sneezing, sharing needles

   C sneezing, kissing, sharing needles

   D sneezing, kissing, coughing

(c) The influenza virus genome contains 14 001 bases.

   (i) What type of nucleic acid is present in the genome of the influenza virus particle as it infects a cell? [1]

   (ii) This nucleic acid is converted into mRNA inside the host cell nucleus. What is the correct name for this conversion? [1]

      A translation   **B** copy error

      C transcopying   **D** transcription

   (iii) This process has an error rate of 1 base in 10 000. Calculate the average number of mutations that might be expected in a virus particle. Give your answer to 1 decimal place. [2]

(d) Some of the molecules coded for by the virus genome form the virus antigens that are recognised by the host's immune system. Explain how a mutation in the genome could lead to the host suffering from influenza more than once. [4]

(e) The influenza virus infects ciliated epithelial cells of the respiratory system. Suggest why this might lead to a secondary infection by bacteria. [4]

(Total for Question 1 = 14 marks)

2 The following graph shows the changes in population size of bacterial cultures grown in the presence of three antibiotics, A, B and C. In each case the antibiotic was added 7 hours after the experiment started.

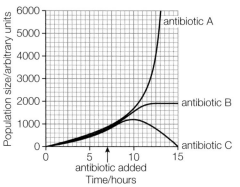

(a) (i) Identify which antibiotic is bactericidal and use evidence from the graph to justify your decision. [2]

   (ii) Identify which antibiotic is bacteriostatic and use evidence from the graph to justify your decision. [2]

(b) In previous experiments, the bacteria used had shown a response to antibiotic A that was similar to the response shown to antibiotic B. Suggest and explain a reason for the shape of the curve for antibiotic A. [4]

(c) Describe a technique that could demonstrate the effectiveness of antibiotics on bacteria. [5]

(Total for Question 2 = 13 marks)

3 MRSA is a strain of the bacterium *Staphylococcus aureus*. MRSA can survive treatment with several antibiotics. An infection with MRSA is difficult to treat.

(a) Describe what is meant by the term *evolutionary race*. [2]

(b) Explain how some strains of bacteria have become able to survive treatment with antibiotics. [4]

(c) Describe the techniques that have been adopted to overcome the problem of antibiotic-resistant bacteria. [3]

(Total for Question 3 = 9 marks)

4 (a) The following list gives some examples of how immunity can develop in a mammal.

   P: Antibodies are transferred into the blood of a baby, from its mother, in the days before birth.

   Q: T killer cells are produced by the body when it is infected by a virus.

   R: Polio virus, which has been made incapable of replicating, is given to babies to stimulate the production of memory cells.

   S: Anti-venom contains antibodies produced in an animal. This anti-venom can be injected to give protection against snake bite venom.

   Complete the following table by writing a letter from the list above to match the type of immunity described in each case. [4]

Immunity	Active	Passive
natural		
artificial		

(b) Vaccination minimises the spread of disease in a population by developing herd immunity. State which of these is an approximate percentage of a population which needs to be vaccinated to ensure herd immunity. [1]

    A 10%   **B** 50%   **C** 85%   **D** 97.5%

(c) An investigation was carried out into the production of antibodies by lymphocytes. Combinations of T cells and B cells were grown in four different cultures: W, X, Y and Z. The culture medium was checked for the presence of antibodies after several days. Results are shown in the table below.

Culture	T cells present	B cells present	Antigen present	Antibody production
W	✓	✓	✗	no
X	✓	✗	✓	no
Y	✗	✓	✓	no
Z	✓	✓	✓	yes

(i) State **two** conclusions about antibody production that you can draw from these results. [2]

(ii) Explain why no antibodies were produced in cultures X and Y. [2]

(iii) Suggest **two** other components of the culture medium and explain why each is needed. [4]

(Total for Question 4 = 13 marks)

5 Immune responses involve communication between cells by means of chemicals.

(a) Which chemical is used by T helper cells to cause B cells to differentiate into active B cells? [1]

    A cytokinesis   **B** cytokines

    C cytoplasm   **D** cytokinetics

(b) A strain of mice is unable to produce one of these chemicals called TNF (TNF-deficient mice). These mice were selected to investigate the importance of chemical communication in preventing tuberculosis. State **two** symptoms of tuberculosis. [2]

(c) TNF-deficient mice and normal mice were both exposed to tuberculosis bacteria. After several weeks, the numbers of some types of white blood cell in the lungs were counted. The results are shown in the graph below.

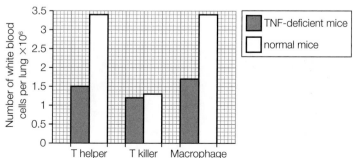

(i) Describe the effects of TNF on the numbers of these white blood cells in the lungs. [2]

(ii) Describe the role of T killer cells in a normal immune response and give a reason for the results for T killer cells in this investigation. [3]

(iii) Using the information in the graph, explain why TNF-deficient mice are more likely to die of tuberculosis. [3]

(d) Drugs that inhibit TNF production are commonly used to reduce inflammation in people with rheumatoid arthritis. Predict possible consequences of using TNF inhibitors for people who are infected with tuberculosis but have no symptoms. [2]

(Total for Question 5 = 13 marks)

6 A new technique for vaccinating people involves injecting them with DNA. Viruses have proteins on their coats that are coded for by their DNA. The genes for producing viral proteins can be isolated and inserted into loops of DNA (plasmids). Plasmids can enter human cells which will then produce the viral proteins. The proteins will become part of the surface membrane of the human cell. The immune system recognises these proteins as foreign and responds by producing antibodies and T killer cells.

The process is summarised for one protein in the following diagram.

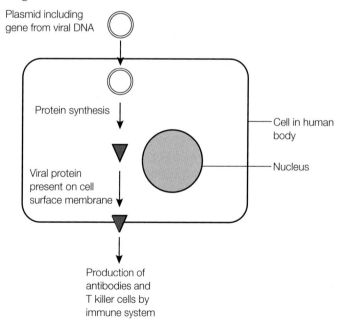

(a) Explain why the response of the immune system to the viral proteins is an example of active immunity. [2]

(b) Explain how active immunity provides immunity against future infections by the virus. [2]

(Total for Question 6 = 4 marks)

TOPIC 6 MICROBIOLOGY, IMMUNITY AND FORENSICS

6C DECOMPOSITION AND FORENSICS

Rainforests, with their dense, leafy canopies, are very productive plant communities, so they have a high level of leaf fall. But not all of the leaves end up as leaf litter. Much plant material is eaten by animals, including birds, monkeys and insects, before it reaches the ground. In the warm moist climate of a rainforest, decomposers work fast. In spite of the high productivity of the plants, the forest floor does not disappear under a layer several metres deep of old leaves and fruits, because fungi, bacteria, many different arthropods and worms break down the plant material and return the nutrients to the soil to sustain the continual growth. These decomposers play an important role everywhere, removing dead material and waste, and making sure materials such as carbon and nitrogen cycle continuously between the living and the non-living parts of the ecosystem.

In this section, you will be looking at the way microorganisms work to decompose organic material and the role they play in the carbon cycle in nature. You will also be looking at how microorganisms, along with larger organisms including flies and beetles, work together to bring about the decomposition of bodies after death. You will discover how forensic scientists use their knowledge and understanding of these biological processes to help them understand what has happened when there is an unexpected or violent death. This will include the DNA technology now widely used in identifying both innocent and guilty people when a crime is investigated. You will also discover how the same technology is used to classify living organisms and illuminate the relationships between species.

MATHS SKILLS FOR THIS CHAPTER

- **Recognise and make use of appropriate units in calculations** (*e.g. calculating numbers of DNA fragments from a PCR process*)

- **Translate information between graphical, numerical and algebraic forms** (*e.g. use of a cooling curve to estimate time of death*)

What prior knowledge do I need?

Chapter 2B (Book 1: IAS)

- The structure of DNA
- The genetic code

Chapter 4B (Book 1: IAS)

- The principles of classification
- Different definitions of a species

Chapter 6A

- The growth patterns of bacteria
- Factors affecting the growth rate of bacterial colonies

What will I study in this chapter?

- The role of microorganisms in the decomposition of organic material and the recycling of carbon
- How to determine the time of death of a mammal by examining the extent of decomposition, stage of succession, body temperature, degree of muscle contraction and forensic entomology
- The use of the polymerase chain reaction in the amplification of DNA
- The use of gel electrophoresis to separate DNA fragments of different lengths
- The value of DNA profiling in identification of people and of species of animals and plants, and in determining genetic relationships between organisms

What will I study later?

Chapter 7B

- The role of ATP in the contraction of muscles

Chapter 8C

- How DNA sequencing can be used to identify specific pathogens
- The importance of microarrays in recombinant DNA technologies

LEARNING OBJECTIVES

■ Know the role of microorganisms in the decomposition of organic matter and the recycling of carbon.

As you saw in **Chapter 6B**, some microorganisms cause infectious diseases in animals (including people) and plants. However, the majority of microorganisms play a very positive role in the ecosystems of the natural world. Many bacteria and fungi are decomposers. They break down organic material to produce simple inorganic molecules such as carbon dioxide and water. They release inorganic nitrogen (in the form of ammonia, nitrites and nitrates) which returns to the soil in the **nitrogen cycle**, and they release sulfur compounds which return to the soil or water. They also play a big part in the carbon cycle, which you studied in **Section 5C.3**.

DECOMPOSITION IN NATURE

Decomposing bacteria play a vital role. Plants constantly take minerals such as nitrates from the soil and compounds such as carbon dioxide from the air to build macromolecules in their cells. These are transferred to animals through the food chains and food webs that link all living organisms. Fortunately, this is not a one-way process, so the resources of the Earth can be reused. The decomposers make sure that the chemical constituents of life are continually recycled within ecosystems (see **fig A**).

Living organisms die. Animals also produce both faeces and urine. Decomposers are important because they prevent the bodies of plants and animals and all the excrement from all of the animals from covering the surface of the Earth. Bacteria and fungi are the main groups of organisms which make sure this does not happen.

▲ **fig A** The actions of decomposing microorganisms break down organic material, including any cellulose in undigested plant material. They release carbon dioxide into the atmosphere, and nitrates and other mineral ions into the soil.

The decomposers feed on faeces and dead bodies, digesting them and using the nutrients for respiration, to build their own cells, and for reproduction. The microorganisms also release waste products, and these provide nutrients in a form that plants can use once more. In **Section 6C.3**, you will look at the role these

microorganisms play in the decomposition of the human body. They play a similar part for every organism after death.

The carbon cycle is part of this nutrient recycling. You can remind yourself of the carbon cycle in nature by looking back at **Section 5C.3 fig A**. One reason bacteria are so important in the carbon cycle is because some microorganisms produce the enzyme cellulase. This enzyme breaks down the cellulose in plant cell walls to produce sugars. These sugars can then be used as food by the decomposers and a wide variety of other organisms.

The decomposers (bacteria and fungi) release 60×10^9 tonnes of carbon per year into the atmosphere. Carbon dioxide is also released into the atmosphere by combustion when anything that has been living is burned. This includes wood, straw or fossil fuels made from animals and plants which lived millions of years ago. At the moment, combustion only releases $6–7 \times 10^9$ tonnes of carbon per year into the atmosphere, but this figure is rising. If you compare these two figures you can see the importance of microorganisms in the global carbon cycle.

▲ **fig B** The dead body of this tree is slowly being broken down by the action of decomposers. Some of the fungi are clearly visible, but the billions of bacteria are too small for you to see without a microscope.

CONDITIONS FOR RECYCLING NUTRIENTS

The chemical reactions that take place in microorganisms, like those in many other living things, work faster in warm conditions. However, if the temperature gets too hot they stop altogether, because the reactions are controlled by enzymes which denature with the increased temperature. The reactions also stop if conditions are too cold.

EXAM HINT

Remember to relate the conditions for decomposition back to the activity of enzymes.

Most microorganisms grow better in moist conditions, which make it easier to dissolve their food and also prevent them from drying completely. So the decay of dead plants, animals and faeces occurs far more rapidly in warm, moist conditions than it does in cold, dry ones. A leaf takes about a year to decompose in a temperate climate, but can take only 6 weeks to decay in a tropical rainforest. The majority of decomposers respire like any other organism, so decay occurs more rapidly when there is plenty of oxygen available.

People use the processes of decomposition in the treatment of sewage (a mixture of waste from the human body and water) and the production of compost. However, it is in the natural world where the role of the decomposers is the most important, and the cycling of resources plays a vital role in maintaining the fertility of our soil and the health of our atmosphere. The processes that remove materials from the soil are balanced by processes that return materials in a stable community of plants and animals. In other words, the materials are constantly cycled through the environment. The time that microbes and detritus feeders take to break down the waste products and the dead bodies of organisms in ecosystems is equivalent to the time taken for most of the energy originally captured by green plants in photosynthesis to be transferred to other organisms or back into the environment.

CHECKPOINT

1. ▶ Explain why the recycling of nutrients in ecosystems is so important.

2. Describe the roles of microorganisms in the recycling of carbon in ecosystems.

3. Sketch the carbon cycle (if necessary look back to **Section 5C.3 fig A** to refresh your memory) and indicate where microorganisms are involved in the process.

SKILLS ▶ CRITICAL THINKING

SUBJECT VOCABULARY

nitrogen cycle the recycling of nitrogen between living things and the environment by the actions of microorganisms

2 FORENSIC SCIENCE AND THE TIME OF DEATH

People die; sometimes they are killed unlawfully or murdered. The murder rate varies from country to country. For example, on the Caribbean island of Jamaica, approximately 60 people per 100 000 population are murdered each year, whereas in the Republic of Maldives only 2.6 people per 100 000 population are killed in this way. Unsurprisingly, murder is a crime which has a high profile in the media, and it is the subject of many detective stories. To secure a prosecution for murder, a number of facts need to be verified:

• who died and who killed them

• what was used to commit the crime

• when the victim died

• where the murder took place

• why the murder took place; that is, establish the motive for the crime.

Forensic science is the application of science to crime scenes. It is important when we need to investigate an unexplained death. It can tell us when someone died, what they died of, whether a crime was committed and even the identity of the person who died. Forensic science also gives us the tools to help us solve crimes, including murders. We can use the technology available to identify a criminal or release an innocent person, but only if the evidence is gathered correctly (see **fig A**).

▲ **fig A** People are stopped from entering crime scenes so specially trained officers can collect evidence. In this topic you will look at the science behind some of the techniques they use to identify both the innocent and the guilty.

INVESTIGATING TIME OF DEATH

When a body is found, it is not easy to estimate the time of death. If the person has been dead for more than 48 hours it becomes even more difficult. Forensic evidence is usually combined with the evidence of witnesses and circumstantial evidence to produce a best estimate of the time of death.

A number of changes occur in the body of any mammal when it has died and we can use these changes to help estimate the time of death. For example, normal human body temperature is around 36–37 °C. It is kept warm by the energy transferred during metabolic reactions such as cellular respiration. After death, the metabolic reactions which have warmed the body begin to slow and eventually stop. At the same time, energy is transferred from the surface of the body into the surroundings by radiation, conduction and the evaporation of water, which cools the body. The body reactions do not all stop immediately when a mammal dies, so although the body temperature starts to fall immediately after death, it stays almost stable for a time before decreasing steadily to room temperature. As a result of these changes, the temperature of a body will give some indication of how long the person has been dead.

A number of other factors affect how quickly the body temperature drops.

• **FAT LEVELS** The outside of the body cools much more rapidly than the inside and the amount of body fat affects the rate of cooling because it acts as insulation. As a result, a fatter person will cool more slowly after death than a thin person.

• **MODE OF DEATH** People who die somewhere cold, for example in cold water or on an exposed mountain in winter, will cool more rapidly than someone who dies on land in a hot climate or a heated room.

• **SURROUNDING TEMPERATURE** Bodies cool faster in cold countries or if they are placed in fridges or freezers than they do in hot countries or in heated rooms.

• **SURFACE AREA:VOLUME RATIO** Smaller people will cool faster because they have a larger surface area : volume ratio than larger people.

• **CLOTHING AND COVERING** A body which has clothes on to insulate it, or a body wrapped in bed coverings or rolled in carpet, will be insulated and so will cool more slowly than a body without any coverings.

Soon after the point of death the body of a mammal starts to cool in a fairly predictable manner despite all these variations. From analysing the cooling curve for bodies under known conditions it is possible to estimate the time of death, as you can see in **fig B**.

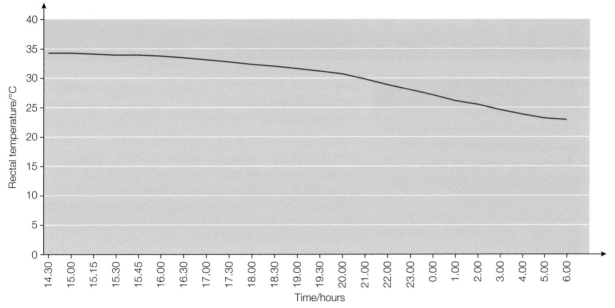

▲ **fig B**　Cooling curve for a human male, body length 1.6 m, weight 70.76 kg, external temperature 10 °C, time of death 13:30. This man had died from heart failure. *Data from Professor Len Nokes.*

EXAM HINT

As with any graph, you may need to demonstrate your understanding and maths skills by describing or explaining the shape of the graph. You may also need to calculate intercepts to the graph or extrapolate from the graph.

RIGOR MORTIS

When someone dies, their heart stops pumping blood around their body and the brain dies within minutes. When you are alive, your brain cells respire actively and aerobically all the time. They do not contain much stored ATP, fat or glycogen, so they are affected very quickly by a lack of oxygen. However, some tissues, such as the muscle cells, have large stores of ATP and glycogen and can continue to respire anaerobically for some time after death. ATP is needed to maintain the muscles in a relaxed state. After death, the muscle cells use all of the ATP and the muscle fibres become permanently contracted. The body becomes rigid and this effect is known as **rigor mortis**. On average, rigor mortis starts about 2–4 hours after death and takes between 6 and 8 hours to take full effect. It begins in the muscles in the face and neck, progresses down the body, and spreads steadily to the larger muscles of the body. Rigor is seen clearly in all mammals, and in other animals too.

EXAM HINT

You may need to link the occurrence of rigor mortis to the structure and action of muscle fibres, which you will study in **Topic 7**.

The main factor which determines how quickly rigor mortis begins is the amount of ATP stored in the muscles at the time of death. The level of ATP in the muscles varies from person to person, depending on their genetics, their level of fitness and the level of activity before death. For example, rigor mortis usually starts very quickly in drowning victims, because they have used up all their muscle ATP trying to stay on the surface.

The temperature of the person when they die and the temperature of their surroundings also affect how quickly rigor mortis begins. Rigor mortis is not permanent; it usually passes between 36 and 48 hours after death, but it can last considerably longer. The relaxation of the muscles as rigor mortis passes happens because muscles soften as enzymes released from the lysosomes begin to break down the tissue. Using some basic mathematics, it is possible to predict when the ATP store was 'full' and hence the time of death. Putting these data together with temperature changes can greatly improve the prediction of time of death.

CHECKPOINT

1.　(a) Explain why the body of a mammal cools down after death.

　　(b) Why is the cooling rate slower in the first hours after death?

　SKILLS　CRITICAL THINKING

　▶ (c) Explain how factors such as external temperature, whether the body is wet or dry, and whether a body is wrapped or exposed will affect the rate of cooling after death.

2.　Explain why rigor mortis alone is of limited use in determining the time of death.

SUBJECT VOCABULARY

forensic science the application of scientific techniques to the investigation of a crime
rigor mortis temporary muscle contraction causing the body to become rigid after death

In **Section 6C.2**, you looked at how forensic scientists can use body temperature and the degree of muscle contraction in a body to help them calculate the time of death. These are particularly useful in the first 48 hours or so after death. However, some bodies are found much longer than 48 hours after death. In that situation, forensic experts use other measurements to help them calculate when the person died. The state of decay of a body is a very important tool in this process.

Most bodies follow a similar pattern as they decay. When the gut movements stop, the catabolic enzymes of the digestive system start breaking down the walls of the gut and then the surrounding cells. As body cells die from lack of oxygen, the lysosomes within them rupture and release lysozymes, the enzymes which break down the cells. These processes make the body a more suitable habitat for the organisms responsible for further decay.

THE STAGES OF SUCCESSION

A newly dead body is like a piece of freshly exposed soil or rock in that it is a new available habitat. The principles of succession that you studied in **Section 5B.2** also apply to the succession of species on a mammalian body after death. An understanding of this succession allows forensic specialists to estimate how long a body has been present very accurately (see **fig A**).

The first stage of succession on a body involves the colonisers. The first colonisers are anaerobic bacteria, which do not need oxygen and grow rapidly in the lactic acid-rich environment of the muscles after death. These bacteria, which in life are confined to certain areas of the body such as the gut, reproduce freely and take control. The enzymes break down cells so the bacteria spread. However, they are quickly joined by several fly species. The best known of these are the blowflies. These insects are extremely sensitive to the smell of dead organisms, and can arrive on a body within minutes of death. They are attracted to the moisture and smell around all the natural holes of the body (e.g. eyes, mouth, nose, etc.), as well as any open cuts or wounds. At first, the body attracts these flies as a site on which to lay eggs. The maggots (larval form of a fly) hatch and immediately begin feeding on the tissues, breaking them down. They dig deeper

into the flesh. Eventually, the maggots pupate, turn into flies and immediately mate and start the cycle again. The soft tissues of the body gradually liquefy and adult flies can feed on this too.

Beetles also start to lay eggs on the carcass (decaying body); their larvae feed on maggots rather than eat the body itself. Then parasitic wasps arrive to lay their eggs in the fly and beetle larvae. Gradually, as the body is digested, it becomes dry and the early colonisers begin to disappear. Different species such as cheese flies and coffin flies move in. Eventually, the body is too dry for maggots and a number of beetle species with strong, chewing mouthparts move in, for example, carcass beetles, ham beetles and hide beetles. They feed on the remains of the muscles and the connective tissues. At the very end, mites and moth larvae will feed on the hair until only dry bones are left.

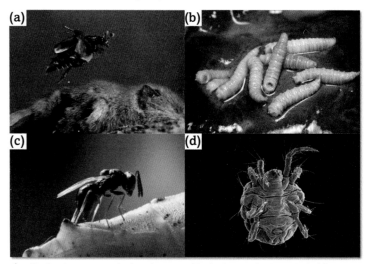

▲ **fig A** Organisms which colonise and help decay a body include (a) carcass beetles, (b) blowfly larvae, (c) parasitoid wasps and (d) mites.

The stages of decay follow a regular pattern, but the speed at which they occur is variable and depends on a number of factors.

- **TEMPERATURE** The warmer the body, the faster the rate of decay because the speed of all the chemical reactions increases as the temperature rises. For example, in Zimbabwe, an entire elephant can be reduced to a skeleton in just 7 days. On the other hand, a body kept in a freezer will decay very slowly.

- **THE LEVEL OF EXPOSURE** A buried body will decay more slowly than a body left in the open air, because the body on the surface is more available to flies and other decomposers. The temperature underground is lower and more stable. A body hidden in a house will be less exposed to insects, but usually warmer than a body outside. A body in a plastic bag outside will decay more slowly than a body exposed to the air.

Fig B shows you the transformation that results from the action of the decomposers.

▲ **fig B**   The processes of decay convert a once-living mammal, through stages of decomposition, into dry remains of hair and bone.

FORENSIC ENTOMOLOGY

Many of the organisms that colonise a body after death are insects, and the study of insect life relating to crime has developed into the science known as **forensic entomology**.

Forensic entomologists know the detailed life cycles of blowflies and other insects that colonise dead bodies, and they also know how the life cycles vary in different countries and with different environmental conditions such as temperature. They can use this evidence to help estimate how long a body has been in the place it is found. For example, if a body found outside had no evidence of blowfly activity, this would tell the forensic team that the body had almost certainly been there for a maximum of 24 hours because blowflies often find a body within minutes of death!

Blowflies lay their eggs in the natural openings of a body but will also lay eggs in knife and gunshot wounds. When the maggots hatch, they feed on the body and grow until they form a pupa and a new adult fly hatches. You can see the complete life cycle in **fig C**.

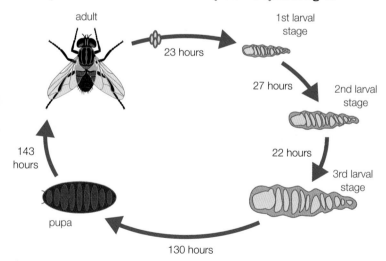

▲ **fig C**   The time taken for the blowfly life cycle to be completed depends on the temperature. The times shown in this diagram are at 21 °C. In colder weather, the maggots grow more slowly, and in very hot conditions they complete their life cycle faster.

Forensic entomologists collect eggs, maggots and pupae when a body is discovered. If possible, the forensic entomologists will grow live specimens on to the adult stage so a precise identification of the species of fly or beetle can be made. This makes calculating time of death even more accurate because each species has a life cycle of a different length.

EXAM HINT

You will need to apply your knowledge to unfamiliar contexts. A question might provide a range of information about a crime scene and ask you to deduce what effect certain factors had on the body decomposing. Ensure you read the information fully and carefully before considering each factor individually. You then need to consider how the different factors may have interacted.

Evidence from forensic entomologists is important in police investigations and in court. For example, a couple were found dead in their home in New South Wales, Australia. They had both been shot. The body of the woman was found in bed. It was more decomposed than the body of the man found in the kitchen, and the maggots were more advanced. Police gathered evidence from reported phone calls and sightings suggesting that the woman had been alive until the Saturday evening. There was a suspect, but he had a strong alibi for the Saturday night. However, when the evidence from the forensic entomologist was considered, a very different picture emerged. The minimum age of the oldest maggots taken from the woman was four days. This put the death at least one whole day earlier than the circumstantial evidence. The maggots showed she had been killed on the Friday night. When the witness evidence was checked again, it was discovered that people had made mistakes (they often do). Maggots do not lie: it was Friday when they saw the woman alive. And the prime suspect had no alibi for the Friday night. Faced with the evidence, he confessed. And why was one body more decomposed, with bigger, stronger maggots than the other? The woman had gone to bed with an electric blanket on and this was where she was shot. The electric blanket kept her body warm and so the chemical reactions of both decay and the development of the maggots were faster.

DID YOU KNOW?

Moving forensic science forward

A big research project conducted in Canada in the 1990s looked at the insect succession in buried carcasses. It used large mammals as human models, even dressing them in human clothes before burying them in the type of shallow grave often used by murderers to hide their victims. The results show some interesting differences in body temperature between exposed and buried bodies (see **fig D**). These are the sort of data that are useful to forensic experts. It is not based on observations from humans but it is still a very useful model.

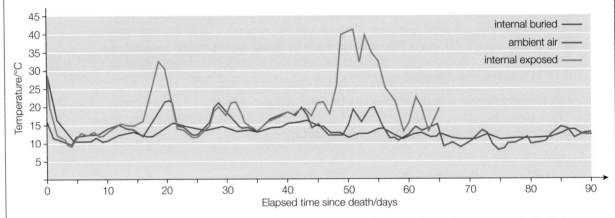

▲ **fig D**   Internal temperature of buried and exposed animal carcasses compared to the ambient air temperatures in the Cariboo Region of Canada. Measurements were taken using electronic probes in the animal carcasses and weather station data.

CHECKPOINT

SKILLS   INTERPRETATION

1. ▶ Make a flowchart to describe the main stages of decomposition of a mammalian body showing the organisms you would expect to be involved.

2. The conditions in which a body is kept after death have a major effect on the speed at which decomposition occurs. Select **two** factors and explain how they affect decomposition rates.

3. Explain how the process of succession is helpful to forensic scientists in determining the time of death.

4. (a) Using **fig D**, describe the pattern of temperature changes in both an exposed and an unexposed body in relation to the ambient air temperature and explain the patterns you see.

SKILLS   CREATIVITY

   ▶ (b) These data were collected using large domestic animals in human clothes as mimics of the human body. Suggest **two** advantages and **two** disadvantages of using non-human mammals in this type of investigation.

SUBJECT VOCABULARY

forensic entomology the study of insect life relating to crime

In the last 50 years, one of the areas of biology that has developed fastest is the study of genetics, genomics and proteomics, highlighted by our ever-growing knowledge of the human genome. You learned about DNA and the genome in **Topic 2** and **Topic 3 (Book 1: IAS)**. Here, you will find out how we are learning to analyse DNA and use the results.

WHAT IS THE GENOME?

The genome is the total of all the genetic material in an organism. In prokaryotes, the DNA is in the cytoplasm, in the main chromosome and in the plasmids. In eukaryotes, the DNA is in the nucleus of the cell and in the mitochondria; in green plant cells, there is also DNA in the chloroplasts. The DNA that constitutes the human chromosomes consists of many millions of base pairs. However, the genes themselves (the coding regions of DNA that determine the protein structures that support the entire organism) are only about 2% of that DNA. These coding regions are known as the **exons**. The large, non-coding regions of DNA are removed from messenger RNA before it arranges itself on the ribosomes and is translated into proteins. They are known as the **introns**.

ANALYSING THE DNA

In **Section 2B.3 (Book 1: IAS)**, you discovered the principles of the structure of DNA: the double helix made up of sugar–phosphate backbones with purine and pyrimidine bases, held together along the length of the molecule by hydrogen bonds. In recent years, our ability to analyse the structure of the molecule has developed tremendously. We have analysed the entire human genome, not just once but thousands of times. In **DNA sequencing/gene sequencing** we analyse individual strands of DNA or individual genes, giving us the pattern of bases that codes for a particular protein in the cell. We can also use it to identify different species of living organisms. In **DNA profiling** we analyse the patterns in the non-coding areas of DNA (the introns) and use them to identify individuals (for example, in a court of law or in paternity cases).

Our successful analysis of DNA at both levels depends on one process: the **polymerase chain reaction (PCR)**.

THE POLYMERASE CHAIN REACTION

Traditional DNA profiling needs at least 1 μg (microgram) of DNA, which is the equivalent of the DNA from about 10 000 human cells. In a crime investigation, there may only be a very small DNA sample available. The polymerase chain reaction (PCR) adapts the natural process that replicates DNA in the cell. PCR makes it possible to produce enough DNA to sequence a genome or produce a DNA profile from very small traces of biological material. When a very small sample of DNA is increased using PCR to produce a large enough sample for analysis, we say it has been **amplified**.

DEVELOPING PCR

When scientists were trying to develop a way of amplifying very small amounts of genetic material, they had a particular problem. The DNA sample needed to be heated to around 90–95 °C to separate the two strands and make them available for replication. These high temperatures completely denature the DNA polymerase enzymes from most organisms, so the scientists could not replicate the separated DNA strands.

Kary Mullis is the scientist who solved the problem. He decided to try using enzymes from a bacterium (*Thermus aquaticus*) that lives in hot springs to develop a technique for replicating DNA artificially in the laboratory. The enzymes in this bacterium have evolved to survive in extreme conditions. Mullis hypothesised that they would be robust enough to resist the high temperatures and frequent temperature changes needed to separate DNA strands and lead to replication. He was right.

HOW PCR WORKS

The DNA sample that is to be amplified is mixed with the enzyme Taq (*Thermus aquaticus*) DNA polymerase, primers (small sequences of DNA that must join to the beginning of the separated DNA strands before copying can begin), a good supply of the four different nucleotides, and a suitable buffer for the reaction. The mixture is placed in a PCR machine. It is heated to 90–95 °C, which causes the DNA strands to separate as the hydrogen bonds between them break. The mixture is then cooled to 50–60 °C so that the primers bind (anneal) to the single DNA strands. Finally, the mixture is heated to 75 °C, which is the optimum temperature for the Taq DNA polymerase enzyme to build the complementary strands of DNA (see **fig A**).

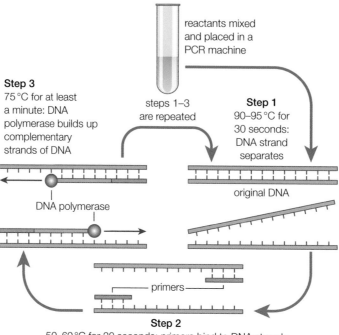

Step 3
75 °C for at least a minute: DNA polymerase builds up complementary strands of DNA

steps 1–3 are repeated

Step 1
90–95 °C for 30 seconds: DNA strand separates

reactants mixed and placed in a PCR machine

original DNA

DNA polymerase

primers

Step 2
50–60 °C for 20 seconds: primers bind to DNA strands

▲ **fig A**   The main stages of the polymerase chain reaction.

These steps are repeated around 30 times, producing about 1 billion copies of the original DNA. On average, the whole process takes about 3 hours, and much of that time involves heating and cooling the reaction mixture in the PCR machine. The amplified sample gives forensic scientists the material they need to produce a DNA profile of a suspected criminal.

CHECKPOINT

1.   Describe how DNA can be amplified in the polymerase chain reaction.
2.   Explain why the use of enzymes from the bacterium *Thermus aquaticus* was such a breakthrough.
3.   Suggest why PCR has been an important development in the investigation of crime.

SUBJECT VOCABULARY

exons the coding regions of DNA (the genes)
introns the large, non-coding regions of DNA that are removed before messenger RNA is translated into proteins
DNA sequencing/gene sequencing the analysis of the individual base sequence along a DNA strand or an individual gene
DNA profiling the identification of repeating patterns in the non-coding regions of DNA
polymerase chain reaction (PCR) the process used to amplify a sample of DNA (to make more copies of it very rapidly)
amplified the process by which DNA is replicated repeatedly (using the polymerase chain reaction) to produce a much bigger sample

We can identify individuals and species by patterns in their DNA. This important discovery has had enormous implications in many areas of science, including forensic science and paternity testing. The process is known as DNA profiling.

INTRONS AND SATELLITES

The human genome contains between 20 000 and 25 000 genes. The chromosomes are made up of hundreds of millions of base pairs. Yet less than 2% of the genome codes for proteins. Over 90% of the DNA is made up of the introns: repetitive coding regions between the genes. Their function is still not fully understood, but we do know that some of these sequences can code for small interfering RNA molecules (siRNA) that interact with mRNA and prevent the production of certain proteins, and that they are inherited in the same way as the active genes. Introns are the regions of the chromosomes that are used in DNA profiling.

Within the introns are short sequences of DNA that are repeated many times to form **micro-satellites** and **mini-satellites**. In a mini-satellite, a 10–100 base sequence is repeated 50 to several hundred times. A micro-satellite has 2–6 bases repeated 5 to 100 times. The same mini- or micro-satellites appear in the same positions on each pair of homologous chromosomes. However, the number of repeats of each satellite varies because different patterns will be inherited from each parent.

There are many different introns and a huge variation in the number of repeats so the probability that any two individuals have the same pattern of DNA is very unlikely, unless they are identical twins. However, the more closely related two individuals are, the more likely it is that similarities will be apparent in their DNA patterns.

LEARNING TIP

Remember that it is the non-coding DNA (introns) that is used for DNA profiling. This is because the introns show variation in the number of repeated sequences. Coding DNA is almost always identical as it has to code for specific proteins.

HOW IS A DNA PROFILE PRODUCED?

The strands of DNA from a sample are cut into fragments using special enzymes known as **restriction endonucleases**. These enzymes cut the DNA at particular points in the intron sequences. There are many different restriction enzymes, each type cutting a DNA molecule into fragments at different specific base sequences known as **recognition sites**. Using restriction enzymes that cut either side of mini- and micro-satellite units leaves the repeated

sequences intact. The technique produces a mixture of different-sized DNA fragments, depending on the number of repeated sequences, which consist largely of mini- and micro-satellite sequences (see **fig A**).

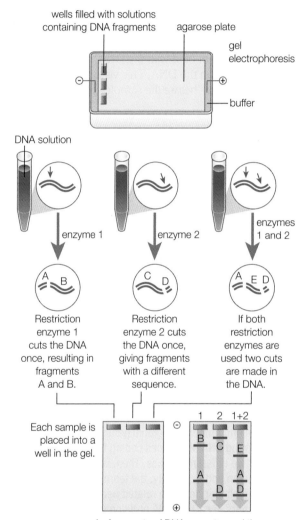

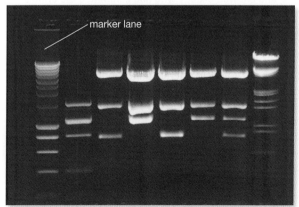

▲ **fig A**   A summary of the initial stages of DNA profiling.

GEL ELECTROPHORESIS

Scientists need to separate and identify the fragments of DNA produced by the restriction endonucleases. This process starts with gel electrophoresis, which is a variation of chromatography. DNA fragments are placed in wells in an agarose gel medium in a buffering solution to keep the pH constant throughout the process. Known DNA fragments are added in a separate well to make it easier to identify the unknown fragments.

The gel contains a dye (e.g. EtBr, ethidium bromide) that binds to the DNA fragments as they move through it. This dye fluoresces when placed under short-wave ultraviolet (UV) light, so it can reveal the band pattern of DNA. A visible dye is also added to the DNA samples. It does not bind with the DNA, but moves through the gel slightly faster than the DNA. This visual signal means the current can be turned off before the samples get to the end.

When an electric current passes through the apparatus, the DNA fragments move towards the positive anode because of the negative charge on the phosphate groups in the DNA backbone. The fragments move at different rates according to their size and charge. Once the electrophoresis is complete the plate is placed under short-wave UV light. The DNA fluoresces and shows up clearly so the scientists can identify it.

This is the original method of DNA profiling, which needs a relatively large sample of DNA. It reveals large DNA fragments containing a minimum of 50 base pairs (mini-satellites). The resulting DNA profile (fingerprint) looks similar to a supermarket barcode. By using extensions of this technique, scientists can now identify smaller regions of DNA and specific genes.

DID YOU KNOW?

Inventing DNA profiling

In the 1970s, Alec Jeffreys at Leicester University, UK, showed that when he used restriction endonuclease enzymes to cut the different regions of the chromosomes into smaller pieces, the repeating patterns in these units varied among individuals. Comparing seal and human myoglobin genes, he discovered groups of repeating DNA sequences coding for the same polypeptide in two different species. Then, working with DNA from their lab technician and her family, the team noticed the repeating units were more similar between related people than between complete strangers. By 1985, DNA profiling (sometimes referred to as DNA fingerprinting) had been created. In 1987, it was used for the first time in a murder investigation. It proved one man innocent, and showed exactly who had committed the crimes.

USING DNA PROFILES

IN FORENSIC SCIENCE

As you have seen, forensic science is the application of science to the processes of law. In forensic science, to develop a DNA profile in criminal investigations, we use gene probes to identify **short tandem repeats**. These are micro-satellite regions that are widely used in DNA identification. The DNA profile will be more accurate if more micro-satellites are used. Family members are more likely than unrelated individuals to have a number of micro-satellites in common, but the more sites that are examined the more likely it is that a different combination will have been

inherited from the two different parents. Statistically, the chance of two people matching on 11 or more sites is so small that it is considered to be reliable evidence in court.

IN FAMILIAL TESTING

Sometimes there is doubt about who is the father of a child. This is almost always in the context of problems in human relationships. DNA profiling can be used to prove or disprove family relationships. The mother of the child is almost always known; it is paternity that has to be taken on trust. Thus, most DNA tests are to test paternity. Specific micro-satellite markers are used to make the matches.

Another instance where DNA profiling is important is for inheritance claims. If someone claims to be a member of the family, with a claim on inheritance, DNA evidence can prove things one way or the other.

Familial DNA testing is also important in immigration cases. If someone has immigrated to a country and wants their children to join them, they may have to prove that the children are really theirs. Sometimes birth certificates are not available, but DNA testing can quickly prove whether people are genuinely biologically related.

IN THE IDENTIFICATION AND CLASSIFICATION OF ORGANISMS

The DNA patterns which are revealed in DNA profiling are unique to individuals, but we can use the similarity of patterns to identify relationships between individuals and even between species. Scientists are now using these new technologies to help identify and classify different species of animals and plants, and to understand the genetic relationships between them.

EXAM HINT

To identify links between species the coding parts of DNA are used. Genes coding for proteins that are universal are best (e.g. RNA polymerase or cytochrome oxidase).

DNA BARCODES

If scientists raise an organism in the lab, they usually know what species it is. But it may be difficult to identify organisms found in the field. The Consortium for the Barcode of Life (CBOL), the International Barcode of Life project (IBOL) and the European Consortium for the Barcode of Life (ECBOL) are large groups of scientific organisations that are developing DNA barcoding as a global standard for species identification (see **fig B**). This involves looking at short genetic sequences from a part of the genome common to particular groups of organisms. For example, a region of the mitochondrial cytochrome oxidase 1 gene (CO1), containing 648 bases, is being used as the standard barcode for most animal species. To date, we have effectively used this sequence in identifying fish, bird and insect species, including butterflies and flies. It is not possible to use this region to identify plants because it evolves too slowly to give sufficient differences between species. However, botanists have identified two gene regions in chloroplasts that have been approved for use to produce a standard barcode for plants, to be used in the same way as the animal barcodes. It is important that every specimen used to produce the definitive barcodes is preserved for reference. DNA profiling like this can be used both to identify species and to show the links between them.

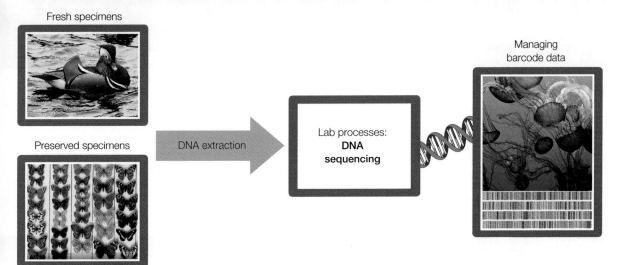

▲ **fig B**   The production of a DNA barcode: eventually scientists hope to be able to use these reduced DNA profiles to identify and classify all the different species of animals and plants.

Barcoding will not replace traditional taxonomy but it will support it. Scientists hope that field instruments will be developed that they will be able to use to analyse genes and identify species in their natural habitats. This will make it so much easier to identify plants with no flowers or fruits, immature animals and larval forms of insects. For example, quick identification of invasive species will make it much easier to deal with the threat.

The CBOL project recognises that it will take a long time to barcode all the species of living organisms, but it is progressing rapidly. Hundreds of thousands of species are now on the databases, with the numbers increasing all the time. Fortunately, the tests needed to obtain the barcode from the DNA are both fast and relatively cheap. Within the next 20 years, it should be possible to identify and barcode all known plant and animal species based on DNA analysis. Fungi and bacteria may be more difficult.

CHECKPOINT

1.   Explain the term *DNA profiling*.

2.   Describe how gel electrophoresis is used to separate DNA fragments produced by restriction enzymes in the production of a DNA profile.

3. ▶ (a) Explain why DNA profiling is useful for identifying different individuals in forensic investigations.

   (b) Individuals have been wrongly accused of serious crimes partly because at some point another member of their family has committed a minor crime. Discuss how this might be possible and how it can be avoided.

4. ▶ Explain how profiling small areas of the DNA of animals and plants can be used to identify species and show the relationships between them.

SKILLS   REASONING

SKILLS   INTERPRETATION

SUBJECT VOCABULARY

micro-satellite a section of DNA with a 2–6 base sequence repeated 5 to 100 times
mini-satellite a section of DNA with a 10–100 base sequence repeated 50 to several hundred times
restriction endonucleases enzymes used to cut up strands of DNA at particular points in the intron sequences
recognition site specific base sequences where restriction endonucleases cut the DNA molecule
short tandem repeats micro-satellite regions that are widely used in DNA identification of suspects; statistically, the chance of two people matching on 11 or more sites is so small that it is considered to be reliable evidence in court

COLD CASES

SKILLS ANALYSIS, INTERPRETATION, CREATIVITY
CONTINUOUS LEARNING, COMMUNICATION

Before the development of new technologies such as PCR, forensic scientists needed relatively large samples of DNA to make a DNA profile to use in criminal investigations. Now we can amplify tiny samples of DNA. As a result, detectives are now revisiting old cases, and looking again for DNA evidence to produce profiles which are helping to catch the people who committed crimes many years ago.

POPULAR SCIENCE MAGAZINE

▲ **fig A** DNA fingerprinting or profiling is used in genetics, forensics, drug discovery, biology and medicine. Modern forensic science can use tiny scraps of evidence, such as the tissue trapped behind the fingernails of a victim, to prove the guilt or innocence of a suspect.

In 1984, a talented young British biochemist called Helena Greenwood was living in the United States with her husband Roger, working on improving methods of DNA analysis. One day, when Helena was at home alone, David Frediani broke in. He held her at gunpoint and assaulted her. She persuaded him not to kill her by suggesting she wouldn't tell anyone what had happened. As soon as Frediani left the house, she gave the police his description and they found him quickly. A trial date was set for 1985, and Frediani was allowed out on bail.

Afraid that Frediani might return, Helena and Roger moved to another city. Three weeks before Frediani's trial, Roger found Helena dead in their garden, strangled and beaten. Evidence from Fridiani's credit cards showed he had been in the area just before the murder, so he was the prime suspect – but the police had no forensic evidence linking him to Helena's body. Frediani stood trial and was imprisoned for the original assault but, although there were no other suspects for Helena's murder, there was no proof that Frediani was involved either.

Fourteen years later, in 1999, the San Diego police reopened a number of unsolved murders to see if new forensic techniques such as PCR and DNA profiling could help them. Helena had fought her attacker, and skin fragments were found beneath her fingernails when her body was examined. These samples had been saved with the rest of the evidence. PCR gave the forensic team the tool they needed to look again and see if they could find Helena's murderer.

The DNA sample was amplified using PCR, and a DNA profile was produced using, in part, techniques that Helena had been working on at the time of her murder. The DNA fingerprint from the skin under the fingernails of the murdered woman matched one man – David Frediani! He must have thought he was safe, but more than 15 years after he killed Helena Greenwood, Frediani was tried, found guilty and sentenced to life imprisonment.

SCIENCE COMMUNICATION

This story has been published in a number of magazines and on websites about cold cases.

1 How might the publication of stories like this one help deter criminals?

2 A book was written about Helena's story called *Pointing from the Grave: A True Story of Murder and DNA*. Based on what you know from this brief summary of the cold case investigation, discuss ways in which this is a good title for a book explaining Helena's case and the DNA technology used to solve it.

BIOLOGY IN DETAIL

Now you are going to think about the science in this resource.

3 (a) Describe how the time of death may be determined using the following techniques. In each case, suggest any factors which must be considered before relying on a result.

   • Body temperature

   • Rigor mortis

   • Succession stage

   (b) Explain how the time of death will still be important in a cold case such as the second investigation into Helena's death, even though it will not be reinvestigated as the evidence will be long gone.

4 DNA analysis has had a major impact on forensic science.

   (a) Explain why this is the case.

   (b) Carry out some research and discover how DNA profiling is moving forward and becoming even more useful in forensic science.

ACTIVITY

Because scientists can now amplify tiny traces of DNA, they are looking back and reopening cold cases. These are unsolved crimes that happened a long time ago but for which the technology is now available to solve them.

• Use the internet to help you investigate a cold case which has been solved using DNA profiling.

• **Either:** Write an article for a popular magazine

   Or: Produce a short piece for a TV current affairs programme about the case you have researched. Make sure you explain clearly to your audience how the process works and how DNA technology allows us to bring criminals to justice long after they hope the crime has been forgotten. Keep it simple but get the science correct.

1 (a) The polymerase chain reaction (PCR) has a number of stages.

1 heat to 95 °C

2 repeat many times

3 cool to 55 °C

4 heat to 75 °C

Which is the correct sequence for this procedure?

A 4–3–2–1

B 1–2–3–4

C 1–3–4–2

D 1–4–3–2 [1]

(b) Samples from PCR can be used in many ways. Which of the following would require the DNA from just **one** person?

1 to identify the possible gene variants of a particular gene

2 to identify a criminal

3 to test for paternity

4 to see if there is a correlation between a gene variant and a disease

A 1

B 2

C 3

D 4 [1]

(c) PCR can be used to amplify samples of DNA.

(i) Explain what is meant by the term *amplify*. [2]

(ii) If a PCR process starts with one fragment, calculate how many fragments should be present after 15 cycles. Show your working. [2]

(iii) Suggest **three** reasons why the theoretical value may not be achieved. [3]

(Total for Question 1 = 9 marks)

2 Often only small amounts of DNA can be collected from a crime scene. This DNA can be amplified using PCR.

(a) (i) What is the name of the enzyme used in PCR?

A endonuclease

B invertase

C polymerase

D transcriptase [1]

(ii) After PCR, DNA fragments can be separated to create a DNA profile. What is the name of the process that could be used to separate DNA fragments to create a profile?

A amniocentesis

B electrophoresis

C endocytosis

D chromatography [1]

(b) Following a burglary, a DNA profile was created using a small sample of blood left behind on a broken window. This DNA profile was then compared with DNA profiles from four suspects: S1, S2, S3 and S4. These DNA profiles are shown in the diagram below.

(i) Comment on which of the suspects is most likely to have left the blood sample on the broken window pane. Refer to the theory used in DNA profiling and explain how you came to this conclusion. [3]

(ii) Suggest why evidence from DNA profiles may not be absolutely conclusive. [3]

(iii) Explain how DNA profiling could be useful to scientists classifying animals and plants. [3]

(Total for Question 2 = 11 marks)

3 (a) Haemoglobin consists of two polypeptides, α-globin and β-globin. Before separating out fragments of DNA by electrophoresis, the DNA must be cut into short lengths. Digesting normal human DNA with the enzyme HPa I produces a fragment which is 7.6 kbp long and contains the β-globin gene. Digesting DNA from a person who has sickle cell anaemia produces a fragment 13 kbp long.

What type of enzyme is HPa I? [1]

A protease

B ligase

C restriction endonuclease

D lysin

(b) (i) Describe how these fragments could be separated using gel electrophoresis. [4]

(ii) The DNA sample is often treated with a protease during preparation for electrophoresis. Suggest why this step is required. [2]

(c) (i) Describe what is meant by the term *gene probe*. [2]

(ii) Explain why a gene probe is needed and describe
how it can be used to screen members of a family
with a genetic disorder such as muscular dystrophy. [5]

(Total for Question 3 = 14 marks)

4 (a) During the polymerase chain reaction the temperature
of the mixture is altered. List the temperatures involved
and describe the process occurring at each temperature. [6]

(b) Explain the purpose of adding primers to the PCR mix. [3]

(c) A student attempted to separate some fragments of DNA
using electrophoresis. Her method is written below.

*We set up a gel plate with six wells. We placed the
DNA samples in the wells at the anode. The plate was
switched on for two minutes. After this the gel plate
was moved to a bath of purified water to make the
banding pattern visible.*

Suggest **two** changes to the method described and
explain how each change will improve the method. [4]

(Total for Question 4 = 13 marks)

5 (a) Body temperature can be used to calculate how long a
person has been dead. List **three** factors that affect how
quickly the body cools. [3]

(b) (i) After committing murder, the murderer placed
some blowfly maggots on the face of the victim.
Suggest and explain why the murderer did this. [4]

(ii) Suggest why the murderer's scheme may not work. [2]

(Total for Question 5 = 9 marks)

TOPIC 7 RESPIRATION, MUSCLES AND THE INTERNAL ENVIRONMENT

CHAPTER

7A CELLULAR RESPIRATION

Clostridium botulinum is a bacterium that produces one of the most powerful poisons known: botulinus toxin. A single microgram can kill a person. The bacteria are not common, but they can grow in food that has been preserved badly. Yet botulinus toxin is widely used in the beauty business. Botox injections containing 5 ng of toxin temporarily paralyse muscles and thus remove wrinkles from the face.

Clostridium botulinum can only respire *without* oxygen. They are known as obligate anaerobes and they are killed by high levels of oxygen. Although most organisms, including people, can respire without oxygen for a short time, most living things need oxygen to keep them alive long-term.

In this chapter, you will find out about the main stages of the process of respiration that take place in eukaryotic cells. You will discover how the anaerobic processes of glycolysis in the cytoplasm are linked to the aerobic processes of Krebs cycle and oxidative phosphorylation in the mitochondria. You will learn how ATP is produced in oxidative phosphorylation, the importance of chemiosmosis, and the differing yields of ATP.

You will also consider how scientists investigate the rate of respiration in organisms using a respirometer, and look at some of the techniques used to investigate where ATP is formed in the cell.

Finally, you will look at how anaerobic respiration takes place in the cytoplasm of animal cells, and consider the effect of lactate on muscle tissues and how they contract.

MATHS SKILLS FOR THIS CHAPTER

- **Use fractions, ratios and percentages** (*e.g. in considering the production of ATP during aerobic respiration*)
- **Estimate results** (*e.g. production of ATP in the process of cellular respiration*)
- **Use appropriate number of significant figures** (*e.g. calculating percentage efficiency of anaerobic respiration in mammalian muscles*)
- **Construct and interpret frequency tables and diagrams, bar charts and histograms** (*e.g. anaerobic exercise and oxygen debt*)
- **Translate information between graphical, numerical and algebraic forms** (*e.g. oxygen debt*)
- **Plot two variables from experimental or other data** (*e.g. investigating the rate of respiration*)
- **Calculate rate of change from a graph showing a linear relationship** (*e.g. investigating the rate of respiration*)

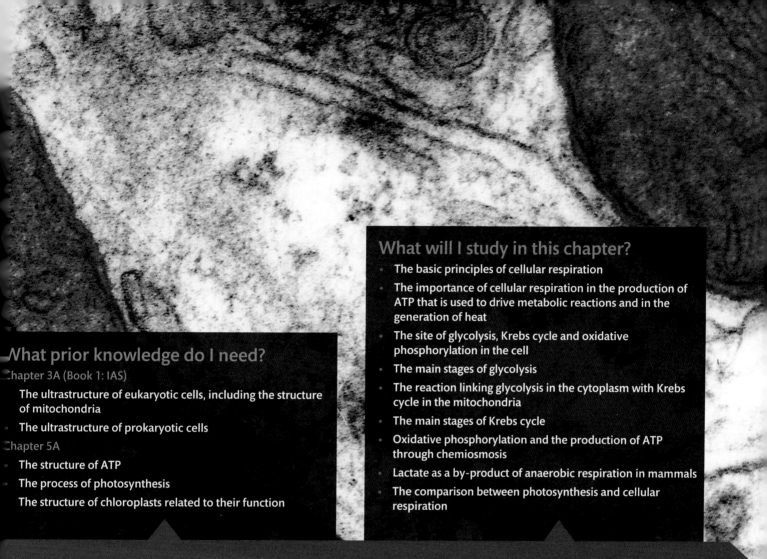

What will I study in this chapter?

- The basic principles of cellular respiration
- The importance of cellular respiration in the production of ATP that is used to drive metabolic reactions and in the generation of heat
- The site of glycolysis, Krebs cycle and oxidative phosphorylation in the cell
- The main stages of glycolysis
- The reaction linking glycolysis in the cytoplasm with Krebs cycle in the mitochondria
- The main stages of Krebs cycle
- Oxidative phosphorylation and the production of ATP through chemiosmosis
- Lactate as a by-product of anaerobic respiration in mammals
- The comparison between photosynthesis and cellular respiration

What prior knowledge do I need?

Chapter 3A (Book 1: IAS)

- The ultrastructure of eukaryotic cells, including the structure of mitochondria
- The ultrastructure of prokaryotic cells

Chapter 5A

- The structure of ATP
- The process of photosynthesis
- The structure of chloroplasts related to their function

What will I study later?

Chapter 7B

- The importance of ATP in the biochemistry of muscle contraction

Chapter 7C

- The importance of ATP and cellular respiration in the maintenance of a stable equilibrium in the body, in the transmission of nerve impulses, in the functioning of the kidney and many other processes

LEARNING OBJECTIVES

■ Understand the overall reaction of aerobic respiration as splitting of the respiratory substrate to release carbon dioxide as a waste product and reuniting hydrogen with atmospheric oxygen with the release of large amounts of energy.

■ Understand that respiration is a multi-step process, with each step controlled and catalysed by a specific intracellular enzyme. (Names of specific enzymes are not required.)

The cells of all organisms need energy to break and make bonds during the chemical reactions that lead to growth, reproduction and the maintenance of life. Autotrophic organisms make their own food, usually by photosynthesis; heterotrophic organisms eat and digest other organisms. The energy in the chemical bonds of the food is transferred to the bonds in ATP (adenosine triphosphate) during cellular respiration. This provides the energy for all other metabolic reactions all the time (see **fig A**).

EXAM HINT

Many students forget that plants need to respire. Photosynthesis is not an alternative; all living things need to respire.

▲ **fig A**   When a mammal is asleep, the body continues to use lots of energy for breathing, blood circulation, excretion, growth, repair and maintaining the body temperature.

WHAT IS CELLULAR RESPIRATION?

Cellular respiration is the process by which the energy from food molecules is transferred to ATP (see **Section 5A.1**). The substance that is broken down is called the **respiratory substrate**. The main respiratory substrate used by cells is glucose.

Aerobic respiration is the type of cellular respiration that takes place in the presence of oxygen. The overall reaction of aerobic respiration involves breaking down the respiratory substrate to release carbon dioxide as a waste product and reuniting hydrogen with atmospheric oxygen to form water, with the release of large amounts of energy. The volume of oxygen used and the volume of carbon dioxide produced change depending on the level of activity of the organism, the type of food being respired and other external factors such as temperature (see **Section 7A.2**).

Aerobic respiration of glucose is usually summarised as follows:

$$C_6H_{12}O_6 + 6O_2 \rightarrow 6CO_2 + 6H_2O \; (+ \text{ATP}) \; \Delta H \approx -2880 \, \text{kJ}$$

$$\text{glucose} + \text{oxygen} \rightarrow \text{carbon dioxide} + \text{water} \; (+ \text{ATP})$$

$$\Delta H \approx -2880 \, \text{kJ}$$

EXAM HINT

Remember that this is a very simplified version of respiration. It is not one simple reaction; it is a complex, multi-step process.

ATP provides readily transferable energy for all cellular reactions. In **Section 5A.1**, you learned that the third phosphate bond of ATP molecules can be broken in a hydrolysis reaction, catalysed by the enzyme ATPase (see **fig B**). This provides energy for other chemical reactions in a cell. The result is adenosine diphosphate (ADP) and a free inorganic phosphate group (P_i). For every mole of ATP hydrolysed about 30.5 kJ of energy is made available. Some of this energy is transferred to the environment, warming it up, but the rest is available for any biological activity that requires energy. The breakdown of ATP into ADP and P_i is reversible. The phosphorylation of ADP to ATP is also catalysed by ATPase and requires 30.5 kJ of energy.

$$-30.5 \, \text{kJ}$$
$$\text{ATP} \Longleftrightarrow \text{ADP} + P_i$$
$$+30.5 \, \text{kJ}$$

▲ **fig B**   The reversible hydrolysis of ATP which makes energy available to cells. The energy needed to synthesise ATP from ADP and P_i comes from cellular respiration. (See **Section 5A.1 fig B** for more details.)

ATP cannot be stored in the body in large amounts. The raw materials to make ATP are almost always available, so the compound is made when it is needed. However, once all the raw materials have been used, cellular respiration cannot continue and no more ATP is made. This is observed when rigor mortis sets in after death (see **Section 6C.2**). The contracting proteins of the muscles cannot work and the muscles become rigid after cellular respiration stops and ATP production ends.

AN OUTLINE OF AEROBIC RESPIRATION

The simple equation given above for aerobic respiration does not show that the complete process involves many steps. It involves a complex series of reactions, which are outlined in **fig C**.

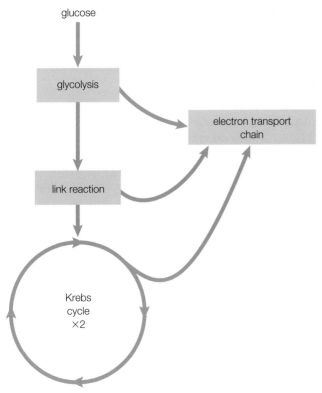

▲ **fig C**   A simplified model of the many-stepped process of aerobic respiration.

Aerobic respiration occurs in two distinct phases. The first stage is called glycolysis and does not require oxygen (see **Section 7A.2**). This is where the respiratory substrate begins to break down and the molecules are prepared for entry into the second stage of the process. A small amount of ATP is produced. The second set of reactions is called the Krebs cycle and needs oxygen (see **Section 7A.3**). The link reaction moves the products of glycolysis into the Krebs cycle and the electron transport chain.

Each step of the process of glycolysis is controlled by a specific intracellular enzyme, so many different enzymes are involved. The rate of the reaction is controlled by inhibiting these enzymes, usually by other chemicals in the reaction chain.

Most organisms depend on aerobic respiration, which means they need the presence of oxygen to allow the respiratory process to occur and provide them with sufficient energy to survive. They may be able to cope with a temporary lack of oxygen, but only for a very short time. Some organisms (called facultative anaerobes) can survive without oxygen; they can rely on anaerobic respiration if necessary. There are a few groups of organisms (called obligate anaerobes) that cannot use oxygen at all and may even be killed by it (see **fig D**).

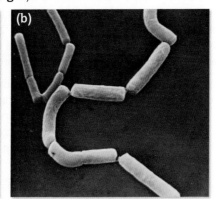

▲ **fig D**   (a) If hummingbirds are deprived of oxygen, they will die very rapidly as their cells cannot obtain enough ATP. (b) If bacteria *Clostridium perfringens*, which cause gas gangrene in wounds, are given oxygen when they are actively dividing, they will die because oxygen is toxic to them.

WHERE DOES CELLULAR RESPIRATION TAKE PLACE?

Glycolysis is the first part of the respiratory pathway and is not associated with any particular cell organelle. The enzymes controlling glycolysis are found in the cytoplasm. All the other stages in aerobic respiration (the link reaction, Krebs cycle and the electron transport chain involved in producing ATP) occur inside the mitochondria.

As you learned in **Section 3A.2 (Book 1: IAS)**, mitochondria are relatively large organelles with a complex internal structure. They have a double membrane and the inner one is formed in many folds called cristae (see **fig E**). The matrix of the mitochondrion contains the enzymes of the Krebs cycle, and the cristae carry the **stalked particles** associated with ATP synthesis. The number of mitochondria in a cell tells you how active the cell is. Cells with very low energy requirements generally contain very few mitochondria; for example, fat storage cells. Cells that are very active have large numbers of mitochondria in their cytoplasm; for example, muscle and liver cells.

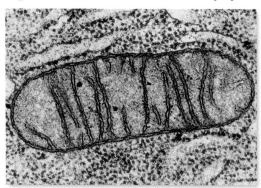

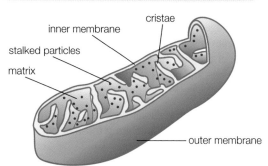

▲ **fig E** A mitochondrion: the powerhouse of the cell. Different processes of aerobic respiration occur in different places within these important organelles.

THE HYDROGEN ACCEPTORS

Simple representations of cellular respiration suggest that ATP is produced as a direct result of the breakdown of glucose, but this is not the case. Most of the ATP produced during cellular respiration is made through a series of oxidation and reduction reactions in the electron transport chain. **Reduction** is the addition of electrons to a substance and in the cell this results from the addition of hydrogen or the removal of oxygen. **Oxidation** is the removal of electrons from a substance and any compound that has oxygen added, or hydrogen or electrons removed, is oxidised.

During cellular respiration, hydrogen is removed from compounds and received by a **hydrogen acceptor** which is, therefore, reduced. This happens several times during the reactions of respiration, as you will learn later. The hydrogen is split to give a proton and an electron and the electron then passes along the electron transport chain. A series of linked oxidation and reduction (redox) reactions occurs. Each redox reaction releases a small amount of energy which is used to drive the synthesis of a molecule of ATP (see **Section 7A.4**).

The most common hydrogen acceptor in cellular respiration is **NAD (nicotinamide adenine dinucleotide)**. NAD is a coenzyme: a small molecule that assists in enzyme-catalysed reactions. When NAD accepts hydrogen atoms from a metabolic pathway, it becomes reduced to form **reduced NAD**. **FAD (flavin adenine dinucleotide)** is another coenzyme hydrogen carrier. It accepts hydrogen from reduced NAD and forms reduced FAD. Each time this happens, a molecule of ATP is created in the process.

FINDING OUT ABOUT CELLULAR RESPIRATION

We have gradually developed our understanding of the process of respiration over the years. At the beginning, scientists based their research on whole animals and plants. Now, we continue the work at the level of very small cell fragments. We can work at this level because of the evidence made available by technology such as the electron microscope. We need to understand the factors that affect the rate of respiration and exactly how and where the reactions occur within the cell. Two examples are given below.

INVESTIGATING FACTORS AFFECTING THE RATE OF RESPIRATION (USING WHOLE ORGANISMS)

It is not always easy to demonstrate the rate of cellular respiration without sophisticated biochemical techniques designed to measure the rate in isolated cell organelles. However, in a school lab, a **respirometer** can give some valuable information about the rate of cellular respiration. This instrument measures the uptake of oxygen (the quantity used) or the output of carbon dioxide (the quantity produced) by whole organisms.

A basic respirometer consists of a closed chamber into which no air can enter and which contains one or more living organisms, such as germinating seeds (see **fig F**). You can use a chemical (usually soda lime or potassium hydroxide) to absorb the carbon dioxide produced by respiration. Therefore, any changes in volume will be caused by the uptake of oxygen by the organisms. As the organisms use oxygen, the pressure reduces and so the fluid in the manometer moves towards the tube containing the organisms. Using the syringe, you can measure the volume of gas you need to return the manometer to normal. You can then use this measurement to calculate the intake of oxygen per minute, which gives you an approximate respiration rate for the organisms.

EXAM HINT

When you carry out **Core Practicals 15** and **16** (see **Sections 7A.2** and **7A.5**) you will investigate factors affecting the rate of aerobic or anaerobic respiration using a respirometer, taking into account the safe and ethical use of organisms. Make sure you understand well how respirometers work as your understanding of the experimental method may be assessed in your examination.

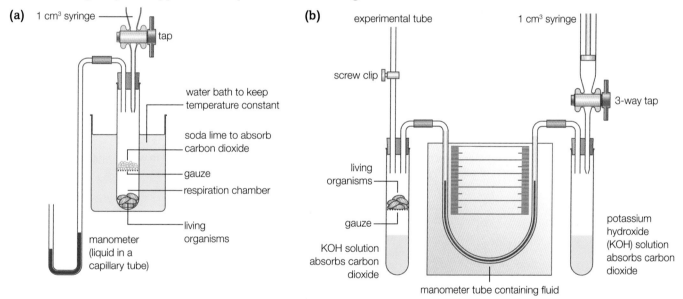

▲ **fig F**   (a) Any movement of the liquid in the tube of a respirometer indicates an uptake of oxygen by the organisms in the respiration chamber. You can measure the distance moved over time to measure the rate of respiration. (b) Some respirometers are more sophisticated than others.

You can change the external conditions (e.g. temperature) and measure the effect on the rate of respiration by recording changes in the uptake of oxygen. This simple apparatus has obvious limitations but you can use it to get an overall impression of the rate of respiration of organisms under different conditions.

INVESTIGATING THE SITE OF ATP SYNTHESIS (USING CELL FRAGMENTS)

You can investigate respiration at the cellular level in a number of ways.

- You can break open cells and centrifuge the contents to obtain a fraction containing just mitochondria. If these are kept supplied with glucose and oxygen, they will produce ATP.
- Using high-resolving electron microscopes, you can see that the surface of the inner membrane of the mitochondrion is covered in closely packed stalked particles. These provide a greatly increased surface area, which is an ideal site for enzymes to work.

- You can separate the stalked particles and the small pieces of membrane associated with them from the rest of the mitochondrial structure. Scientists have demonstrated that ATP synthesis only occurs here. This and other evidence proves that the stalked particles are vital for the formation of ATP.

CHECKPOINT

SKILLS ▶ CREATIVITY

1. Explain why cellular respiration is such an important reaction.

2. ▶ $C_6H_{12}O_6 + 6O_2 \rightarrow 6CO_2 + 6H_2O + ATP$
   This equation is sometimes given as the equation of aerobic respiration. Explain the strengths and limitations of representing aerobic respiration in this way.

3. (a) Explain how respirometers are limited in what they can tell us about cellular respiration.

SKILLS ▶ DECISION MAKING

   ▶ (b) Evaluate the two pieces of apparatus in **fig F** to explain which you think would deliver more reliable evidence and why.

4. Describe the kind of evidence that is needed to identify the sites of the various stages of cellular respiration in a mitochondrion.

SUBJECT VOCABULARY

cellular respiration the process by which food is broken down to yield ATP (a source of energy for metabolic reactions)

respiratory substrate the substance used as a fuel and oxidised during cellular respiration

aerobic respiration the form of cellular respiration that occurs in the mitochondria in the presence of oxygen

stalked particles structures on the inner mitochondrial membrane where ATP production occurs

reduction the addition of electrons to a substance (e.g. by the addition of hydrogen or removal of oxygen)

oxidation the removal of electrons from a substance (e.g. by the addition of oxygen or removal of hydrogen)

hydrogen acceptor a molecule which receives hydrogen and becomes reduced in cell biochemistry

NAD (nicotinamide adenine dinucleotide) a coenzyme that acts as a hydrogen acceptor

reduced NAD NAD which has received a hydrogen atom in a metabolic pathway

FAD (flavin adenine dinucleotide) a hydrogen carrier and coenzyme; in cellular respiration, FAD receives hydrogen to form reduced FAD driving the production of ATP

respirometer a piece of apparatus used for measuring the rate of respiration in whole organisms or cultures of cells

LEARNING OBJECTIVES

■ Understand the roles of glycolysis in aerobic and anaerobic respiration, including the phosphorylation of hexoses, the production of ATP by substrate level phosphorylation, and the production of reduced coenzyme, pyruvate and lactate.

■ Understand what happens to lactate after a period of anaerobic respiration in animals.

In this section, you will look at the different stages of the biochemistry of cellular respiration separately, and then consider the overall process. This will make it easier to understand what is happening inside the cells, but remember that in living cells this is one continuous process. The first stage is **glycolysis**. In glycolysis, a 6-carbon (6C) glucose molecule is broken into two molecules of the 3-carbon (3C) compound pyruvate in a series of 10 reactions. The **pyruvate ions** produced by glycolysis can either be used in aerobic respiration or anaerobic respiration.

GLYCOLYSIS

Glycolysis means 'sugar breakdown'. It takes place in the cytoplasm of the cell. The main stages of glycolysis are shown in **fig A**.

6C sugar (glucose)

ATP

ADP

phosphorylated 6C sugar

2 × 3C sugar

NAD

reduced NAD

ADP

ATP

2 × pyruvate

▲ **fig A** The main stages of glycolysis result in the production of pyruvate ions, a small quantity of ATP and reduced NAD. These reactions all happen within the cytoplasm of the cell.

Glucose is a 6-carbon (6C) sugar, a hexose. The glucose needed for glycolysis may come directly from the blood or it may be produced by the breakdown of glycogen stores in muscle and liver cells (see **Section 1A.3 (Book 1: IAS)**).

The first step in glycolysis uses ATP to provide the energy to phosphorylate the 6C sugar glucose, adding two phosphate groups. This phosphorylation makes the sugar more reactive and

also makes it unable to pass through the cell membrane, so it becomes trapped within the cell.

The phosphorylated sugar is then broken down to give two molecules of a 3-carbon sugar. Each of these molecules is then converted through several steps into a molecule of pyruvic acid. This is found in solution as pyruvate ions. During these reactions, a small amount of ATP is produced as follows.

• Two hydrogen atoms are removed from the 3C sugars and collected by NAD, forming reduced NAD. This occurs in the cytoplasm of the cell. The reduced NAD then passes through the outer mitochondrial membrane into the electron transport chain, which is explained in detail in **Section 7A.4**.

• A small amount of ATP is also made directly from the energy transfer when the 3C sugar is converted to pyruvate. The phosphorylation of the sugar at the beginning of glycolysis is reversed when the final intermediate compound is converted to pyruvate. The phosphate group which is released is used to convert ADP to ATP.

• If there is plenty of oxygen the pyruvate will enter the mitochondria and be used in the aerobic reactions of the Krebs cycle. If oxygen levels are low, the pyruvate remains in the cytoplasm and is converted into either **ethanol** (in plants and yeast) or **lactate** (in mammals) with no additional ATP produced. This is **anaerobic respiration** in the cytoplasm (see **fig B**).

EXAM HINT

This production of a small amount of ATP in glycolysis is known as substrate level phosphorylation.

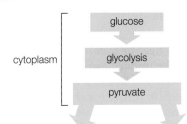

cytoplasm

glucose

glycolysis

pyruvate

aerobic respiration	anaerobic respiration
• in mitochondria	• in cytoplasm
• complete oxidation	• incomplete oxidation
• waste products: H_2O, CO_2	• waste products: lactic acid or ethanol and CO_2
• net energy: 31 ATP	• net energy: 2 ATP

▲ **fig B** The alternative routes for the products of glycolysis produce very different amounts of ATP.

ANAEROBIC RESPIRATION

Anaerobic respiration occurs in the absence of oxygen. It occurs in most types of organism, but the final products differ between types of organism.

> **EXAM HINT**
>
> It is more important to learn the number of carbon atoms of the different compounds than the names of the compounds themselves.

ANAEROBIC RESPIRATION IN MAMMALS

If you do high-intensity exercise, your muscles do not receive enough oxygen to meet their needs. When this happens, the products of glycolysis cannot continue to the aerobic stages of cellular respiration. The muscles have to respire anaerobically.

In anaerobic respiration in mammals, the pyruvate from glycolysis is converted to lactic acid. This is another 3C compound that dissociates to form lactate ions and hydrogen ions. Anaerobic respiration only produces two molecules of ATP per glucose molecule respired. In contrast, up to eight ATP molecules are produced in glycolysis when pyruvate can continue into the Krebs cycle. This very low production of ATP in anaerobic respiration is because some of the reduced NAD is used to reduce pyruvate to lactate rather than entering the electron transport chain. The lactate moves out of the cells into the blood:

$$C_6H_{12}O_6 \rightarrow 2C_3H_6O_3 \; (+ \; ATP) \quad \Delta H \approx -150\,kJ$$
$$\text{glucose} \rightarrow \text{lactic acid} \; (+ \; ATP) \quad \Delta H \approx -150\,kJ$$

The levels of lactate and hydrogen ions increase during anaerobic respiration in the muscles and this causes the pH to fall and the muscle tissue becomes acidic. It used to be thought this reduced the ability of the muscles to contract, so the contractions lost their force and eventually stopped completely. Modern research suggests this is not the case; muscle contraction does not seem to be affected by lactate ions or the fall in pH caused by hydrogen ions. It appears that the movement of lactate and hydrogen ions into the blood from the muscles lowers the pH of the blood, which as a result affects the central nervous system (see **Section 8A.1**). This may reduce nervous stimulation from the central nervous system that reduces, and eventually stops, muscle contraction. Scientists think this is a protective adaptation to give the muscles time to recover and return to aerobic respiration, which helps to raise the pH of the blood again.

When exercise stops, the levels of lactate in the blood remain high. The lactate is toxic so it must be oxidised back to pyruvate to enter the Krebs cycle and be respired aerobically, producing carbon dioxide, water and ATP. The lactate is carried to the liver in the blood. It is converted back to pyruvate and respired in the liver cells. Oxygen is needed to oxidise the pyruvate made from the accumulated lactate. This is why you continue to breathe deeply for some time after you have finished exercising. You can see these effects in **fig C**.

> **EXAM HINT**
>
> Do not overlook the importance of anaerobic respiration. The conversion of pyruvate to lactate requires hydrogen from reduced NAD. This reoxidises the NAD so it is available to continue the conversion (oxidation) of 3C sugars to pyruvate. This is the step where ATP is produced in glycolysis so it allows continued production of a small amount of ATP.

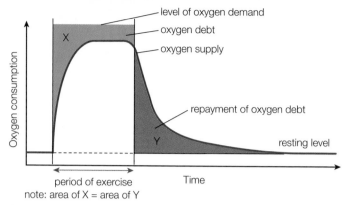

▲ **fig C** This graph shows the difference between the oxygen demand during exercise and the available oxygen supply to the muscles.

Sprint athletes may run up to 95% of a race relying on the anaerobic respiration of their muscles. Long-distance runners have to maintain a much higher level of aerobic respiration because their muscles could not continue to work for the length of time needed to finish the race if lactate levels were not kept to a minimum. Training allows athletes to get more oxygen to their muscles faster as a better blood supply develops, and to tolerate higher levels of lactate before the muscles become tired. When your body is exposed repeatedly to high lactate levels, more lactate transporter molecules develop in the mitochondrial membranes. As a result, lactate is processed to pyruvate more quickly when oxygen is available.

ANAEROBIC RESPIRATION IN PLANTS AND FUNGI

Yeast is well known for anaerobic respiration, with ethanol and carbon dioxide as the main waste products. This is the basis of many processes in biotechnology, such as baking leavened bread. However, anaerobic respiration also occurs in plants, in particular in the root cells in soils containing high levels of water. When plant cells respire anaerobically, they also produce ethanol:

$$C_6H_{12}O_6 \rightarrow 2C_2H_5OH + 2CO_2 \; (+ \; ATP) \quad \Delta H \approx -144\,kJ$$
$$\text{glucose} \rightarrow \text{ethanol} + \text{carbon dioxide} \; (+ \; ATP) \quad \Delta H \approx -144\,kJ$$

DISCOVERING THE GLYCOLYSIS PATHWAY

We took many years to understand the pathways of glycolysis and the closely associated processes of lactic and alcoholic fermentation. From the beginning, yeast rather than plants has been used to investigate glycolysis (see **fig D**). It is easy to grow, it reproduces rapidly, and there are no ethical issues about using a fungus. In addition, yeast contains all the enzymes of glycolysis.

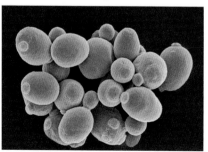

▲ **fig D** This scanning electron micrograph shows yeast, the main experimental organism for investigating glycolysis and anaerobic respiration.

PRACTICAL SKILLS   CP15

During glycolysis, the 3-carbon sugars produced by splitting glucose are reduced to form pyruvate ions. Enzymes known as **dehydrogenases** catalyse reduction (see **Sections 2B.1** and **2B.2 (Book 1: IAS)** and **Section 7A.3**).

This process is the basis of **Core Practical 15: Use an artificial hydrogen carrier (redox indicator) to investigate respiration in yeast**. There are a number of different artificial hydrogen carriers which receive the hydrogen ions produced during glycolysis and change colour when they are reduced. They are called **redox indicators**. Examples include:

- methylene blue: changes from blue to colourless when it is reduced
- tetrazolium chloride (TTC): changes from colourless to red when it is reduced
- dichlorophenolindophenol (DCPIP): changes from blue to colourless when it is reduced.

You can add one of these redox indicators to actively respiring yeast mixtures and use them to investigate the effect of different temperatures, or different substrate concentrations, on the rate of respiration. You can calculate the rate of respiration simply by measuring the time it takes to observe the appropriate change in colour. **Fig E** shows you the indicator methylene blue.

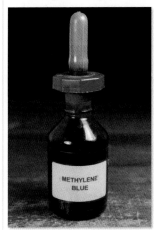

> ⚠ Safety Note: Eye protection is needed for this activity and there should be no skin contact with the indicators used.

▲ **fig E**   When the indicator methylene blue changes from blue to colourless you know it is completely reduced. You can use this to measure the respiration rate in yeast.

CHECKPOINT

SKILLS ▶ INTERPRETATION

1. ▶ Produce an annotated diagram of glycolysis labelling the important biochemistry of each step.
2. Summarise how the anaerobic respiration of glucose releases useful energy.
3. Using the information in **fig B**, calculate the percentage efficiency of anaerobic respiration in mammalian muscles compared to aerobic respiration.
4. Explain why breathing rate and heart rate continue to be raised after exercise.

SUBJECT VOCABULARY

glycolysis the first stage of cellular respiration, which occurs in the cytoplasm and is common to both aerobic and anaerobic respiration

pyruvate ions the end-product of glycolysis

ethanol an organic chemical with the formula C_2H_5OH produced as a result of anaerobic respiration (fermentation) in fungi and some plant cells

lactate ions of a 3-carbon compound (lactic acid) which is the end-product of anaerobic respiration in mammals

anaerobic respiration the form of cellular respiration that occurs in the cytoplasm when there is no oxygen present

dehydrogenases enzymes that remove hydrogen from a molecule (during oxidation reactions)

redox indicators chemicals which are different colours when they are oxidised and reduced; they can act as artificial hydrogen carriers to investigate respiration in yeast

7A │ 3 THE KREBS CYCLE

LEARNING OBJECTIVES

■ Understand the role of the link reaction and the Krebs cycle in the complete oxidation of glucose and formation of carbon dioxide (CO_2) by decarboxylation, ATP by substrate level phosphorylation, reduced NAD and reduced FAD by dehydrogenation (names of other compounds are not required) and that these steps take place in mitochondria, unlike glycolysis which occurs in the cytoplasm.

When there is plenty of oxygen available, the pyruvate produced as the end-product of glycolysis is passed into the mitochondria. In the mitochondria it enters the **Krebs cycle** through the **link reaction**. The Krebs cycle is a series of biochemical steps that leads to the complete oxidation of glucose, producing carbon dioxide, water and relatively large amounts of ATP.

Like glycolysis, the Krebs cycle is a process with many steps. Each individual step is controlled and catalysed by a specific intracellular enzyme. The reactions of the cycle occur in the matrix of the mitochondrion. ATP is produced in the stalked particles on the inner mitochondrial membranes in the presence of oxygen. You will learn about the Krebs cycle but you do not need to learn the detailed biochemical steps that occur. **Fig A** gives you an idea of just how complex the Krebs cycle is, even without all the names and enzymes.

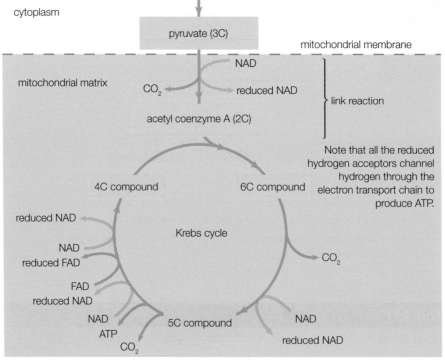

▲ **fig A**   The main stages of the Krebs cycle

THE LINK REACTION

This is the name we sometimes give to the reaction that links glycolysis to the Krebs cycle. The 3-carbon (3C) compound pyruvate crosses through the mitochondrial membrane from the cytoplasm into the mitochondria. An atom of carbon and a molecule of oxygen are removed from pyruvate (decarboxylation), resulting in the formation of a carbon dioxide molecule and a 2-carbon compound. This 2C compound then joins with coenzyme A to form the compound **acetyl coenzyme A (acetyl CoA)**. At the same time, the pyruvate is oxidised, losing hydrogen to NAD (dehydrogenation), resulting in reduced NAD. The reduced NAD is used later in the electron transport chain to produce ATP (see **Section 7A.4**). The energy contained in the acetyl CoA is released in the Krebs cycle. Enzymes called **decarboxylases** remove carbon dioxide and enzymes called **dehydrogenases** remove hydrogen. In summary:

$$\text{pyruvate (3C)} + \text{CoA} + \text{NAD} \rightarrow \text{acetyl CoA (2C)} + CO_2 + \text{reduced NAD}$$

THE KREBS CYCLE

We can summarise the key reactions of the Krebs cycle.

- The 2C acetyl group from acetyl CoA combines with a 4C compound to form a 6C compound. At this point it has entered the Krebs cycle.

- The 6C compound now follows a cyclical series of reactions. During this cycle, the compound is broken down in a number of stages to give the original 4C compound. Two more molecules of carbon dioxide are removed in the process and are released as a waste product.

- The 4C compound then combines with another 2C acetyl CoA molecule and the cycle begins again.

For each molecule of pyruvate that feeds into the Krebs cycle, three molecules of reduced NAD, one of reduced FAD and one of ATP are produced. The reduced NAD and reduced FAD then enter the electron transport chain (see **Section7A.4**). For each molecule of glucose that enters the glycolytic pathway, the Krebs cycle is completed *twice* because the 6C glucose molecule produces two 3C pyruvate molecules, each of which pass through the Krebs cycle.

CHECKPOINT

1. Summarise the differences between the Krebs cycle and glycolysis.

> **SKILLS** ▶ CREATIVITY

2. ▶ The Krebs cycle makes energy for the cell. Explain how this statement is incorrect.

3. Investigate the work of Krebs and write a short description of how he built up his model of the cycle of reactions that occur in the mitochondria.

SUBJECT VOCABULARY

Krebs cycle a series of biochemical steps that leads to the complete oxidation of glucose, resulting in the production of carbon dioxide, water and relatively large amounts of ATP
link reaction the reaction needed to move the products of glycolysis into the Krebs cycle
acetyl coenzyme A (acetyl CoA) the 2C compound produced in the link reaction which feeds directly into the Krebs cycle, combining with a 4C organic acid to form a 6-carbon compound
decarboxylases enzymes that remove carbon dioxide from a molecule
dehydrogenases enzymes that remove hydrogen from a molecule (during oxidation reactions)

DID YOU KNOW?

Understanding the Krebs cycle

Hans Krebs first suggested his ideas for the now famous cycle in 1937 (see **fig B**). It was the result of Krebs' and others' intelligent reasoning and experimentation in the years beforehand.

▲ **fig B** Sir Hans Krebs moved from Germany to Cambridge University in the UK. He discovered both the Krebs cycle and the ornithine cycle, which you will study later in **Chapter 7C**.

In the period 1910–20, several biochemists, including T. Thunberg, L.S. Stern and F. Batelli, used a blue dye that loses its colour when reduced to show that dehydrogenases were involved in transferring hydrogen atoms from particular organic acids known to occur in the cells of finely chopped animal tissue.

In 1935, Albert Szent-Györgyi produced a sequence of enzymatic reactions showing the oxidation of several organic acids from succinic acid, which we now know is part of the Krebs cycle.

Krebs (see **fig B**) then conducted a series of experiments to show that only certain organic acids were oxidised by cells, and that certain inhibitors could stop the oxidations. After much work he suggested the sequence we now know as the Krebs cycle. His most important discovery was that the 2C molecule and the 4C molecule combined to form 6C citric acid. This was the 'missing link' that allowed him to show that the process was a cycle. Krebs also showed that all the reactions he suggested could occur at a fast enough rate to account for the use of pyruvate and oxygen in the tissue which was already known. This suggested that his pathway was the main, if not the only, pathway for the oxidation of food molecules. He won the Nobel Prize in Physiology and Medicine in 1953 for his work, which changed perceptions of cell biology for ever.

LEARNING OBJECTIVES

■ Understand how ATP is synthesised by oxidative phosphorylation associated with the electron transport chain in mitochondria, including the role of chemiosmosis and ATP synthase.

Oxidative phosphorylation is the final process of aerobic respiration. In this process, reduced NAD or FAD from glycolysis and the Krebs cycle is used with oxygen to make ATP. The process involves an **electron transport chain**, which is a series of electron carrier molecules along which electrons from reduced NAD or FAD are transferred. At the same time, the remaining hydrogen ions (protons) are used in **chemiosmosis** to supply the energy needed to synthesise ATP.

Hydrogen atoms are removed from the compounds in glycolysis and the Krebs cycle, and hydrogen atoms in the end combine with oxygen atoms to form water, but it is mainly electrons that are passed along the carrier system. This is why the system is called the electron transport chain. The hydrogen ions remain in solution. You can think of the various parts of the electron transport chain as being at different energy levels. The first member of the chain is the highest level, and subsequent steps have lower levels. Each electron is passed down from one energy level to another, driving the production of ATP (see **fig A**). ATP production is called oxidative phosphorylation because ADP is phosphorylated in a process that depends on the presence of oxygen. The electron transport chain model describes the sequence of reactions by which living organisms make ATP.

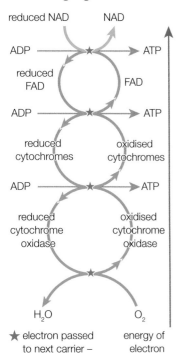

reduced NAD   NAD

ADP ──────★───→ ATP

reduced FAD   FAD

ADP ──────★───→ ATP

reduced cytochromes   oxidised cytochromes

ADP ──────★───→ ATP

reduced cytochrome oxidase   oxidised cytochrome oxidase

H_2O O_2

★ electron passed to next carrier – oxidation is loss of electrons

energy of electron

◀ **fig A** These are the main components of the electron transport chain. As the carriers become reduced and then oxidised again, energy is released to drive the production of molecules of ATP.

There are four main electron carriers involved.

- The coenzymes NAD and FAD both act as hydrogen acceptors for hydrogen released in the Krebs cycle. One molecule of ATP is produced when the FAD is reduced and accepts hydrogen from the reduced NAD, which becomes oxidised in the process.
- **Cytochromes** are protein pigments with an iron group (rather similar to haemoglobin) which are involved in electron transport. They are reduced by electrons from reduced FAD and reduced NAD which are consequently oxidised again. A molecule of ATP is produced at this stage.
- **Cytochrome oxidase** is an enzyme that receives the electrons from the cytochromes and is reduced as the cytochromes are oxidised. A molecule of ATP is also produced at this stage.
- Oxygen is the final hydrogen acceptor in the chain. When the oxygen is reduced, water is formed and the chain is at an end.

As a result of each molecule of hydrogen passing along the electron transport chain from reduced NAD, sufficient energy is released to make three molecules of ATP. When the hydrogen enters the chain from reduced FAD, only two molecules of ATP are produced.

WHERE IS ATP ACTUALLY MADE?

Glycolysis occurs in the cell cytoplasm and the other stages of respiration occur in the mitochondria. The link reaction and Krebs cycle occur in the matrix of the mitochondria. The electron transport chain and ATP production occur on the inner membrane of the mitochondria, which is folded up to form the cristae, producing a large surface area. The surface of the cristae is covered with closely packed stalked particles which seem to be the site of the ATPase enzymes.

THE CHEMIOSMOTIC THEORY OF ATP PRODUCTION

In 1961, Peter Mitchell first described the link between the electrons that are passed down the electron transport chain and the production of ATP. He called it the **chemiosmotic theory**. The theory explains what happens to the hydrogen ions (protons) that remain after the electrons are passed along the electron transport chain, and how the movement of the hydrogen ions is linked to the production of ATP.

Mitchell proposed that hydrogen ions are actively transported into the space between the inner and outer mitochondrial membranes using the energy provided as the electrons pass along the transport chain (see **fig B**). The active transport of the hydrogen ions across the inner membrane results in a different hydrogen ion concentration on each side of the inner membrane. The membrane space has a higher concentration of hydrogen ions than the matrix, so there is a concentration gradient across the membrane. As a result of the different hydrogen ion concentrations, there is also a pH gradient. And because positive hydrogen ions are concentrated in the membrane space, there is an electrochemical gradient too.

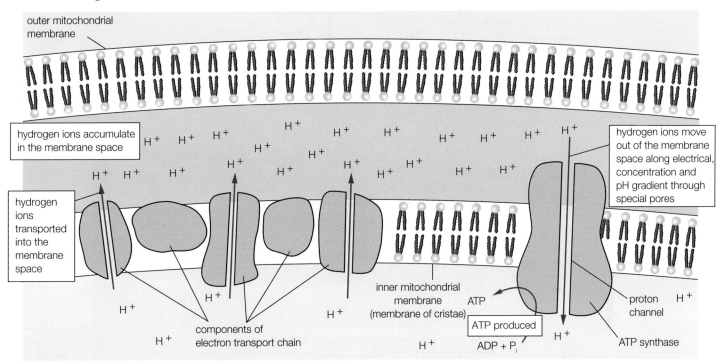

▲ **fig B**   The chemiosmotic model for the production of ATP. Hydrogen ions are moved into the space between the inner and outer mitochondrial membranes by the electron transport chain. They move back into the mitochondrial matrix *down* the concentration, electrical and pH gradients through special pores (holes) associated with an ATPase, driving the production of ATP.

All of these factors mean that the hydrogen ions tend to move back into the matrix. The only way they can move back into the matrix is through special pores (holes). These pores are on the stalked particles and have an ATP synthase enzyme associated with them. As the hydrogen ions move along their electrical, concentration and pH gradients through these pores, their energy is used to drive the synthesis of ATP. This is a process found in all living things.

HOW MUCH ATP IS GAINED?

It is important to remember that the two stages of respiration we have looked at work together. Glycolysis continually feeds into the Krebs cycle, and the control of the whole process depends on various enzymes and the levels of some of the substrates and products of the reactions.

Cellular respiration has evolved to produce energy in the form of ATP for use in the cells. The fact that the process is the same in almost all living organisms suggests (a) that it evolved at a very early stage in the development of organisms on Earth, and (b) that it is a very effective method of producing available energy. If this were not the case, alternative successful life forms with a different system of respiration would have evolved long ago. But exactly how much ATP is gained during the oxidation of one molecule of glucose in its journey along the respiratory pathways?

The easiest way to understand this is to consider the whole process and where the ATP is produced (see **fig C**). For many years, scientists said that the average amount of ATP produced from one

EXAM HINT

Here is another opportunity for your examiners to test IAS knowledge at IAL. You may need to relate the structure and properties of membranes (see **Section 2A.1–4 (Book 1: IAS)**) to the movement of ions across the membrane.

glucose molecule in aerobic respiration was 36 molecules of ATP, assuming that glucose enters the cycle and that oxidation is complete. The actual total was taken as 38 molecules of ATP, but it takes two molecules of ATP to transport the reduced NAD molecules produced in glycolysis through the mitochondrial membrane, leaving 36 available for the body cells. If this is compared with the two molecules of ATP that result when the breakdown of a glucose molecule is completely anaerobic, the importance of the oxygen-using process becomes very clear.

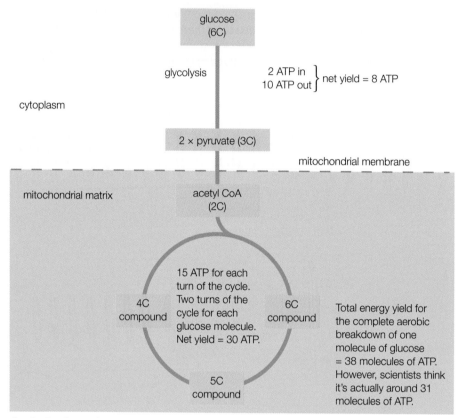

▲ **fig C** This model showing the 38 molecules of ATP gained by the complete oxidation of a molecule of glucose in cellular respiration has been accepted for a long time. Modern scientists now realise that the figures depend on many things and are quite variable.

These figures make the assumption that the ATP is always produced in whole numbers. Over the past 50 years, our understanding of the processes of the electron transport chain has increased and the figures have become less certain. Scientists now think that ATP production may not always be in whole numbers. Currently, the best estimates from scientists are that the oxidation of two molecules of reduced NAD supplies enough energy to make five molecules of ATP. Similarly, they now think that the oxidation of two molecules of reduced FAD produces about three molecules of ATP. This gives an overall production of around 31 molecules of ATP for the process of aerobic cellular respiration.

In addition, the proton gradients in the mitochondria are not only used to produce ATP. They can also be used to drive the active transport of several different molecules and ions through the inner membrane into the matrix. And reduced NAD can be used as a reducing agent for many different reactions. So, although the functional production of ATP is usually quoted as 36 (or 38 or 31), the amount of ATP resulting from one molecule of glucose going through complete oxidation is probably rarely over 30.

CHECKPOINT

1. ▶ Using **fig C** to help you, draw a large, fully labelled diagram summarising the process of aerobic cellular respiration, starting with glucose.

2. Explain how the oxidation of glucose results in the formation of ATP.

3. (a) Explain why aerobic respiration produces so much more ATP than anaerobic respiration.

   (b) Explain why the quoted ATP yield of aerobic respiration is so variable.

4. Describe Mitchell's chemiosmotic theory of ATP production and state why it is so important to our understanding of cellular respiration.

SUBJECT VOCABULARY

oxidative phosphorylation the oxygen-dependent process in the electron transport chain where ADP is phosphorylated

electron transport chain a series of electron-carrying compounds along which electrons are transferred in a series of oxidation/reduction reactions, driving the production of ATP

chemiosmosis the process that links the electrons that are passed along the electron transport chain to the production of ATP, by the movement of hydrogen ions through the membrane along electrochemical, concentration and pH gradients

cytochromes members of the electron transport chain; they are protein pigments with an iron group (similar to haemoglobin) which are reduced by electrons from reduced FAD, which is reoxidised, with the production of a molecule of ATP

cytochrome oxidase an enzyme in the electron transport chain which receives the electrons from the cytochromes and is reduced as the cytochromes are oxidised, with the production of a molecule of ATP

chemiosmotic theory the model developed by Peter Mitchell to explain the production of ATP in mitochondria, chloroplasts and elsewhere in living cells

LEARNING OBJECTIVES

■ Understand what is meant by the term *respiratory quotient (RQ)*.

There is not always sufficient glucose and glycogen to provide all the ATP needed in a cell. When this is the case, other substances can be respired, in particular other carbohydrates and fats. It is easy to understand that, for example, disaccharide sugars can be broken down and enter the respiratory pathway, but where do the other substances fit in?

RESPIRATORY SUBSTRATES

Different **respiratory substrates** are used at different times in the cells of the body. Lipids are an excellent source of energy. In **Section 1A.4 (Book 1: IAS)**, you learned that they can be hydrolysed by lipases to give fatty acids and glycerol. The glycerol is phosphorylated and enters the glycolytic pathway as GALP. The fatty acids pass through a series of reactions which remove 2-carbon sections of acetyl CoA. Hydrogen atoms are removed in the process and are transferred along the electron transport chain to form ATP even before the acetyl CoA enters the Krebs cycle. The fatty acid then goes through the reaction series again to remove the next 2-carbon fragment. A single fatty acid can produce a large amount of ATP; for example, a single molecule of stearic acid (an 18-carbon chain) produces about 180 molecules of ATP. Thus, the complete oxidation of fat produces a large amount of useable energy. This is why fat is a very important energy source for the body, particularly for active tissues such as heart muscle, liver and kidneys.

Protein is not usually used as a respiratory substrate. It is broken down only if supplies of both carbohydrates and lipids are very low, that is to say only if the body senses it is starving. The amino acids which make up the peptide chains must be deaminated (the amino groups must be removed) before the rest of the molecule can be used for cellular respiration.

Fig A shows how a variety of respiratory substrates enter the respiratory pathway to provide ATP for the cells.

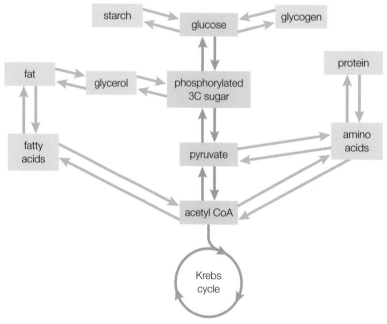

▲ **fig A** A summary of the complex pathways by which different respiratory substrates enter into glycolysis and the Krebs cycle.

RESPIRATORY QUOTIENTS

The amounts of oxygen used and carbon dioxide produced during cellular respiration change depending on the level of activity of the organism, the type of food being respired and other factors. We can produce what is known as the **respiratory quotient (RQ)** by measuring the amounts of carbon dioxide produced and oxygen used by an organism in a given time period.

$$\text{respiratory quotient} = \frac{\text{carbon dioxide produced}}{\text{oxygen used}}$$

The RQ helps us understand what types of foods are being oxidised in the body of an organism at a particular time. In theory, carbohydrates give an RQ of 1, fats give an RQ of 0.7 and protein gives an RQ of 0.9. Under normal conditions, protein is not often used to provide energy, so an RQ of around 1 suggests that a lot of carbohydrate is being used in cellular respiration. An RQ of less than 1 indicates that a combination of carbohydrate and lipid is being respired.

If the RQ of an organism is greater than 1, then anaerobic respiration may be occurring, with relatively little oxygen being used compared with the carbon dioxide produced. You will often observe very low RQ values in photosynthetic organisms because much of the carbon dioxide produced is used in making new sugars and so it cannot be measured.

PRACTICAL SKILLS CP16

For **Core Practical 16: Use a simple respirometer to determine the rate of respiration and RQ of a suitable material (such as germinating seeds or small invertebrates)**, you can use the respirometer described in **Section 7A.1 fig F (b), page 165**. There are a number of important things to remember when using this apparatus. You need to understand these and may need to apply them in examination questions.

1   The potassium hydroxide in the test tubes absorbs carbon dioxide from the air. You can use sodium hydroxide to do the same thing.
2   As the organisms in the left-hand tube respire, they use oxygen from the air and the pressure in the tube falls. The fluid level in the manometer tube nearest to the organisms rises because of the pressure difference between the left-hand tube and the right-hand tube.
3   Any carbon dioxide produced is absorbed by the potassium hydroxide solution.
4   Use the syringe connected to the right-hand tube to return the manometer levels to their original point. You can calculate the volume of oxygen used by observing the volume of gas you need to squeeze from the syringe to return the manometer levels to their original level.
5   If you repeat the experiment using water rather than potassium hydroxide, you can also calculate the amount of carbon dioxide produced by the difference in the movement of the manometer fluid.

> **!**
>
> Safety Note: The apparatus is very fragile and the syringe should be moved slowly and carefully. Skin contact with the solutions and organisms should be avoided.

CHECKPOINT

1.   Explain what is meant by the terms *respiratory substrate* and *respiratory quotient*.
2.   Use **fig A** to help you explain why some respiratory substrates produce more ATP than others.
3.   The respiratory quotients (RQs) of four different organisms are measured.
     Organism A has an RQ of 1.0.
     Organism B has an RQ of around 0.7.
     Organism C has an RQ of 0.85.
     Organism D has an RQ of 1.2.

     Suggest explanations for these very different measurements.

SKILLS ▶ INNOVATION

SKILLS ▶ ANALYSIS

SUBJECT VOCABULARY

respiratory substrate the substance used as a fuel and oxidised during cellular respiration
respiratory quotient (RQ) the relationship between the amount of carbon dioxide produced and the amount of oxygen used when different respiratory substrates are used in cellular respiration

1 (a) An electron from a glucose molecule would travel through the following processes:

1 Krebs cycle

2 glycolysis

3 electron transport chain

4 link reaction

In which order would the electron travel through these processes? [1]

A 2–4–1–3

B 2–1–3–4

C 1–3–2–4

D 1–4–3–2

(b) In cellular respiration, glucose is broken down to produce ATP. During this process, hydrogen atoms are removed from the glucose molecule and attach to which of the following substances? [1]

A water

B oxygen

C nitrogen

D NAD

(c) The Krebs cycle is a part of aerobic respiration and is an example of a metabolic pathway.

(i) Explain why the Krebs cycle is described as a metabolic pathway. [2]

(ii) State precisely where in the cell the Krebs cycle occurs. [2]

(d) The diagram below shows some of the stages that occur in the Krebs cycle.

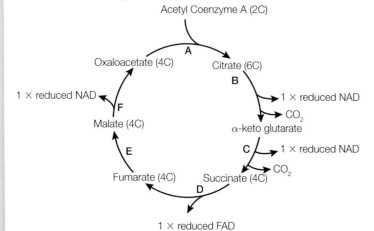

Oxidoreductase enzymes are involved in some of the reactions in the Krebs cycle. Using the letters A to F and the information given in the diagram, identify the letters of **all** the stages that involve an oxidoreductase enzyme. [1]

(Total for Question 1 = 7 marks)

2 (a) What are the products of anaerobic respiration? [1]

A lactate in plants and ethanol in mammals

B ethanoic acid in plants and lactate in mammals

C ethanol in plants and lactate in mammals

D lactate in plants and ethanoic acid in mammals

(b) State where in the cell anaerobic respiration takes place. [1]

(c) The diagram below shows some of the stages of anaerobic respiration in a muscle cell.

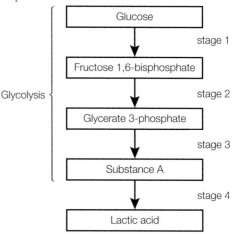

(i) What is the name of substance A? [1]

A glycerate bisphosphate

B acetate

C pyruvate

D ethanol

(ii) Which stage uses ATP? [1]

A stage 1

B stage 2

C stage 3

D stage 4

(iii) Which stage produces ATP? [1]

A stage 1

B stage 2

C stage 3

D stage 4

(d) Explain why lactic acid is produced in actively respiring animal cells. [4]

(Total for Question 2 = 9 marks)

3 (a) The diagram below represents some of the stages of respiration in a yeast cell.

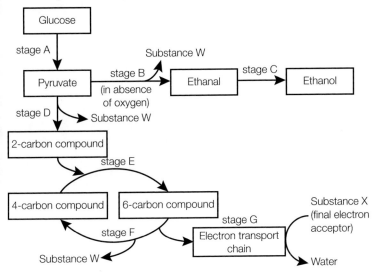

(i) What are the correct names for substances **W** and **X**?   [1]

    A carbon monoxide and oxygen

    B hydrogen and oxygen

    C carbon dioxide and hydrogen

    D carbon dioxide and oxygen

(ii) Which letter represents the stage during which most ATP is produced?   [1]

    A stage A

    B stage B

    C stage F

    D stage G

(iii) State **two** letters that represent stages during which no ATP is made.   [1]

(iv) State **two** letters that show where substrate level phosphorylation takes place.   [2]

(b) During glycolysis the coenzyme NAD is reduced. Describe the ways in which the reduced NAD can be used in the yeast cell.   [3]

(Total for Question 3 = 8 marks)

4 (a) The photograph below shows a mitochondrion as seen using an electron microscope.

(i) Name the parts labelled **B** and **C**.   [2]

(ii) State the **letter** that represents the location of the electron transport chain.   [1]

(b) Antimycin A inhibits aerobic respiration. It binds to one of the electron carriers in the electron transport chain. An experiment was carried out to investigate the effect of antimycin A on the respiration of yeast cells.

- Yeast cells were mixed with a buffer solution containing ADP, phosphate ions and glucose.
- The mixture was placed in a water bath at 30 °C.
- It was incubated for 30 minutes.
- During this time, the oxygen content of the suspension was measured.
- The experiment was repeated with antimycin A added to the suspension 5 minutes after the start of the incubation.
- The results are shown in the table below.

Time of incubation/mins	Oxygen content of suspension/ arbitrary units	
	Without antimycin A	With antimycin A added after 5 minutes
0	6.4	6.4
5	3.7	3.7
10	2.4	3.7
15	1.6	3.7
20	0.9	3.7
25	0.5	3.7
30	0.5	3.7

(i) Comment on why the oxygen content of the suspension of cells without antimycin A did not reach zero.   [2]

(ii) Explain why the oxygen concentration of the suspension did not decrease after antimycin A was added.   [2]

(iii) Explain what effect the addition of antimycin A will have on the production of ATP.   [3]

(Total for Question 4 = 10 marks)

CHAPTER 7B MUSCLES, MOVEMENT AND THE HEART

Arabian horses are renowned around the world for their endurance and stamina. Thoroughbreds are faster over short distances, but Arabians are superior as soon as the distances get longer. Why is this? Every horse needs muscles to move, and those muscles need a good supply of food and oxygen. What scientists are discovering is that the balance of muscle fibre types in these different horses is very different. The blood supply to the muscles is different too: thoroughbreds have a high aerobic and anaerobic capacity in their muscles, but Arabians use fat for energy very efficiently in their muscles and this enables them to continue racing for longer.

In this chapter, you will study the basic structure of the musculoskeletal system and how the different tissues work together when you move, including the antagonistic pairs of skeletal muscles and the synovial joints. By looking at the detailed structure and biochemistry of the muscles you will discover how they contract and why some people are sprinters while others run marathons.

You will consider how the heart muscle differs from the skeletal muscle, and find out how your heart rate is controlled. Your body operates most effectively within narrow boundaries and you will look at the principles of homeostasis, the way this 'steady state' is maintained. Finally, you will be considering how the heart and breathing rates are controlled so your body can respond to both exercise and stress in everyday life.

MATHS SKILLS FOR THIS CHAPTER

- **Construct and interpret frequency tables and diagrams, bar charts and histograms** (*e.g. impact of exercise on heart rate and heart volume, anaerobic exercise and oxygen debt*)

- **Translate information between graphical, numerical and algebraic forms** (*e.g. impact of exercise on heart rate and heart volume, oxygen debt*)

- **Use appropriate number of significant figures** (*e.g. calculating percentage efficiency of anaerobic respiration in mammalian muscles*)

- **Plot two variables from experimental or other data** (*e.g. investigating the rate of respiration, ventilation rates*)

- **Calculate rate of change from a graph showing a linear relationship** (*e.g. investigating the rate of respiration*)

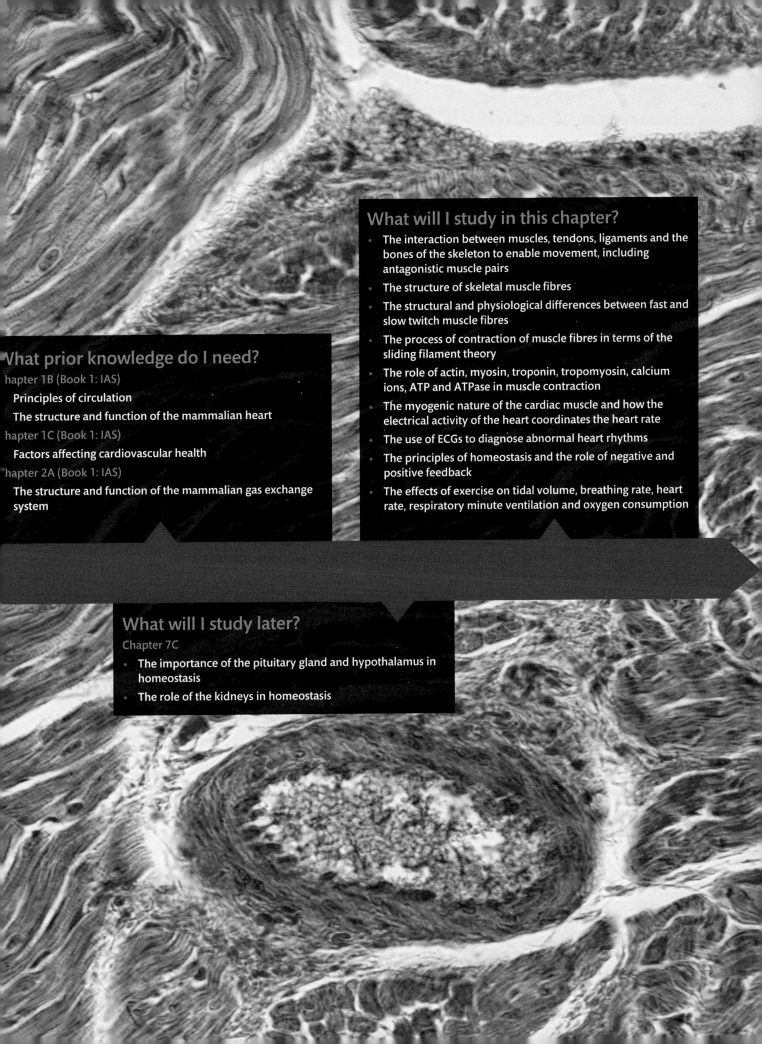

What will I study in this chapter?

- The interaction between muscles, tendons, ligaments and the bones of the skeleton to enable movement, including antagonistic muscle pairs
- The structure of skeletal muscle fibres
- The structural and physiological differences between fast and slow twitch muscle fibres
- The process of contraction of muscle fibres in terms of the sliding filament theory
- The role of actin, myosin, troponin, tropomyosin, calcium ions, ATP and ATPase in muscle contraction
- The myogenic nature of the cardiac muscle and how the electrical activity of the heart coordinates the heart rate
- The use of ECGs to diagnose abnormal heart rhythms
- The principles of homeostasis and the role of negative and positive feedback
- The effects of exercise on tidal volume, breathing rate, heart rate, respiratory minute ventilation and oxygen consumption

What prior knowledge do I need?

Chapter 1B (Book 1: IAS)

Principles of circulation

The structure and function of the mammalian heart

Chapter 1C (Book 1: IAS)

Factors affecting cardiovascular health

Chapter 2A (Book 1: IAS)

The structure and function of the mammalian gas exchange system

What will I study later?

Chapter 7C

- The importance of the pituitary gland and hypothalamus in homeostasis
- The role of the kidneys in homeostasis

■ Know the way in which muscles, tendons, the skeleton and ligaments interact to enable movement, including antagonistic muscle pairs, extensors and flexors.

Imagine you are moving (walking, running or jumping) and then try to imagine doing any of those things without your skeleton and the tissues associated with it. Your bony skeleton supports your whole body. You can move around as a result of the way the bones of your skeleton and other skeletal tissues work together. The properties of your skeletal tissues vary, because they are adapted to perform their very different functions effectively.

THE SKELETAL TISSUES

The main tissue of the skeleton is **bone**. Bone is strong and hard. It is made of bone cells which are fixed firmly in a matrix of collagen and calcium salts. Bone is particularly strong under compression (squashing) forces. Bone needs to be both strong and hard, but it must also be as light as possible to reduce the weight you have to move about. Compact bone is dense and heavy. This is the bone found in the long bones of your body. Spongy bone has a much more open structure so it is much lighter. It is found in large masses of bone such as your pelvis and the head of the femur.

Cartilage is a hard but flexible tissue. It is made up of cells called **chondrocytes** within an organic matrix which consists of varying amounts of collagen fibrils. Cartilage is elastic (stretches and returns to its original size) and able to withstand compressive (squeezing) forces. It is a very good shock absorber and is found between bones such as the vertebrae in your spine and in the joints. There are two main types of cartilage found in the skeleton.

- Hyaline cartilage is found at the ends of bones (and in the nose, air passageways and parts of the ear).
- White fibrous cartilage has bundles of densely packed collagen in the matrix. It has great tensile strength but is less flexible than the other forms of cartilage. It forms the discs between your vertebrae and is found between the bones in the joints.

Tendons are made up almost entirely of **white fibrous tissue**. This consists of bundles of collagen fibres which form a strong but relatively inelastic tissue. You learned the structure of collagen fibres in **Section 1A.5 (Book 1: IAS)**. It is ideal for joining muscles to bones. One end of the tendon is attached to a muscle and the other end attached either directly to a bone or to the fibrous cover of the bone, as you can see in **fig A**. This makes a secure attachment for muscles to bone, and provides a little shock absorption if the joint is given a sudden stretch. However, if tendons stretched a lot, much of the work done by the muscles would be wasted because they would stretch the tendons without moving the bones!

Ligaments hold the bones together in the correct alignment. They form a capsule around the joint and hold the bones together inside the joint itself. Ligaments need to be elastic so the bones of the joint can move when they need to. They are made of yellow elastic tissue which gives an ideal combination of strength with elasticity. Some ligament capsules are very loose, so the joint can move quite freely. Others are very tight because different joints need different properties. The differences in the properties of the ligaments come from the varying amounts of collagen and white fibrous tissue in the mixture.

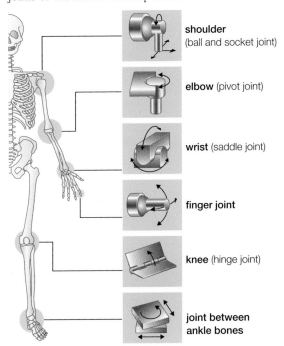

Flexor and extensor muscles work antagonistically to operate the joint.

- extensor muscle (quadriceps)
- flexor muscle (biceps femoris)
- tendons attach muscle to bone
- femur
- patella (knee cap)
- ligaments attach bone to bone
- cartilage acts as shock absorber between bones
- tibia
- fibula

▲ **fig A**   The tissues of the skeletal system are shown here at the knee. They work together to hold the bones together, to stop them wearing away (becoming thinner and weaker with use) and to make movement possible.

EXAM HINT

Some students confuse the roles of tendons and ligaments. It is important to remember that ligaments hold bone to bone and stretch to allow movement. Tendons hold muscle to bone and do not stretch.

JOINTS, MUSCLES AND MOVEMENT

You need the joints of your body to allow movement and locomotion. The ends of the bones at a joint are shaped to move smoothly over each other. The way in which the two bones meet varies according to the type of movement required, as you can see clearly in **fig B**. For example, the hip and shoulder have ball and socket joints which give you very free movement, whereas the hinge joints of the knees restrict possible movement.

shoulder (ball and socket joint)

elbow (pivot joint)

wrist (saddle joint)

finger joint

knee (hinge joint)

joint between ankle bones

▲ **fig B**   The skeleton has several different types of joint. They each make a different type of movement possible.

The bones in a joint form two solid masses moving over each other while experiencing severe forces. If the joint consisted simply of bone on bone, the ends would soon become thinner and weaker through rubbing together. To prevent this, the joint is lined with a replaceable layer of rubbery cartilage. This makes it possible for the joint to articulate (move) smoothly. The most mobile joints also produce a liquid lubricant known as **synovial fluid**; it fills the joint cavity and ensures easy friction-free movement (see **fig C**).

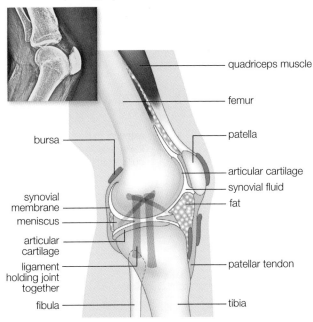

- quadriceps muscle
- femur
- patella
- articular cartilage
- synovial fluid
- fat
- bursa
- synovial membrane
- meniscus
- articular cartilage
- ligament holding joint together
- patellar tendon
- fibula
- tibia

▲ **fig C**   The knee is a synovial joint. The lubrication provided by the synovial fluid gives smooth, pain-free movement in this important joint.

HOW DO WE MOVE?

You move by the action of muscles on bones. Each of your skeletal muscles is attached by tendons to two different bones, extending across at least one joint. When muscles contract they pull on a bone and so it moves relative to another bone. However, when muscles relax they do not push in a corresponding way. They simply stop contracting and can be pulled back to their original shape. This is why the muscles of the skeleton are found in pairs. One pulls the bone in one direction, the other pulls it back to its original position. The muscles which extend a joint are called **extensors** and the muscles that bend or flex a joint are called **flexors**. Because they work in direct opposition to each other, these muscles are known as **antagonistic pairs**. You can see a clearer picture of how movement happens in **fig D**.

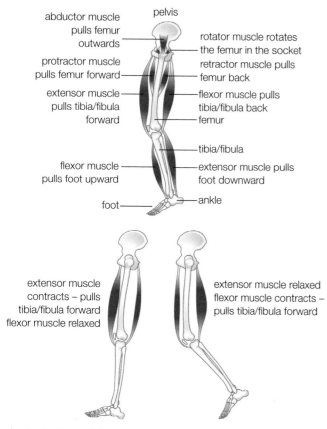

abductor muscle
pulls femur
outwards

pelvis

rotator muscle rotates
the femur in the socket

protractor muscle
pulls femur forward

retractor muscle pulls
femur back

extensor muscle
pulls tibia/fibula
forward

flexor muscle pulls
tibia/fibula back

femur

tibia/fibula

flexor muscle
pulls foot upward

extensor muscle pulls
foot downward

foot

ankle

extensor muscle
contracts – pulls
tibia/fibula forward
flexor muscle relaxed

extensor muscle relaxed
flexor muscle contracts –
pulls tibia/fibula forward

▲ **fig D** Some of the antagonistic pairs of muscles involved in the movement of the leg.

CHECKPOINT

1. Joints like the knee, the hip and the shoulder have a synovial membrane and produce synovial fluid. Many other joints do not. Describe synovial fluid and explain why it is so important in some joints but not in others.

2. Make a table to summarise the properties of the main skeletal tissues.

SKILLS ▶ ANALYSIS

3. ▶ Describe how muscles, tendons, the skeleton and ligaments interact to enable movement.

SUBJECT VOCABULARY

bone the strong, calcium-rich tissue which is the main component of the vertebrate skeleton

cartilage hard but flexible skeletal tissue that often acts as a shock absorber and prevents wear in joints

chondrocytes the cells that form cartilage

tendons inelastic tissue which joins muscles to bones

white fibrous tissue inelastic connective tissue made up of bundles of collagen fibres

ligaments elastic tissue which forms joint capsules

synovial fluid fluid which lubricates the most mobile joints

extensors muscles that extend (stretch or open) a joint

flexors muscles that flex (close or bend) a joint

antagonistic pairs muscles which work in opposition to each other, pulling in opposite directions

LEARNING OBJECTIVES

■ Know the structure of a mammalian skeletal muscle fibre.

Muscle is a specialised tissue which is very similar throughout the animal kingdom. In this section, you will learn about mammalian muscle. Muscles are mostly made of protein. They consist of large numbers of very long cells known as muscle fibres. These muscle fibres are held together by connective tissue. They can contract (shorten) to do work. When they relax, they can be pulled back to their original length. Muscles have a good blood supply. This provides them with the glucose and oxygen they need for cellular respiration (see **Chapter 7A**). This supplies the ATP that muscles need for contraction. The rich blood supply also makes it easy to remove the carbon dioxide and other waste products that result. Muscles respond to stimulation from the nervous system and to chemical stimulation from hormones such as adrenaline.

In mammals, the muscle tissues can represent as much as 40% of the body weight of the animal. There are three main types of muscle, each specialised for a particular function:

- skeletal muscle
- smooth muscle
- cardiac muscle.

SKELETAL MUSCLE

Skeletal muscle/striated muscle/voluntary muscle is the muscle attached to the skeleton and involved in locomotion. It is under the control of the voluntary nervous system, and its appearance under the microscope is striated (with stripes). It contracts rapidly, but also fatigues or tires relatively quickly. You will mainly study striated muscle. The microscopic structure of striated muscle that you can see under the electron microscope gives us lots of clues about how the tissue contracts.

The muscle fibres are made up of many **myofibrils** lying parallel to each other. Each myofibril is made up of individual units of the muscle structure, called **sarcomeres**. The structure of the sarcomeres is made of the proteins **actin** and **myosin**. The cytoplasm of the myofibrils is called the **sarcoplasm**. It contains many mitochondria which supply the energy that the muscle needs to contract. A network of membranes present in the whole system is called the **sarcoplasmic reticulum**, which stores and releases calcium ions (see **fig A**). You will learn more about how skeletal muscle contracts in **Section 7B.4**.

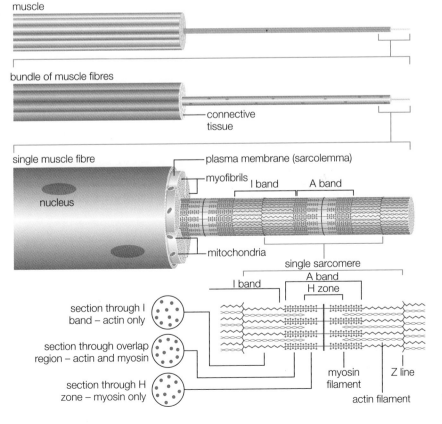

fig A The detailed structure of skeletal muscle.

SMOOTH MUSCLE AND CARDIAC MUSCLE

The body contains two other types of muscle.

- **Smooth muscle/involuntary muscle** is not striped and is under the control of the involuntary nervous system. It is the muscle found in the gut where it is involved in moving the food along. It is also the muscle in blood vessels. It contracts and fatigues slowly.
- **Cardiac muscle** is only found in the heart. It is striated, and the fibres are joined by special cross-connections. Cardiac muscle contracts spontaneously, which means it does not need to be stimulated by either nerves or hormones, and it does not fatigue. Your heart beats continually through your life without tiring.

DID YOU KNOW?

Properties of skeletal muscle

Muscles respond to electrical stimulation. Scientists use this property to investigate the way muscles work. If you give a variety of different electrical stimuli to a calf muscle (gastrocnemius) from a frog you can record the effects on a revolving drum (kymograph). **Fig B** shows the results from the stimulation of a single muscle fibre. If you use a whole muscle, the results are more confusing because different fibres have different thresholds (levels at which they start to respond) and contract with different strengths.

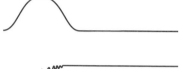

A single stimulus causes a single contraction or twitch of the muscle fibre. It is an 'all-or-nothing' response. This means that if the stimulus is below a certain level, nothing happens. If it is above the threshold level, the muscle fibre twitches. However big the stimulus, the size of the twitch is still the same.

When a whole series of rapid stimuli are given the muscle fibre becomes fully contracted and as short as possible and simply stays like this. It is known as tetanus. This is the normal situation in a muscle when you are lifting an object or indeed standing up and maintaining your posture against gravity. Single twitches are relatively rare.

▲ **fig B**   Recordings of the responses of a single muscle fibre to different stimuli.

A single stimulus causes a single contraction or twitch (small sudden movement) of the muscle fibre. It is an 'all-or-nothing' response. This means that if the stimulus is below a certain level, nothing happens. If it is above the **threshold level**, the muscle fibre twitches. But the size of a single twitch is always the same, however big the stimulus. Single twitches are relatively rare in whole muscles.

When a series of rapid stimuli is given, the muscle fibre becomes fully contracted and as short as possible and stays like this. This is known as **tetanus** and it is the normal situation in a muscle when you are lifting an object or standing up and maintaining your posture against gravity; many fibres are in tetanus.

A muscle cannot remain in tetanus continuously. Eventually it fatigues and cannot contract any more; this is the point at which there is no more ATP or calcium.

CHECKPOINT

1. Summarise the similarities and differences between the three main types of muscle.
2. ▶ A whole muscle responds differently to a stimulus from a single fibre. Suggest an explanation for this.

SKILLS ▷ CREATIVITY

SUBJECT VOCABULARY

skeletal muscle/striated muscle/voluntary muscle muscle with a striped appearance under the microscope; it moves the bones of the skeleton under voluntary control
myofibril a very long contracting fibre in skeletal muscle cells
sarcomere basic unit of muscle structure
actin one of the proteins which form the contracting mechanism of muscle cells
myosin one of the proteins which form the contracting mechanism of muscle cells
sarcoplasm the cytoplasm of muscle cells
sarcoplasmic reticulum the equivalent of the endoplasmic reticulum, found in muscle cells
smooth muscle/involuntary muscle muscle which is under the control of the involuntary nervous system
cardiac muscle muscle which makes up the heart
threshold level the level of stimulus that triggers a response
tetanus the state of a muscle which is fully contracted and remains so for some time; the disease tetanus got its name because the muscles of the body go into tetanus and cannot relax, so you cannot breathe and this causes death

7B 3 DIFFERENT TYPES OF MUSCLE FIBRE

LEARNING OBJECTIVES

■ Understand the structural and physiological differences between fast and slow twitch muscle fibres.

The cells of the skeletal muscles have certain features in common. They usually contain many mitochondria. As you learned in **Section 7A.1**, mitochondria are the site of aerobic respiration and they play an important role in supplying the active muscle cells with ATP. The muscle cells also contain **myoglobin**. This is a protein similar to haemoglobin, but it has one protein chain instead of four. It has a much higher affinity (attraction) for oxygen than haemoglobin, so it readily accepts oxygen from the blood. Myoglobin acts as an oxygen store in the muscles. Most muscles also have a rich blood supply to bring oxygen to the tissue which is respiring tissue. You learned about the structure of haemoglobin in **Section 1A.5 (Book 1: IAS)**, and you looked at the oxygen dissociation curve of haemoglobin in **Section 1B.2 (Book 1: IAS)**. In **fig A** you can see the oxygen dissociation curves for haemoglobin and myoglobin. This will help you understand why myoglobin is so effective at supplying oxygen to the muscles. For example, you can see that haemoglobin is 50% saturated with oxygen when the partial pressure of oxygen is 26 torr. Myoglobin is still 50% saturated with oxygen at partial pressures of oxygen of only 2.8 torr. When oxygen levels fall, haemoglobin gives up its oxygen which can then be picked up by myoglobin in the muscles.

EXAM HINT

Ensure you link back to what you learned in **Section 1B.2 (Book 1: IAS)** and give full explanations of the role of the respiratory pigments. The myoglobin has a higher affinity for oxygen so it will hold onto the oxygen until all the oxygen in oxyhaemoglobin has been used. It will then release its oxygen enabling continued aerobic respiration.

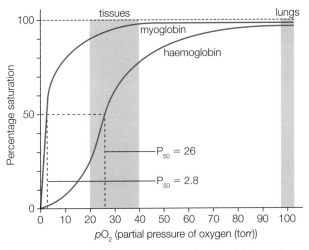

▲ **fig A** The oxygen dissociation curves for myoglobin and haemoglobin. Myoglobin has a higher affinity for oxygen, so it will take oxygen from haemoglobin as blood passes through the muscles.

TYPES OF SKELETAL MUSCLE FIBRE

There are two types of skeletal muscle fibre in mammals, which give very different levels of performance. Most muscles contain a mixture of the two types of fibre. The balance will affect both the performance and the colour of the muscle.

- **Slow twitch muscle fibres** are adapted for steady action over a period of time. They contract relatively slowly and can stay in tetanus for a long time. They are used to maintain your body posture, and during long periods of activity. These slow twitch fibres have a rich blood supply, lots of mitochondria and plenty of myoglobin. These adaptations mean that they can maintain their activity without needing to respire anaerobically for any length of time. The rich blood supply and high levels of myoglobin mean they are a deep red colour. Slow twitch muscle fibres need glucose as a fuel. This is supplied by their big network of blood vessels so they can continue to produce ATP for as long as oxygen is available.

EXAM HINT

Remember to link all the features of each type of muscle fibre to their function. The extra myoglobin in slow twitch fibres can release oxygen to enable continued aerobic respiration. That releases much more ATP than anaerobic respiration.

- **Fast twitch muscle fibres** contract very rapidly, making them well suited for sudden, rapid brief activity. They often function anaerobically (without oxygen) using glycolysis (see **Section 7A.2**) so they often fatigue quickly. Fast twitch fibres are supplied with few blood vessels compared to slow twitch fibres. They have low levels of myoglobin for storing oxygen and also contain a small number of mitochondria compared to slow twitch fibres. As a result, fast twitch fibres look much paler in colour. However, fast twitch fibres contain rich glycogen stores, which can be converted to glucose for both aerobic and anaerobic respiration. They also contain relatively high levels of creatine phosphate, which can be used to form ATP from ADP. These adaptations mean many more myofibrils are packed into fast twitch fibres than slow twitch fibres, because little space is taken up with mitochondria. They cannot produce high levels of ATP over a long period of time, but they are capable of very fast, powerful contractions for a brief period.

Most people have roughly equal amounts of slow twitch muscle fibres and fast twitch fibres in their muscles, but in some people the proportions can vary quite dramatically (see **fig B**). For example, long-distance runners, cyclists, swimmers and other endurance athletes usually have very high proportions of slow twitch fibres. In contrast, weightlifters and sprinters, who need the maximum strength from their muscles in short rapid activity, usually have an unusually high proportion of fast twitch fibres in their muscles.

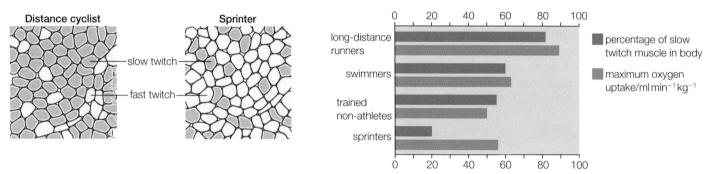

▲ **fig B** Scientists have found that athletes in different sports generally have different distributions of slow and fast twitch fibres in their muscles.

Why do we see these differences in the muscles of different types of athlete? Part of the answer is training. The *number* of muscle fibres you possess does not change, but the *size* and *type* of fibre can alter in response to exercise. So if you practise sprinting, you will develop more fast twitch muscle fibres. If you do endurance training, then your slow twitch muscle fibres will increase in number.

There are a variety of different genes which affect the basic components of our muscles, and you can then enhance this natural tendency with further training. Most of us have about 50% of each type of muscle fibre, but some people have around 75% fast twitch and others have 75% slow twitch. These differences make a difference to sporting potential. For example, someone born with a high proportion of fast twitch muscle fibres may be a good sprinter but they are unlikely to make a top-class marathon runner. Someone with more than the average number of slow twitch muscle fibres is unlikely to be a successful weightlifter, but may make a very good endurance athlete.

Scientists have also discovered 'superfast' twitch fibres, which contract even more quickly and strongly than usual. Athletes with these superfast fibres have an even bigger natural advantage if they train to run fast in sprint races, for example.

SUBJECT VOCABULARY

myoglobin a pigment with a high affinity for oxygen found in the muscle

slow twitch muscle fibres muscle fibres which contract and fatigue slowly; they contain many mitochondria, a lot of myoglobin and a rich blood supply

fast twitch muscle fibres muscle fibres which contract very rapidly and strongly and fatigue quickly; they have relatively low levels of myoglobin and low numbers of mitochondria

CHECKPOINT

1. (a) Describe the roles of mitochondria and myoglobin in muscle fibres.
   (b) Describe how mitochondria and myoglobin vary between fast twitch muscle fibres and slow twitch muscle fibres.

   SKILLS CREATIVITY

2. ▸ Chickens are birds which spend much of their time walking around on the ground. If they are startled or frightened, they fly up almost vertically to escape. However, chickens don't fly for long. When a chicken is cooked, the breast meat is pale and the leg meat is much darker. Explain these observations in relation to your knowledge about different types of muscle fibre.

3. Draw up a table to compare fast twitch muscle fibres and slow twitch muscle fibres.

LEARNING OBJECTIVES

■ Understand the process of contraction of skeletal muscle in terms of the sliding filament theory, including the role of actin, myosin, troponin, tropomyosin, calcium ions (Ca^{2+}), ATP and ATPase.

As you saw in **Section 7B.2 fig A**, muscle fibres consist of many myofibrils lying parallel to each other. Each myofibril is made up of sarcomeres, the individual units of the muscle structure, and sarcomeres consist mainly of two proteins, actin and myosin, which overlap to form the characteristic stripes of striated muscle. Observations of micrographs show that whether a muscle fibre is contracted or relaxed, the dark A bands remain the same length. However, the light I bands and the H zone become shorter when a muscle fibre contracts and return to normal length when it relaxes again. This suggests that the two types of filaments slide over each other during contraction. This is the basis of the **sliding filament theory** (see **fig A**).

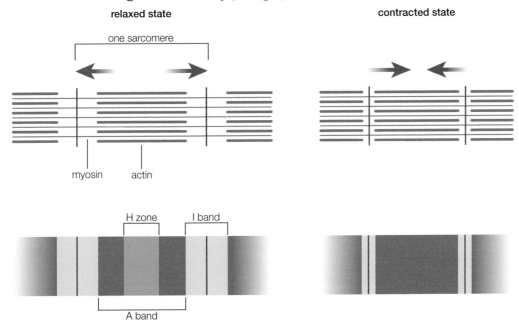

relaxed state

contracted state

one sarcomere

myosin   actin

H zone   I band

A band

▲ **fig A**   The sliding filament theory. As the fibres slide over each other, the light bands appear much smaller.

To understand how the sliding filament theory explains the way in which muscles contract, it is important to understand the structure of actin and myosin. A myosin molecule is made up of two long polypeptide chains twisted together; each one ends in a large, globular head which has ADP and inorganic phosphate molecules attached to it. In some circumstances, the head can act as an ATPase enzyme. A myosin filament is made up of lots of these molecules held together, with their heads sticking out from the filament. An actin filament is made up of two chains of actin monomers joined together like beads on a necklace. The shape of the actin molecule produces binding sites for myosin at regular intervals, in which the globular heads of the myosin molecules can fit. However, another long chain protein molecule called **tropomyosin** wraps around the double actin chains. In a relaxed muscle, the tropomyosin chain covers the myosin binding sites. The tropomyosin also has molecules of another protein, **troponin**, which are attached at regular intervals along the chain. **Fig B** shows some of the main features of these proteins.

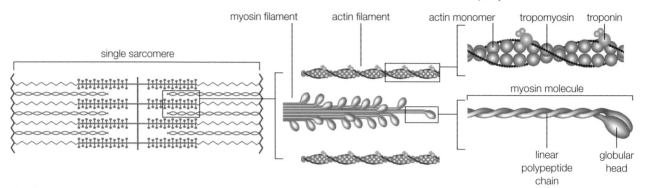

▲ **fig B** Understanding the structure of actin, myosin and the other associated protein molecules helps explain how the muscles contract.

ACTIN–MYOSIN INTERACTIONS

A ratchet mechanism has been proposed for the contraction of the myofibrils. This is summarised in **fig C**. Calcium ions are released from stores in the sarcoplasmic reticulum and attach to binding sites on the troponin molecules, changing their shape. As a result, the troponin molecules pull on the tropomyosin molecules, moving them away from the actin binding sites of the myosin molecules. The shape of the myosin molecule now allows it to attach to the actin, to form cross-bridges of **actomyosin**. This changes the molecular shape and pulls the actin filament across the myosin, and this increases the interlocking region and shortens the sarcomere. The bridges then break and the process is repeated between 50 and 100 times per second. The combined effect of this happening in each sarcomere is a shortening of the whole myofibril. The shortening of many myofibrils together results in the contraction of a muscle.

> **LEARNING TIP**
>
> You can imagine the myosin filament as a rowing boat and the myosin heads as oars. As the 'oars' make contact with the actin and swing they pull the actin filament past the myosin filament, increasing the overlap.

The diagram shows the actin and myosin unit before contraction starts. The myosin heads have ADP and P_i bound closely to them as well.

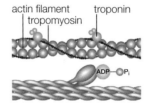

Calcium ions bind to the troponin molecules, changing their shape, so troponin molecules pull on the tropomyosin molecules they are attached to. This moves the tropomyosin away from the myosin binding sites, exposing them ready for action.

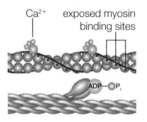

The myosin heads bind to the actin, forming an actomyosin bridge.

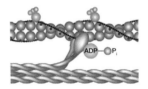

ADP and P_i are released from the myosin head. The myosin changes shape: the head bends forward moving the actin filament about 10 nm along the myosin filament, shortening the sarcomere.

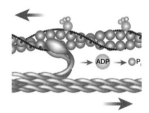

Free ATP binds to the head, causing another shape change in the myosin, so the binding of the head to the actin strand is broken. This activates ATPase in the myosin head, which also needs calcium ions to work. The ATP is hydrolysed, providing the energy to return the myosin head to its original position, primed with ADP and P_i, ready to go again.

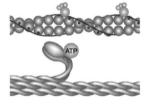

With continued stimulation, calcium ions remain in the sarcoplasm and the cycle is repeated. If not, calcium ions are pumped back into the sarcoplasmic reticulum using energy from ATP. The troponin and tropomyosin return to their original positions and the contraction is complete. The muscle fibre is relaxed.

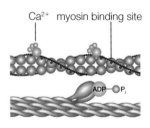

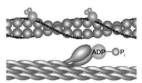

▲ **fig C** The proposed ratchet mechanism for the contraction of the myofibrils.

The ratchet mechanism needs both calcium ions and ATP to work. If we extract and then mix actin and myosin together in the lab, nothing happens. If we add ATP, actomyosin forms and contracts. This shows us that the process must be an active one. As electron micrographs become increasingly clear, they have shown the presence of 'bridges' between the actin and myosin strands. It appears that these bridges are formed and broken down during contraction. ATP is needed to break the actin–myosin bonds. It is also needed to return the calcium ions to the sarcoplasmic stores once contraction ends.

EXAM HINT

Many students think the ATP is used to make the filaments slide during the power stroke of the myosin head. In fact it is used to break the cross-bridges and move the myosin head back to its starting point.

CHECKPOINT

1. Draw and label the appearance of a sarcomere as you would expect it to look under an electron microscope:

   (a) fully contracted

   (b) fully relaxed.

2. Describe the role of calcium ions in the contraction of skeletal muscles.

3. (a) Explain why the presence of ATP is so important for the contraction of striated muscles.

   SKILLS ▸ CREATIVITY

   ▸ (b) Suggest how this explains what happens when rigor mortis sets in after death.

SUBJECT VOCABULARY

sliding filament theory the theory that the actin and myosin filaments overlap during muscle contraction

tropomyosin a long chain protein molecule that wraps around the double actin chains and covers the myosin binding sites in the relaxed state

troponin protein molecules attached to tropomyosin; they change shape when attached to calcium ions, pulling on the tropomyosin and revealing the myosin binding sites on the actin strands

actomyosin the chemical produced when cross-bridges form between actin and myosin during muscle contraction

LEARNING OBJECTIVES

■ Know the myogenic nature of cardiac muscle.

■ Understand how the normal electrical activity of the heart coordinates the heartbeat, including the roles of the sinoatrial node (SAN), the atrioventricular node (AVN), the bundle of His and the Purkyne fibres.

■ Understand how the use of electrocardiograms (ECGs) can aid in the diagnosis of abnormal heart rhythms.

Your heart beats continually throughout your life, with an average of about 70 beats per minute, although in small children the heart rate is much higher. Your heart can respond to need. When you are exercising, your tissues need more oxygen so your heart beats faster and supplies more blood to the tissues. The blood brings the glucose and oxygen which are needed by the rapidly respiring cells and removes the increased waste products. Stress can also raise the heart rate, whereas rest and relaxation can lower it.

HOW IS THE HEARTBEAT CONTROLLED?

In the very early embryo, cells which are destined to become the heart begin contracting rhythmically long before the organ itself forms. Cardiac muscle cells are **myogenic**, which means they contract without any external stimulus. They also have **intrinsic rhythmicity**. An adult heart removed from the body will continue to contract as long as it is bathed in a suitable oxygen-rich fluid. This intrinsic rhythm of the heart is around 60 beats per minute. This is slower than our hearts beat most of the time when we are awake because, as you will see, we have many different ways of controlling the heart to make sure it delivers the amount of blood we need, exactly when we need it.

The intrinsic rhythm of the heart is maintained by a wave of electrical excitation similar to a nerve impulse which spreads through special tissue in the heart muscle (see **fig A**).

The area of the heart with the fastest intrinsic rhythm is a group of cells in the right atrium known as the **sinoatrial node (SAN)**, and this acts as the heart's own natural pacemaker which keeps the heart beating regularly.

- The sinoatrial node establishes a wave of electrical excitation (depolarisation) which causes the atria to start contracting. This initiates the heartbeat.

- Excitation also spreads to another area of similar tissue called the **atrioventricular node (AVN)**.

- The AVN is excited as a result of the SAN but it produces a slight delay before the wave of depolarisation passes into the **bundle of His**, a group of conducting fibres in the septum of the heart. This makes sure the atria have stopped contracting before the ventricles start.

- The bundle of His splits into two branches and carries the wave of excitation on into the **Purkyne tissue**.

- The Purkyne tissue consists of conducting fibres that penetrate down through the septum, spreading around the ventricles.

As the depolarisation travels through the tissue it starts the contraction of the ventricles, starting at the bottom and so squeezing blood out of the heart.

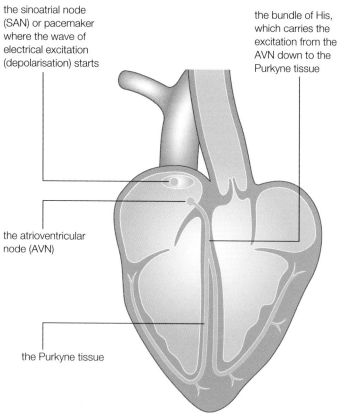

the sinoatrial node (SAN) or pacemaker where the wave of electrical excitation (depolarisation) starts

the bundle of His, which carries the excitation from the AVN down to the Purkyne tissue

the atrioventricular node (AVN)

the Purkyne tissue

▲ **fig A**   The area of the heart with the fastest intrinsic rhythm is a group of cells in the right atrium known as the sinoatrial node. It acts as the heart's own natural pacemaker.

EXAM HINT

Remember that the contraction of the ventricles starts at the apex or bottom of the chambers so that blood is squeezed upwards towards the main arteries.

The speed at which the excitation spreads through the heart, with the hesitation before the AVN stimulates the bundle of His, makes sure that the atria have stopped contracting before the ventricles start. It is these changes in the electrical excitation of the heart that cause the repeating cardiac cycle. These electrical changes can be measured in an **electrocardiogram (ECG)** (see **fig B**).

Because your heart has its own basic rhythm, you do not have to think about it. You don't waste body resources on maintaining a vital but continuous event. However, many people have a faster resting heart rate than this basic rhythm; the average is around 70 beats per minute. This is because lots of other factors, including nerve impulses and hormones, constantly affect the heart rate, as you will learn in **Section 7B.7**.

USING ELECTROCARDIOGRAMS

An ECG is used to investigate the rhythms of the heart by producing a record of the electrical activity of the heart. As you know, the rhythm of the heart results from the spread of a wave of depolarisation (electrical activity) through specialised tissue within the heart muscle itself. This depolarisation in the heart causes tiny electrical changes on the surface of your skin. An ECG measures these changes at the surface of your skin.

To take an ECG, 12 electrodes and leads are attached to your body. This is a completely painless process. Your skin is wiped with ethanol to remove any grease or sweat so the electrodes can make good contact with your skin. Sometimes a special gel is applied to the electrodes to make sure they conduct electricity as effectively as possible. Each electrode records information during the electrocardiogram, effectively giving 12 views of the heart.

An ECG can show you what is happening in a normal, healthy heart. However, it is often used to indicate different heart conditions and to monitor patients with heart disease. An ECG is usually done with the patient lying down and resting, but sometimes it is conducted while a patient is exercising (known as a stress test) because some heart conditions appear only during exercise. You can see some ECGs in **fig B**.

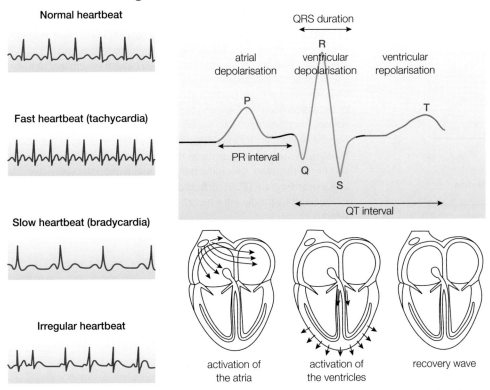

fig B The relationship between the electrical events in the heart and the heartbeat as shown by ECGs. Here you can see both normal and abnormal patterns.

It is possible to collect a wide range of measurements to produce a diagram showing how the electrical activity recorded in an ECG and pressure changes in different chambers of the heart work together during the cardiac cycle (you learned about the cardiac cycle in **Section 1B.4 (Book 1: IAS)**). In **fig C**, you can see how differences in pressure in the regions of the left side of the heart cause the different valves to close. The same process is happening on the right-hand side of the heart at exactly the same time. You can also see how the stages of the cardiac cycle relate to the pressure changes and an ECG showing the electrical activity of the heart.

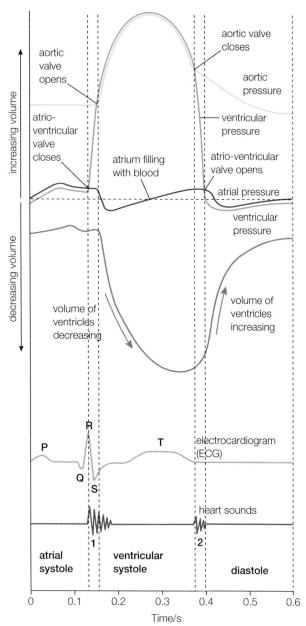

▲ **fig C** Electrical and pressure changes in the heart during the cardiac cycle and the causes of the heart sounds.

CHECKPOINT

1. (a) Describe how the natural intrinsic rhythm of the heart is maintained.

   (b) Explain why the resting heart rate is usually higher than the intrinsic heart rate.

2. Explain how an artificial heart pacemaker, which delivers a regular electric shock to the right atrium, can help maintain a steady heart rate in people when their natural pacemaker is no longer working properly.

   SKILLS **ANALYSIS**

3. ▶ Describe how the events of the heartbeat are shown on an ECG and explain how an ECG can help in the diagnosis of abnormal heart rhythms.

SUBJECT VOCABULARY

myogenic contracts without an external stimulus

intrinsic rhythmicity the intrinsic (internal) rhythm of contraction and relaxation in the cardiac muscle of the heart

sinoatrial node (SAN) a specialised group of cells in the right atrium with the fastest natural intrinsic rhythm; it generates a regular electrical signal and acts as the heart's own natural pacemaker to keep the heart beating regularly

atrioventricular node (AVN) a group of cells stimulated by the wave of excitation from the SAN and the atria; it imposes a delay before transmitting the impulse to the bundle of His

bundle of His a group of conducting fibres in the septum of the heart

Purkyne tissue conducting fibres that penetrate down through the septum of the heart, spreading between and around the ventricles

electrocardiogram (ECG) technology used to investigate the rhythms of the heart by producing a record of the electrical activity of the heart

Mammalian cells are very sensitive to change. Whatever happens in the life of a mammal, for example during high levels of physical exercise or in extremes of external temperatures, the internal conditions of the body must be controlled within a narrow range. This is a dynamic equilibrium and involves matching the supply of oxygen and glucose to the continually changing demands of the body. At the same time, carbon dioxide must be removed and an even temperature and pH maintained. Maintaining a state of dynamic equilibrium through the responses of the body to external and internal stimuli is known as **homeostasis**. The main homeostatic mechanisms in mammals include systems that respond to changes in both external and internal conditions to control the pH, temperature and water potential of the body within narrow limits.

The pH levels of the body must be maintained so that the structures of protein molecules remain stable. This allows enzymes to function at their optimum activity and the structure of cell membranes to be maintained. You saw how the transport of carbon dioxide away from the tissues of the body controls the pH of the blood and body fluids in **Section 1B.2 (Book 1: IAS)**.

The core temperature of the body needs to be stable to maintain the optimum activity of the enzymes that control the rate of cellular reactions. As you learned in **Sections 2A.1** and **2A.2 (Book 1: IAS)**, a stable temperature is very important to maintain the structure and function of the cell membranes, so they can control the movement of substances into and out of the cells. In humans, this temperature is around 37 °C.

The water potential of the body fluids must also remain within narrow limits to avoid osmotic effects that could damage or destroy the cells. You learned about the effects of osmotic changes in **Section 2A.3 (Book 1: IAS)**.

HOMEOSTASIS

Homeostasis involves a high level of coordination and control. Your nervous and chemical control systems interact to maintain a dynamic equilibrium in the body. **Sensors/receptors** detect changes in the body. They send messages to **effectors** that either work to reverse the change or to increase it using a number of different feedback systems. Effectors are usually muscles or glands.

FEEDBACK SYSTEMS

The communication in a homeostatic feedback system may be by hormones (chemical messengers) or by nerve impulses (electrical messages). You will learn more about nerves and hormones in **Topic 8**.

There is often a small overshoot or undershoot (going beyond or not reaching the ideal level) as a feedback system corrects, and that also needs to be corrected, so the levels of most body systems fluctuate (vary between values above and below) slightly around the ideal level in a dynamic equilibrium. In some cases, there are separate mechanisms for controlling changes in different directions; this gives a particularly sensitive response and a great degree of control. An example is the control of blood sugar levels.

NEGATIVE FEEDBACK SYSTEMS

Most of the feedback systems in mammals are **negative feedback systems** (see **fig A**). They provide a way to maintain a condition, such as the concentration of a substance, within a narrow range. The receptors detect a change in conditions; this results in effectors being stimulated to restore the equilibrium. If the concentration goes up, the effectors bring it down again, and vice versa. Examples of negative feedback loops are seen in the production of many hormones and in temperature regulation in mammals (see **Chapter 7C**).

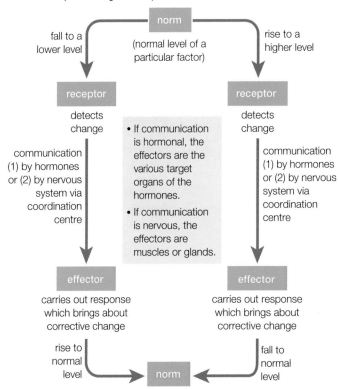

▲ **fig A**   This is a generalised diagram of the type of negative feedback system that is so important in homeostasis.

POSITIVE FEEDBACK SYSTEMS

In **positive feedback systems**, effectors work to increase the effect that has triggered the response. These systems are less common in biological systems and play little part in homeostasis. One example is the contractions of the uterus during labour (giving birth). The pressure of the baby's head on the cervix causes the release of chemicals that increase the contraction of the uterus, so the head is then pushed down even harder (see **fig B**).

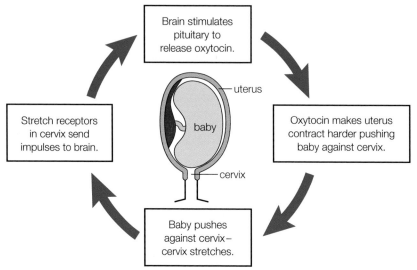

▲ **fig B** The dilation of the cervix and the delivery of a baby. This is one of a small number of examples of positive feedback systems in mammals.

SKILLS   CRITICAL THINKING

CHECKPOINT

1. Define *homeostasis* and explain why it is so important in living organisms.

2. In the past, *homeostasis* was described as the maintenance of a steady internal state. Give the more modern description of homoeostasis and explain why it has been changed.

3. Suggest why there are many negative feedback systems in mammals and relatively few positive feedback systems.

SUBJECT VOCABULARY

homeostasis the maintenance of a state of dynamic equilibrium in the body, despite changes in the external or internal conditions

sensors/receptors specialised cells that are sensitive to specific changes in the environment

effectors systems (usually muscles or glands) that work to reverse, increase or decrease changes in a biological system

negative feedback systems systems for maintaining a condition, such as the concentration of a substance, within a narrow range; receptors detect a change in conditions and, as a result, effectors are stimulated to restore the equilibrium

positive feedback systems systems in which effectors work to increase an effect that has triggered a response

LEARNING OBJECTIVES

■ Understand what is meant by the term homeostasis and its importance in maintaining the body in a state of dynamic equilibrium during exercise, including the role of the hypothalamus in thermoregulation.
■ Be able to calculate cardiac output.
■ Understand how variations in ventilation and cardiac output enable rapid delivery of oxygen to tissues and the removal of carbon dioxide from them, including how the heart rate and ventilation rate are controlled and the roles of the cardiovascular control centre and the ventilation centre in the medulla oblongata.
■ Understand the role of adrenaline in the fight or flight response.

Homeostasis plays an important role during exercise, a situation in which conditions in the body are changing rapidly and demands on the systems are high. When you want to move somewhere fast, your body must respond quickly to supply your muscles with the glucose and oxygen they need and remove excess carbon dioxide. Negative feedback systems are vital in the coordinated response enabling you to exercise effectively, as both your heart and your respiratory system respond.

RESPONDING TO DEMAND

The fine control of the heart rate in mammals demonstrates clearly the importance of having different receptors to detect (a) internal changes, (b) the ways in which the **parasympathetic nervous system** and **sympathetic nervous system** work together in a complementary way, and (c) the interactions between nervous and hormonal control.

In **Section 7B.5**, you learned how the intrinsic rhythm of the heart is controlled by impulses initiated in the sinoatrial node (SAN) that spread through the atrioventricular node (AVN) and the bundle of His to give a regular, coordinated heartbeat. But this intrinsic rhythm cannot cope with changes in demand. For example, during exercise, more oxygen must be carried to the rapidly respiring muscle tissues, and the waste carbon dioxide and lactate which accumulate in the muscle fibres need to be removed. Once the exercise stops, your body must return to normal. The response of the heart to these changes in demand is the result of a number of negative feedback systems.

CHANGING CARDIAC OUTPUT

When your body demands more glucose and oxygen the heart can respond in two ways. The rate at which the heart beats can increase. Also, the **cardiac volume** (amount of blood pumped at each heartbeat) can be increased by a more efficient contraction of the ventricles. The combination of these two factors gives a measure called the **cardiac output**.

$$\text{cardiac output} = \text{cardiac volume} \times \text{heart rate}$$
$$(\text{dm}^3\,\text{min}^{-1}) \qquad (\text{dm}^3) \qquad (\text{beats}\,\text{min}^{-1})$$

In a normal individual at rest, the heart beats about 70 times a minute, and pumps between 4 and 6 dm³ of blood per minute. In

a trained athlete, the resting heart beats more slowly, at around 60 beats per minute. When a fit individual anticipates exercise, the heart rate begins to increase before the exercise begins. The cardiac volume increases more slowly, as it becomes clear from the changes in the body that the exercise is going to continue. Cardiac output during exercise can increase to around 30 dm³ min⁻¹. Without these changes, a high level of physical effort, such as running, whether in a hot or a cold environment, would be impossible (see **fig A**).

▲ **fig A** The demands made on the body of a mammal change when it moves fast. Several feedback systems make sure the mammalian heart responds appropriately, whatever the external temperature.

ADJUSTING THE HEART RATE

Different control systems enable the heart to respond to the varying demands of your body throughout the day.

NERVOUS CONTROL OF THE HEART

Most of the nervous control of your heart is by the autonomic (involuntary) nervous system, so you do not have to think about it. The **cardiovascular control centre** is situated in the medulla oblongata of the brain (see **Section 8B.1**). It plays a major part in controlling changes in the heart rate and the volume of blood pumped with each heartbeat in response to changes in the internal environment.

Chemical, stretch and pressure receptors in the lining of the blood vessels and the chambers of the heart send nerve impulses to the cardiovascular control centre. It responds by sending impulses to the heart along the sympathetic or parasympathetic nerves (see bullets points below). The heart muscle responds to these impulses so controlling the rate at which it beats.

Thus, control of the heart is via the **autonomic (involuntary) nervous system**; it is divided into two parts.

- The sympathetic nervous system is usually *excitatory*; for example, it speeds up the heart rate.
- The parasympathetic system is usually *inhibitory*; for example, it slows down the heart rate.

Most of the body organs are supplied by both types of nerves, giving a level of fine control.

Nerve impulses that travel down the sympathetic nerve from the cardiovascular control centre to the heart release **noradrenaline** to stimulate the SAN. This increases the frequency of the signals from the pacemaker region, so that the heart beats more quickly. Branches of this sympathetic nerve also pass into the ventricles, so they also increase the force of contraction. In contrast, nerve impulses in the corresponding parasympathetic nerve release **acetylcholine (ACh)**, inhibiting the SAN and slowing the heart down (see **fig B** and **fig C**).

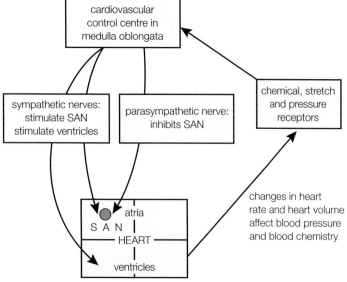

▲ **fig B** The cardiovascular control centre in the medulla oblongata controls the heart rate via parasympathetic and sympathetic nerve stimulation.

THE ROLE OF BARORECEPTORS

Baroreceptors are mechanoreceptors in the carotid arteries in the neck and on the aorta that are sensitive to pressure changes. They are important in the feedback control of the heart rate during exercise (see **fig C**). At rest, they send a steady stream of signals back through sensory neurones to the cardiovascular control centre in the brain. At the beginning of exercise, the blood vessels dilate (vasodilation) becoming wider in response to the hormone **adrenaline**, which is released in anticipation of exercise. As a result of this vasodilation, the blood pressure falls a little. This reduces the stretch on the baroreceptors and they almost stop responding. With this reduced stimulation from the baroreceptors, the cardiovascular control centre immediately sends signals along the sympathetic nerve to stimulate the heart rate and increase the blood pressure again by constricting the blood vessels. When exercise stops, blood pressure in the arteries increases as the heart continues to pump harder and faster than it needs to, and so the baroreceptors are stretched. They respond by sending more sensory nerve impulses to the cardiovascular control centre which consequently sends impulses through the parasympathetic system to slow down the heart rate and cause the blood vessels to become wider. These two actions lower the blood pressure again.

EXAM HINT

The heart rate is altered to deliver more or less blood. This would be ineffective if the blood vessels did not also respond by dilating or constricting in a coordinated response.

The balance of impulses that pass to the heart from the cardiovascular control centre is affected by inputs from a number of different sensory receptors in the main arteries and in the heart itself.

THE ROLE OF CHEMORECEPTORS IN THE AORTA

The walls of the aorta and carotid arteries contain **chemoreceptors** as well as baroreceptors (see **fig D**). These are sensitive to the levels of carbon dioxide in the blood. As carbon dioxide levels go up, the pH of the blood goes down and this is detected by the aortic and carotid chemoreceptors. They send impulses along sensory neurones to the cardiovascular control centre in the medulla oblongata, which increases the number of impulses travelling down the sympathetic nerve to the heart. This results in an increased heart rate, increasing the blood flow to the lungs and so more carbon dioxide is removed from the blood. As blood carbon dioxide levels fall, the blood pH rises. The chemoreceptors respond to this by reducing the number of impulses to the cardiovascular centre. This reduces the number of impulses sent along the sympathetic nerve to the heart and reduces the acceleration of the heart rate so it returns to the intrinsic rhythm. The chemoreceptors also play a role in the control of the breathing rate (see **Section 7B.8**).

We also have a certain amount of conscious control over our heart rate, so nerves from the conscious areas of our brains can also stimulate or inhibit the SAN. Some people can slow their heart rate right down just by concentrating on it.

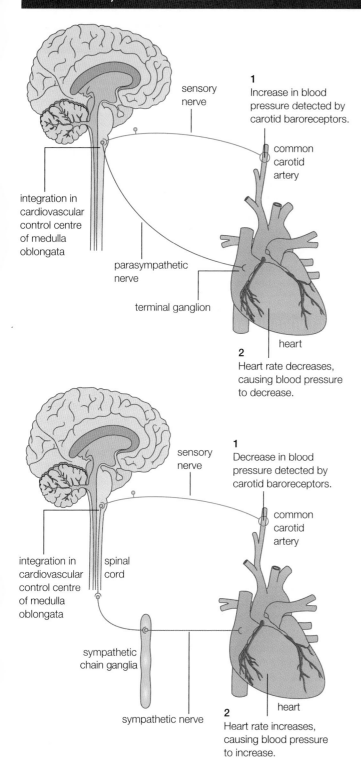

1
Increase in blood pressure detected by carotid baroreceptors.

sensory nerve

common carotid artery

integration in cardiovascular control centre of medulla oblongata

parasympathetic nerve

terminal ganglion

heart

2
Heart rate decreases, causing blood pressure to decrease.

1
Decrease in blood pressure detected by carotid baroreceptors.

sensory nerve

common carotid artery

integration in cardiovascular control centre of medulla oblongata

spinal cord

sympathetic chain ganglia

sympathetic nerve

heart

2
Heart rate increases, causing blood pressure to increase.

▲ **fig C**   A negative feedback system for controlling the heart through the baroreceptors. This is just one of the many complex interactions that make sure the output of your heart matches the demands of your body.

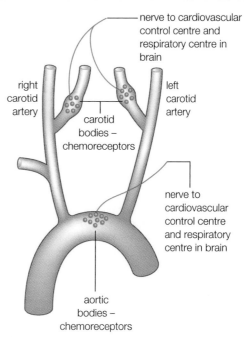

nerve to cardiovascular control centre and respiratory centre in brain

right carotid artery

left carotid artery

carotid bodies – chemoreceptors

nerve to cardiovascular control centre and respiratory centre in brain

aortic bodies – chemoreceptors

▲ **fig D**   The chemoreceptors in the aorta and carotid arteries

HORMONAL CONTROL OF THE HEART

Your heart beats faster when you are nervous, frightened or excited, and if you are anticipating exercise. This happens even if you are sitting still at the time, so the change is not in response to exercise. When you are stressed, the sympathetic nerve stimulates the adrenal medulla to release the hormone adrenaline. You will learn more about hormones and the way they work in **Chapter 7C**. Adrenaline is very similar to the noradrenaline released in the synapses of the sympathetic nervous system. It is carried around the body in the blood and binds to receptors in the target organs, including the SAN. Adrenaline stimulates the cardiovascular control centre in the brain, increasing the impulses in the sympathetic neurones which supply the heart. It also has a direct effect on the SAN by increasing the frequency of excitation and so the heart rate increases. This supplies you with extra oxygen and glucose for your muscles and brain, in case you need to run away or stand and fight. Adrenaline is responsible for the response of the body in stressful situations. This is often called the 'fight or flight' reaction (flight is the act of running away).

ADDITIONAL RESPONSES

The response of the body to exercise or stress is complex. During exercise, impulses from the cardiovascular control centre travel to other effectors as well as the heart. At the same time that the sympathetic system sends many impulses to the heart to speed it up, it sends fewer impulses to many blood vessels. This results in the contraction of the smooth muscles lining the vessels, thus narrowing or closing the vessels. In this way, the blood flow is diverted from areas which are temporarily less important, to provide more blood for the heart and the muscles to use. As you can see from **table A**, the blood supply to the brain is fairly constant, but the amounts of blood flowing elsewhere in the system vary considerably. The changes seen in **fig E** and **table A** are all the result of a combination of nervous and hormonal responses affecting the heart, the blood vessels and a range of body systems.

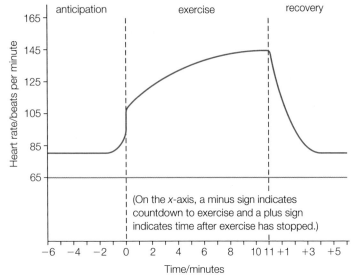

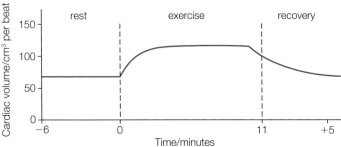

▲ **fig E** The impact of exercise on the heart rate and cardiac volume

STRUCTURE	AT REST		VIGOROUS EXERCISE	
	cm³ min⁻¹	% OF TOTAL	cm³ min⁻¹	% OF TOTAL
heart	190	3.3	740	3.9
liver	1340	23.5	590	3.1
adrenal glands	24	0.4	24	0.1
brain	690	12.1	740	3.9
lung tissue	100	1.8	200	1.0
kidneys	1050	18.4	590	3.1
skeletal muscles	740	13.0	12450	65.9
skin	310	5.4	1850	9.8
other parts	1256	22.0	1716	9.1
total blood flow	5700		18900	

table A Redistribution of blood flow in response to exercise.

CHECKPOINT

1. Draw a clear flow diagram to show the negative feedback system involving the baroreceptors that is used in the response of the heart to exercise.

SKILLS ANALYSIS

2. ▶ Use the data from **fig E** to produce a graph to show what would be expected to happen to cardiac output (in dm³ per minute) during a period of rest, followed by 10 minutes of exercise and another 10 minutes of recovery time.

3. (a) Draw bar charts or pie charts to show the difference in the percentage blood flow to different areas of the body at rest and during exercise using data from **table A**.

(b) Explain why these differences occur.

4. Summarise how nervous and hormonal controls ensure the output of the heart is matched to the demands of the body.

SUBJECT VOCABULARY

parasympathetic nervous system involves autonomic motor neurones which produce acetylcholine as their neurotransmitter and often have a relatively slow, inhibitory effect on an organ system; these neurones have very long myelinated preganglionic fibres that leave the CNS and synapse in a ganglion very close to the effector organ; postganglionic fibres are very short and unmyelinated

sympathetic nervous system involves autonomic motor neurones which produce noradrenaline as their neurotransmitter and often have a rapid response, activating an organ system; these neurones have very short myelinated preganglionic fibres that leave the CNS and synapse in a ganglion very close to the CNS; postganglionic fibres are long and unmyelinated

cardiac volume the volume of blood pumped at each heartbeat

cardiac output a measure of the volume of blood pumped by the heart per minute, calculated by multiplying cardiac volume by heart rate

cardiovascular control centre centre in the medulla oblongata of the brain that receives information from a number of different receptors and controls changes to the heart rate and the cardiac volume through parasympathetic and sympathetic nerves

autonomic (involuntary) nervous system the involuntary nervous system; autonomic motor neurones control bodily functions that the conscious area of the brain does not normally control

noradrenaline neurotransmitter in the sympathetic nervous system and adrenergic synapses of the brain

acetylcholine (ACh) neurotransmitter in the parasympathetic nervous system, the synapses of motor neurones, and cholinergic synapses in the brain

baroreceptors mechanoreceptors in the aorta and carotid arteries that are sensitive to pressure changes

adrenaline hormone produced by the adrenal glands which stimulates the fight or flight response

chemoreceptors sensory nerve cells (or organs) that respond to chemical stimuli

LEARNING OBJECTIVES

- Understand how variations in ventilation and cardiac output enable rapid delivery of oxygen to tissues and removal of carbon dioxide from them, including how the heart rate and ventilation rate are controlled and the roles of the cardiovascular control centre and the ventilation centre in the medulla oblongata.
- Understand what is meant by the term homeostasis and its importance in maintaining the body in a state of dynamic equilibrium during exercise, including the role of the hypothalamus in thermoregulation.

When you exercise, the muscle cells (including the heart muscle) need much more oxygen. They also produce more carbon dioxide and lactic acid. Oxygen is brought into your body, and carbon dioxide is removed, through your lungs. Breathing in and out controls how much air moves into and out of your lungs, and so it controls the oxygen supply and the rate of carbon dioxide removal. The respiratory system has to respond during exercise. Homeostasis is maintained as a result of more negative feedback systems.

COMPONENTS OF THE LUNG VOLUME

In **Section 2A.6 (Book 1: IAS)**, you looked at breathing and gas exchange. To understand in more detail the way breathing is controlled, you need to understand more about how air flows into and out of the lungs.

A certain amount of air is always present in your gas exchange system, filling up the spaces when no air is flowing. Apart from this, the volume of air that enters and leaves the respiratory system is very variable. Most of the components of the lung volume are given specific names (see **fig A**).

- **Tidal volume (VT)** is the volume of air that enters and leaves the lungs at each natural resting breath.

- **Inspiratory reserve volume (IRV)** is the volume of air that you can take in above the normal inspired tidal volume. In other words, this is the extra air that you can take in when you breathe in as deeply as possible after a normal inspiration.

- **Expiratory reserve volume (ERV)** is the volume of air that you can force out above the normal expired tidal volume. This is the extra air you breathe out when you force the air out of your lungs as hard as possible after a normal expiration.

- **Vital capacity (VC)** is the total of the tidal volume and the inspiratory and expiratory reserves. It is the volume of air which you can breathe out by breathing out as hard as you can and then breathing in as deeply as possible.

- **Residual volume (RV)** is the volume of air left in the lungs after the strongest possible expiration.

- **Total lung capacity (TLC)** is the sum of the vital capacity and the residual volume.

- **Inspiratory capacity (IC)** is the volume that can be inspired from the end of a normal expiration; IC = VT + IRV.

EXAM HINT

Be ready to label these volumes and capacities on a trace produced from a spirometer. The trace is a graph so be sure to read the axes carefully.

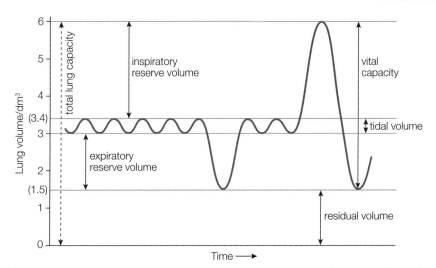

▲ **fig A** This diagram, based on spirometer traces, gives you an idea of the average lung volumes at rest. However, values between men and women, and athletes and non-athletes, can vary considerably.

In a normal person at rest, tidal volume is about 0.5 dm³ or about 15% of the vital capacity of the lungs. The rate of breathing can be expressed as the **ventilation rate**. This is a measure of the volume of air breathed in a minute:

ventilation rate = tidal volume × frequency of inspiration
(dm³ min⁻¹)   (dm³)   (min⁻¹)

The ventilation of the lungs needs to supply all the oxygen that is required by the tissues of the body whatever they are doing, and also remove all the waste carbon dioxide. The ventilation rate is affected by two things: the amount of air you take into your lungs at each breath and the number of times you breathe per minute. For example, the tidal volume can increase from 15% of the vital capacity to 50% during heavy exercise. The frequency of inspiration shows similar increases.

PRACTICAL SKILLS   CP17

To investigate the effect of exercise on characteristics such as tidal volume, ventilation rate and oxygen consumption, we need a way of measuring and recording the air taken in and out of the lungs. To do this, we can use a spirometer (see **fig B**). Spirometers come in a wide variety of shapes and sizes but they all work in a similar way. Some spirometers are portable and can be used during exercise outside, but many are lab-based pieces of equipment. They may be linked directly to a computer or to a simple revolving drum system (a kymograph). We can interpret spirometer traces and use them to calculate how the breathing system responds to exercise.

The subject of the experiment breathes in and out of the airtight chamber making it move up and down, until all the oxygen is used up.

Revolving drum on which a trace is drawn out as the lid moves up and down.

Airtight chamber – in this case a Perspex lid floating on water. The chamber is filled with oxygen at the beginning of the experiment. Attached to the lid of the chamber is an arm with a pen on the end.

Canister of soda lime to remove carbon dioxide from the exhaled air. This is important because carbon dioxide levels affect the rate of breathing and therefore, if carbon dioxide was allowed to accumulate, the investigation would be affected.

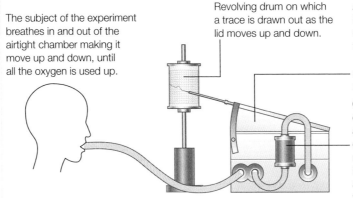

 fig B We can measure the volume of gas inhaled and exhaled under a variety of conditions using a spirometer.

Safety Note: A spirometer should only be set up and filled with oxygen by an experienced teacher or technician who will then supervise its use by students ensuring each uses a new clean mouthpiece.

CONTROL AND REGULATION OF BREATHING

The oxygen needs of the body can change rapidly from the relatively low levels needed at rest to the high levels demanded during strong exercise. The amount of carbon dioxide the body needs to remove changes in a similar way. The ventilation rate adjusts to your activity levels. Both the tidal volume and the frequency of inspiration change to keep the concentration of gases in your blood as close to the ideal level as possible. This control is the result of several feedback systems.

The basic stimulus to inhale and exhale (to breathe in and out) comes from an area of the medulla known as the **ventilation centre**. It involves a feedback system based on the stretching of the bronchi during breathing (see **fig C**).

The ventilation centre contains an **inspiratory centre** which controls breathing in. This is often just referred to as the ventilation centre as it is the main factor in the control of breathing. There is also an **expiratory centre** which controls forced exhalation. Impulses from the ventilation centre travel along sympathetic nerves and cause the intercostal muscles and the diaphragm to contract. This makes us inhale. As the lungs inflate (fill with air), stretch receptors in the walls of the bronchi send nerve impulses to the ventilation centre. The more these receptors are stretched, the more rapidly they send impulses. Eventually, these impulses inhibit the ventilation centre and it stops stimulating the breathing muscles. You stop breathing in and as the muscles relax, you exhale. This gives a basic, deep, slow breathing rhythm, comparable to the resting rhythm of the heart.

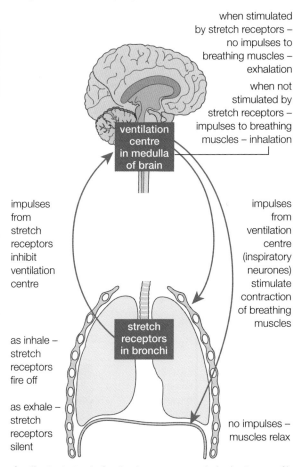

when stimulated
by stretch receptors –
no impulses to
breathing muscles –
exhalation

when not
stimulated by
stretch receptors –
impulses to breathing
muscles – inhalation

ventilation
centre
in medulla
of brain

impulses
from
stretch
receptors
inhibit
ventilation
centre

impulses
from
ventilation
centre
(inspiratory
neurones)
stimulate
contraction
of breathing
muscles

stretch
receptors
in bronchi

as inhale –
stretch
receptors
fire off

as exhale –
stretch
receptors
silent

no impulses –
muscles relax

▲ **fig C** A simple feedback system controls the basic rate of breathing. The breathing muscles are the muscles between the ribs (the intercostal muscles) and the diaphragm.

While we are awake the conscious areas of the brain can take control of breathing. We are all capable of choosing to hold our breath (but only for a certain time), take a deep breath or breathe faster if we want to. However, most of the time, and always when we are unconscious, our breathing is completely under automatic control.

BREATHING AND HOMEOSTASIS

As the demands of the body change, inputs from other receptors interact with the basic respiratory rhythm to give a finely adjusted response to most situations, including stress, exercise and oxygen deprivation.

The main stimulus affecting the breathing rate is the level of carbon dioxide in the blood. An increase in carbon dioxide concentration (and the consequent fall in pH) leads to an increase in both the rate and the depth of breathing (see **fig D**). This is because the diaphragm and the intercostal muscles contract harder and more frequently. A fall in carbon dioxide concentration has the opposite effect.

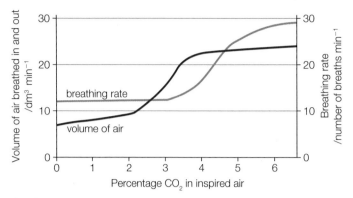

▲ **fig D** This graph shows the effect of carbon dioxide concentration on both the rate and depth of breathing.

As soon as exercise starts, the cortex of your brain consciously recognises movement has begun and sends impulses to stimulate the ventilation centre in the medulla. This stimulates the respiratory muscles and increases the rate and depth of ventilation. Chemoreceptors sensitive to the level of carbon dioxide and the pH of the blood (in the medulla, the carotid bodies in the carotid arteries and the aortic bodies in the aortic arch) send impulses back to the main ventilation centre following rises in carbon dioxide levels. Impulses are then sent out to the breathing muscles (the effectors) so the breathing rate changes. It speeds up in a negative feedback system which removes the extra carbon dioxide and at the same time matches the oxygen needs of the body (see **fig E**). The chemical sensors and the stretch receptors in the muscles and lungs all act on the respiratory centre to maintain increased ventilation rates until the exercise is finished and there is no longer an oxygen debt.

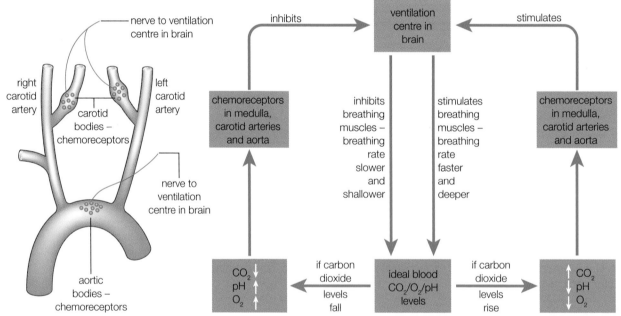

▲ **fig E** This negative feedback system removes carbon dioxide and makes sure your body is supplied with the oxygen it needs, whatever you are doing.

CHECKPOINT

1. Describe how breathing rate in humans can be investigated experimentally.
2. ▶ Produce a diagram to summarise all the different control feedback systems in breathing.
3. Outline the factors that are most likely to affect breathing rate, and explain your answer.

SKILLS   INTERPRETATION

SUBJECT VOCABULARY

tidal volume (VT) the volume of air that enters and leaves the lungs at each natural resting breath

inspiratory reserve volume (IRV) the volume of air that you can take in above the normal inspired tidal volume

expiratory reserve volume (ERV) the volume of air that you can force out above the normal expired tidal volume

vital capacity (VC) the total of the tidal volume and the inspiratory and expiratory reserves

residual volume (RV) the volume of air left in the lungs after the strongest possible expiration

total lung capacity (TLC) the sum of the vital capacity and the residual volume

inspiratory capacity (IC) the volume that can be inspired from the end of a normal expiration;
IC = VT + IRV

ventilation rate a measure of the volume of air breathed per minute, calculated by multiplying tidal volume by breathing frequency

ventilation centre centre in the medulla oblongata of the brain that receives information from a number of different receptors and controls changes to the rate at which you breathe

inspiratory centre main area of the ventilation centre involved in the control of breathing in; breathing out is usually passive

expiratory centre region of the ventilation centre involved in voluntary exhalation (breathing out)

AN ARTIFICIAL PACEMAKER

SKILLS INTERPRETATION, DECISION MAKING, ADAPTIVE LEARNING, CREATIVITY, PERSONAL AND SOCIAL RESPONSIBILITY

The sinoatrial node (SAN) of the heart produces a regular rhythm of contractions throughout life. However if, for any reason, the SAN fails to maintain a regular heart rate, an individual may need an artificial pacemaker inserted. The text below is based on content from a website produced by a national health provider.

HEALTH PROVIDER WEBSITE

Pacemaker implantation is a surgical procedure where a small electrical device called a pacemaker is implanted in your chest. The pacemaker sends regular electrical pulses that help keep your heart beating regularly.

Having a pacemaker fitted can greatly improve your quality of life if you have problems with your heart rhythm, and the device can be lifesaving for some people.

Pacemaker implantation is one of the most common types of heart surgery carried out globally. Rates of implantations of cardiac pacemakers are rising steadily and the rate increases with the age of the population.

How does a pacemaker work?
The pacemaker is a small metal box weighing 20–50 g. It is attached to one or more wires, known as pacing leads, which run to your heart.

The pacemaker contains:

- *a battery* that usually lasts for 6 to 10 years depending on how advanced the device is (more advanced pacemakers tend to use more energy so have a shorter battery life)
- *a pulse generator*
- *a tiny computer circuit* that converts energy from the battery into electrical impulses, which flow down the wires and stimulate your heart to contract.

The rate at which these electrical impulses are sent out is called the discharge rate.

Almost all modern pacemakers work on demand. This means that they can be programmed to adjust the discharge rate in response to your body's needs. If the pacemaker senses that your heart has missed a beat or is beating too slowly, it sends signals at a steady rate. If it senses that your heart is beating normally by itself, it does not send out any signals.

Most pacemakers have a special sensor that recognises body movement or your breathing rate. This allows them to speed up the discharge rate when you are active. Doctors describe this as rate responsive.

Why do I need a pacemaker?
The heart is essentially a pump, made of muscle, which is controlled by electrical signals. These signals can become disrupted for several reasons, which can lead to a number of potentially dangerous heart conditions. Such conditions include:

- *an abnormally slow heartbeat* (bradycardia)
- *an abnormally fast heartbeat* (supraventricular tachycardia) caused by damage to part of the heart called the sinoatrial node
- *heart block* (when your heart beats irregularly because the electrical signals that control it are not transmitted properly)
- *cardiac arrest* (when a problem with the electrical signals in the heart causes the heart to stop beating altogether).

An implantable cardioverter defibrillator (ICD) is a device similar to a pacemaker. This sends a larger electrical shock to the heart that essentially reboots the heart to get it pumping again. Some devices contain both a pacemaker and an ICD.

ICDs are often used as a preventative treatment for people thought to be at risk of cardiac arrest at some point in the future. If the ICD senses that the heart is beating at a potentially dangerous abnormal rate, it will deliver an electrical shock to the heart. This can often help return the heart to a normal rhythm.

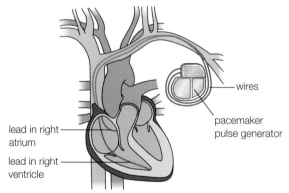

lead in right atrium

lead in right ventricle

wires

pacemaker pulse generator

▲ **fig A** A heart pacemaker can mean the difference between life and death for people with conditions such as sick sinus syndrome. This is when the heart stops for seconds at a time but also has severe and repeated tachycardia. The SAN is diseased and only fires around 40 times a minute.

Source: *http://www.nhs.uk/conditions/pacemakerimplantation/pages/introduction.aspx*

SCIENCE COMMUNICATION

This information on how a heart pacemaker works comes from a website which aims to provide clear and straightforward explanations on medical conditions and procedures for patients and their relatives and friends.

1 (a) In what ways have the writers achieved their aim of producing a clear explanation of how a heart pacemaker works?

   (b) The way this is worded makes it a suitable article for someone about to have a heart pacemaker fitted. Do you agree or disagree with this statement? Why?

   (c) Think about the level of technical language in this piece. In the original, there are no pictures (an image has been added for your benefit). Discuss the impact on the usefulness of the article without the picture.

BIOLOGY IN DETAIL

Now, use your knowledge of the way the heart beats and the natural pacemaker system of the heart to answer the following questions.

2 How do doctors detect the type of heart arrhythmias which might need an artificial pacemaker?

3 Consider each of the three conditions described in the caption to **fig A** and explain why a person with each of these problems would need an artificial pacemaker fitted.

4 What are the similarities and differences between the natural pacemaker system of the heart and an artificial heart pacemaker?

5 Discuss the potential limitations of an artificial pacemaker compared to the natural control of the heart.

6 Investigate why different types of heart pacemakers are needed and what they are used for. Present your findings in a table to make the comparisons easy to see.

ACTIVITY

Coronary heart disease

- Certain arrhythmias of the heart can be treated effectively with an artificial pacemaker. However, some of the most serious heart conditions, affecting millions of people worldwide, cannot be treated with a pacemaker. In 2012, ischaemic heart disease killed about 7.4 million people around the world.

- Using your knowledge of the structure and functions of the heart, along with your knowledge of the blood and the biochemistry of lipids and other molecules, research into this killer disease. Investigate the incidence, the risk factors, the symptoms of disease and how it can be treated and avoided. Produce a 5–10 minute presentation which informs your audience about how the healthy heart works and summarises all the key points about ischaemic heart disease.

1 (a) In the heart there are a number of structures involved in coordinating the contraction of the chambers. In what order are these structures depolarised? [1]

1 bundle of His
2 ventricular muscle cells
3 sinoatrial node
4 atrioventricular node

A 3–4–2–1
B 3–4–1–2
C 4–3–1–2
D 4–2–3–1

(b) Name the region of the human brain involved in control of heart rate. [1]

(c) Heart rate increases during exercise. Describe the mechanisms involved in controlling this increase in heart rate. [6]

(Total for Question 1 = 8 marks)

2 (a) The diagram below shows a mammalian heart.

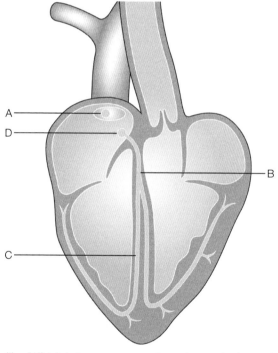

(i) Which letter represents the atrioventricular node? [1]
A
B
C
D

(ii) Which letter represents the Purkyne fibres? [1]
A
B
C
D

(b) Discuss the role of the autonomic nervous system in controlling heart rate. [6]

(c) List **three** differences between cardiac muscle and striated muscle. [3]

(Total for Question 2 = 11 marks)

3 (a) (i) What are the **two** main proteins found in muscle? [1]
A myosin and actin
B actinomyocin and actinase
C collagen and myosin
D actin and collagen

(ii) What is the name of the functional unit in muscles? [1]
A sarcoplasmic reticulum
B sarcoplasm
C sarcolemma
D sarcomere

(b) Describe the process of muscular contraction and how it is controlled. [7]

(c) (i) State how the energy from ATP is used in muscle fibres. [2]

(ii) Suggest what causes muscles to contract and stiffen after death. [3]

(Total for Question 3 = 14 marks)

4 (a) During the cardiac cycle, the atria contract and then the ventricles contract. Explain how this sequence of events is coordinated. [5]

(b) The table below compares the heart rate and stroke volume of a well-trained athlete with a non-athletic person of the same age and size.

	Athlete	Non-athlete
Resting heart rate/bpm	50	72
Heart rate during moderate exercise/bpm	125	185
Stroke volume at rest/cm³	80	70
Stroke volume during moderate exercise/cm³	170	130

(i) Using information in the table, describe the effects that training has on the heart. [2]

(ii) Suggest why the athlete has a lower heart rate at rest. [1]

(iii) Calculate the cardiac output of the athlete at rest. Show your working. [2]

(iv) Calculate the percentage increase in cardiac output of the athlete during moderate exercise. Show your working and give your answer to the nearest whole number. [3]

(Total for Question 4 = 13 marks)

5 The table below shows the blood flow to parts of the body at rest and during vigorous exercise.

Structure	Blood flow at rest		Blood flow during vigorous exercise	
	cm^3 min^{-1}	% of total	cm^3 min^{-1}	% of total
heart muscle	190	3.3	740	3.9
liver	1340	23.5	590	3.1
adrenal glands	24	0.4	24	0.1
brain	690	12.1	740	3.9
lung tissue	100	1.8	200	1.0
kidneys	1050	18.4	590	3.1
skeletal muscles	740	13.0	12 450	65.9
skin	310	5.4	1850*	9.8
other parts	1256	22.0	1716	9.1
total blood flow	5700		18 900	

* When blood flow to the skin was measured just after exercise started it was 150 cm^3 min^{-1}.

(a) (i) State which structure has the largest increase in blood flow during exercise. [1]

(ii) Calculate the percentage increase in blood flow to this structure. Show your working and give your answer to the nearest whole number. [2]

(b) Explain why the heart muscle receives almost four times as much blood during exercise. [2]

(c) Discuss the changes in blood flow to the liver, kidneys and skin. [4]

(Total for Question 5 = 9 marks)

6 (a) Complete the table below showing the functions of components of the joints. [5]

Component of joint	Function
bone	
	holds bones together
	attaches muscle to bone
cartilage	
synovial fluid	

(b) Explain why the tendons are not elastic. [2]

(c) State what is meant by the term *antagonistic* and explain why skeletal muscles are usually arranged in antagonistic pairs. [3]

(Total for Question 6 = 10 marks)

7 The diagram below shows a spirometer. A spirometer can be used to measure aspects of lung function.

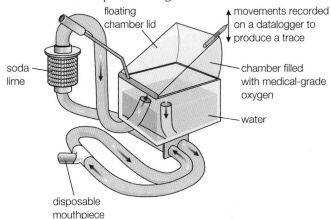

(a) Distinguish between *tidal volume* and *vital capacity*. [2]

(b) Describe **two** precautions you would take when using a spirometer to measure the tidal volume of a student. [2]

(c) The trace below was recorded from a 17-year-old boy at rest.

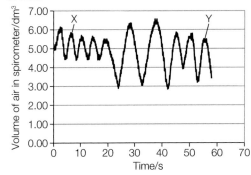

(i) Add a label to the diagram showing a point at which the boy is exhaling. [1]

(ii) Use the trace to calculate the mean tidal volume over the first 20 seconds. Show your working and give your answer to two decimal places. [3]

(iii) The coordinates of point X are: 7 seconds and 5.80 dm^3.

The coordinates of point Y are 55 seconds and 5.50 dm^3.

Use this information to calculate the rate of oxygen consumption. Show your working. [2]

(Total for Question 7 = 10 marks)

TOPIC 7 RESPIRATION, MUSCLES AND THE INTERNAL ENVIRONMENT

Gemsbok are a species of oryx, a type of antelope native to South Africa, which successfully survive in the heat of the African savannah. During the day, their core body temperature may rise as high as 45 °C; but at night, it can fall as low as 35 °C. This reduces their need to pant or sweat and so greatly reduces their water loss. They also have a specialised blood flow: it passes close to the cool nasal passages before it reaches the brain. This cools the blood so that, in spite of the changes in the body temperature, the temperature of the brain of a gemsbok stays the same day and night.

In this section, you will learn about homeostasis and how your body maintains a stable dynamic equilibrium. You will consider how hormones are produced and levels are controlled in the body. You will look at the production of urea in the liver and consider the detailed structure of the kidney and how it controls the concentration of the blood. Your kidneys can produce urine that is extremely dilute or more concentrated than your blood plasma. You will investigate how negative feedback mechanisms involving chemoreceptors and baroreceptors in the brain affect the release of ADH and how this hormone in turn affects the kidney. Finally, you will discover the way the body maintains a stable core temperature during exercise, including the role of the hypothalamus in thermoregulation.

MATHS SKILLS FOR THIS CHAPTER

- **Recognise and make use of appropriate units in calculations** (*e.g. calculating percentage decrease in the concentration of solutes in the blood*)

- **Recognise and use expressions in decimal and standard form** (*e.g. when comparing composition of plasma and glomerular filtrate*)

- **Construct and interpret frequency tables and diagrams, bar charts and histograms** (*e.g. interpret diagrams to show urine production and concentration after drinking water*)

- **Translate information between graphical, numerical and algebraic forms** (*e.g. interpret data on relationship between area of renal medulla and concentration of urine*)

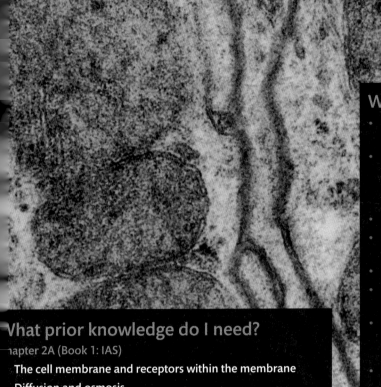

What will I study in this chapter?

- The principles of mammalian hormone production by the endocrine glands
- The mode of action of hormones on their target cells, including receptors, the release of second messengers that activate enzymes and the binding of hormones to transcription factors
- The production of metabolic waste, including urea
- How osmoregulation is brought about in the mammalian body
- The gross and microscopic structure of the kidney
- How urea is produced in the liver and removed from the blood in the kidney by ultrafiltration
- That solutes are selectively reabsorbed in the proximal tubule of the kidney and how the loop of Henle acts as a countercurrent multiplier to increase the absorption of water
- How the pituitary gland and osmoreceptors in the hypothalamus combined with the action of antidiuretic hormone (ADH) bring about negative feedback control of mammalian plasma concentration
- How the kidneys of some desert animals are adapted for life in a very dry environment
- How both behavioural and physiological mechanisms are important in thermoregulation in mammals to maintain their body temperature

What prior knowledge do I need?

Chapter 2A (Book 1: IAS)
- The cell membrane and receptors within the membrane
- Diffusion and osmosis
- Transport across membranes

Chapter 7A
- The production of ATP in cellular respiration

Chapter 7B
- The importance of maintaining pH, temperature and water potential in the body

What will I study later?

Chapter 8A
- The structure and function of the nervous system

Chapter 8B
- How coordination in animals is brought about through nervous and hormonal control

LEARNING OBJECTIVES

■ Understand what is meant by the term homeostasis and its importance in maintaining the body in a state of dynamic equilibrium during exercise, including the role of the hypothalamus in thermoregulation.

■ Understand how the pituitary gland and osmoreceptors in the hypothalamus, combined with the action of antidiuretic hormone (ADH), bring about negative feedback control of mammalian plasma concentration and blood volume.

■ Understand how genes can be switched on and off by DNA transcription factors, including the role of peptide hormones acting extracellularly and steroid hormones acting intracellularly.

In **Chapter 7B**, you started to learn about homeostasis. This is the maintenance of the body in a state of dynamic equilibrium, whatever activities you are doing and whatever the external conditions. Homeostasis is very important because if the internal conditions of the body vary beyond a set of strict limits, the chemical reactions that maintain life cannot occur. Consequently, individual cells and then the whole organism will die. In **Chapter 7B**, you looked at the homeostatic control of your heart and breathing system, especially during exercise. These control systems are largely part of the nervous system. You will study the nervous system in detail in **Chapters 8A** and **8B**. Here, you will learn more about the role that hormones play in maintaining a dynamic equilibrium in your body.

WHAT ARE HORMONES?

Hormones are responsible for one of the main forms of chemical control in animals. They are organic chemicals produced in **endocrine glands** and released into the blood. They travel through the transport system to parts of the body where they cause changes, which may be extensive or highly targeted. Hormones are usually either proteins or peptides (e.g. insulin, antidiuretic hormone) or steroids (e.g. the sex hormones oestrogen and testosterone).

Hormones can have a very rapid effect or they may work over a long period of time. Changes which are caused by the action of hormones include:

- the changes your body experiences during puberty
- the sensations you experience before an interview, a race or an exam
- the way the amount of urine produced by your kidneys varies.

Once a hormone enters the bloodstream, it is carried around in the blood until it reaches its target organ or organs (see **fig A**). The cells of the target organs have specific receptor molecules on the surface of their membranes that bind to the hormone molecules. This leads to a change in the membrane and produces a response.

EXAM HINT

Remember that shapes are important. The shape of the hormone molecule complements the shape of the specific receptor molecule on the surface of a cell. If the cell surface membrane contains the correct molecule, the cell can respond.

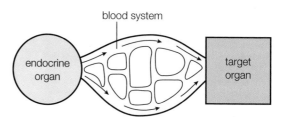

▲ **fig A**   The pathway followed by a hormone from the endocrine gland, where it is produced, to the cells of the target organ or organs.

THE ENDOCRINE GLANDS

Endocrine glands are found around the body, often in association with other organ systems. Several of the glands have more than one function. For example, the ovaries produce ova as well as hormones; the pancreas is both an **exocrine gland** which produces digestive enzymes and an endocrine gland which produces the hormones insulin and glucagon. The glands all have a rich blood supply, with plenty of capillaries within the glandular tissue itself.

The exocrine glands that produce the secretions of the gut release their juices along small tubes or ducts. The endocrine glands do not have ducts and they release their hormones directly into the bloodstream. They include:

- the pituitary gland (see below)
- the hypothalamus (see below)
- the thyroid glands: thyroid hormones are involved in controlling growth and metabolism
- the parathyroid glands: parathyroid hormones are involved in the homeostatic control of calcium metabolism
- the pancreas: the exocrine pancreas produces digestive enzymes and the endocrine pancreas produces insulin and glucagon which are involved in the homeostatic control of blood sugar
- the adrenal glands: produce many hormones including adrenaline, involved in the fight or flight response, and aldosterone, involved in the homeostatic control of the osmotic balance of the body
- the kidneys: produce hormones involved in production of red blood cells and vitamin D metabolism

- the ovaries: produce the female sex hormones
- the testes: produce the male sex hormones.

Most mammals have very similar endocrine glands.

> ### LEARNING TIP
> There are two types of gland. Exocrine glands consist of groups of cells that release a substance (such as an enzyme) into a duct that carries it to where it is needed. Endocrine glands have no duct and release hormones directly into the blood.

HORMONE RELEASE SYSTEMS

The endocrine system interacts very closely with the nervous system (see **Chapter 8B**). Some hormones are released as a result of direct stimulation of the endocrine gland by nerves. For example, the adrenal medulla of the adrenal glands releases adrenaline when it is stimulated by the sympathetic nervous system. The control of hormone release by the nervous system is relatively simple. If the gland is stimulated, hormone is released. If it is not stimulated, no hormone is released. The level of stimulation determines the level of response.

However, many hormones are released from endocrine glands in response to another hormone or chemical in the blood. For example, the pituitary gland in the brain secretes several hormones that directly stimulate other endocrine glands. In addition, changes in chemicals in the blood (e.g. glucose and salt) can stimulate the release of hormones, which consequently act to regulate the levels of those chemicals.

Some hormones are released in response to a chemical stimulus such as another hormone or glucose. In this case, a negative feedback loop controls their secretion (see **fig B**). You looked at the principles of negative feedback in **Section 7B.6**. The presence of the appropriate chemical in the blood stimulates the release of the hormone. As a result of the rise in the hormone levels, the amount of stimulating chemical in the blood drops. Therefore, the endocrine gland receives less stimulation and so the hormone levels drop. This gives a very sensitive level of control that can be adjusted constantly to the needs of the body.

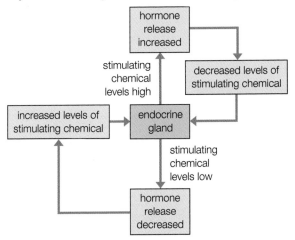

▲ **fig B** A negative feedback loop in a hormonal control system

THE PITUITARY GLAND: CONTROLLING HORMONE RELEASE AND HOMEOSTASIS

The **pituitary gland** in the brain produces and releases secretions that affect most of the other endocrine glands in the body. The pituitary gland has an anterior lobe and a posterior lobe. The pituitary itself is mainly under the control of the **hypothalamus**.

The hypothalamus is the small area of brain directly above the pituitary gland. It performs a variety of functions, although it is relatively small. One of its functions is to monitor the levels of a number of metabolites and hormones in the blood. The hypothalamus controls the activity of the pituitary gland in response to the concentrations of these chemicals. You will see that the interactions between the hypothalamus and the pituitary play a very important role in homeostasis.

How does this control function? **Fig C** shows the close anatomical relationship between the hypothalamus and the pituitary. The hypothalamus contains **neurosecretory cells**. These are nerve cells that produce secretions from the ends of the axons. One group of these cells (neurosecretory cells 1) produce substances that stimulate or inhibit the release of hormones from the anterior pituitary. They are known as either **releasing factors** or **release-inhibiting factors**, depending on what they do. The other group of cells (neurosecretory cells 2) produce secretions that are stored in the posterior pituitary and are released later as hormones.

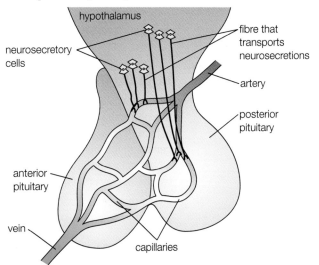

▲ **fig C** The close structural relationship between the hypothalamus and the pituitary is reflected in their function. The hypothalamus produces neurosecretions that control both lobes of the pituitary gland.

Under the control of the hypothalamus, the pituitary gland produces six hormones from the anterior lobe and two from the posterior lobe. Their functions range from controlling the secretions of the thyroid gland to the control of growth, and from sexual development to the control of urine volume. You can see the hormones produced and the role they play in **fig D**. In this section we will focus on **antidiuretic hormone (ADH)** and its homeostatic role in controlling the concentration of the mammalian blood plasma concentration and blood volume.

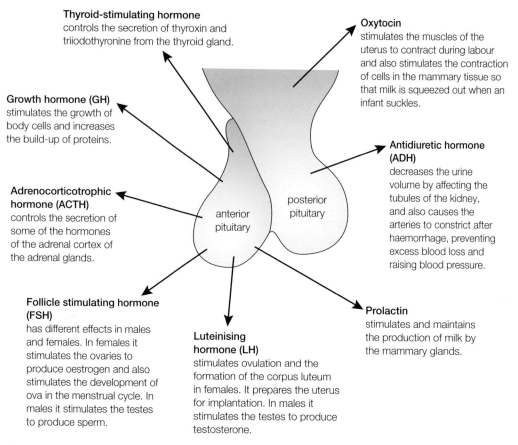

Thyroid-stimulating hormone controls the secretion of thyroxin and triiodothyronine from the thyroid gland.

Oxytocin stimulates the muscles of the uterus to contract during labour and also stimulates the contraction of cells in the mammary tissue so that milk is squeezed out when an infant suckles.

Growth hormone (GH) stimulates the growth of body cells and increases the build-up of proteins.

Antidiuretic hormone (ADH) decreases the urine volume by affecting the tubules of the kidney, and also causes the arteries to constrict after haemorrhage, preventing excess blood loss and raising blood pressure.

Adrenocorticotrophic hormone (ACTH) controls the secretion of some of the hormones of the adrenal cortex of the adrenal glands.

Follicle stimulating hormone (FSH) has different effects in males and females. In females it stimulates the ovaries to produce oestrogen and also stimulates the development of ova in the menstrual cycle. In males it stimulates the testes to produce sperm.

Luteinising hormone (LH) stimulates ovulation and the formation of the corpus luteum in females. It prepares the uterus for implantation. In males it stimulates the testes to produce testosterone.

Prolactin stimulates and maintains the production of milk by the mammary glands.

anterior pituitary

posterior pituitary

▲ **fig D** The hormones of the pituitary gland, especially those from the anterior lobe, function mainly by stimulating other endocrine organs in other parts of the body.

HOW DO HORMONES HAVE THEIR EFFECTS?

Some hormones act by binding to specific receptor sites on the membrane of their target cells. At this point, the hormone needs to affect the target cell in some way to cause the desired change in activity. This is often by switching genes on and off. In **Section 3C.3 (Book 1: IAS)**, you learned how genes can be switched on and off by DNA transcription factors. Different types of hormone act as DNA transcription factors in different ways, and this is how they make changes in the body. There are two main ways in which hormones have their effects:

- the release of a second messenger
- the hormone enters the cell.

RELEASE OF A SECOND MESSENGER

Peptide and protein hormones are not soluble in lipids and therefore cannot cross the cell membrane. Examples include:

- adrenaline, which is involved in the homeostatic response of the heart and breathing system to exercise and stress
- insulin and glucagon, which are involved in the homeostatic control of blood glucose levels
- antidiuretic hormone (ADH), which is involved in the homeostatic control of the blood plasma concentration and blood volume (see **Section 7C.3**).

To have an effect, these hormones have to make changes *inside* the cell. How do they do this?

The hormone molecule binds to a receptor in the cell membrane. This begins a series of membrane-bound reactions that result in the formation of a second chemical messenger inside the cell. This second messenger then activates a number of different enzymes within the cell, altering the metabolism. The most common second messenger is a substance called **cyclic AMP (cAMP)**, which is formed from ATP (see **fig E**). Cyclic AMP binds to other chemicals which pass into the nucleus and act as DNA transcription factors. These changes can cause a number of responses in the cell, including increased cellular respiration, increased contraction of muscle cells, relaxation of smooth muscle in blood vessels, and so on. This is how ADH, the hormone associated with the homeostatic control of blood concentration and volume, has its effects.

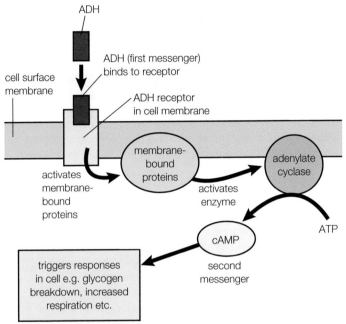

▲ **fig E**   Hormones such as ADH have their effect on cell metabolism via a second messenger system.

THE HORMONE ENTERS THE CELL

Steroid hormones such as oestrogen and testosterone are lipid soluble. They can pass through the membrane and act as the internal messenger themselves. Inside the cell, the hormone binds to a receptor and the hormone–receptor complex passes through the pores of the nuclear membrane into the nucleus. The hormone attached to the receptor acts as a DNA transcription factor, regulating gene expression and switching sections of the DNA on or off (see **fig F**).

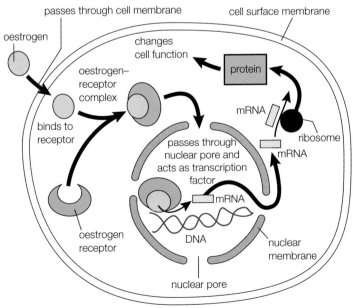

▲ **fig F**   Steroid hormones, such as oestrogen, act directly as DNA transcription factors to have their effects.

CHECKPOINT

1. (a) What is a hormone?

   (b) Give two features of a typical endocrine gland that are directly related to its function.

2. Protein and peptide hormones are not lipid soluble, so they cannot enter a cell directly. Explain how hormones such as ADH and adrenaline still manage to have a wide effect on the biochemistry of the cell.

SKILLS ▷ **CRITICAL THINKING**

3. ▷ 'Steroid hormones act as transcription factors in the nucleus of a cell.' Discuss this statement and point out any inaccuracies.

SUBJECT VOCABULARY

hormones organic chemicals which are produced in endocrine glands, are released into the blood and travel through the transport system to parts of the body where they cause changes, which may be extensive or very targeted; hormones are usually either proteins, parts of proteins such as polypeptides, or steroids

endocrine glands glands without ducts (ductless glands) that produce hormones and release them directly into the bloodstream

exocrine glands glands that produce chemicals (e.g. enzymes) and release them along small tubes or ducts

pituitary gland a small gland in the brain that has an anterior lobe and a posterior lobe and produces and releases secretions that affect the activity of most of the other endocrine glands in the body

hypothalamus a small area of brain directly above the pituitary gland that controls the activities of the pituitary gland and coordinates the autonomic (unconscious) nervous system

neurosecretory cells nerve cells that produce secretions from the ends of their axons; these secretions either stimulate or inhibit the release of hormones from the anterior pituitary, or are stored in the posterior pituitary and then later released as hormones

releasing factors substances that stimulate the release of hormones from the anterior pituitary

release-inhibiting factors substances that inhibit the release of hormones from the anterior pituitary

antidiuretic hormone (ADH) hormone produced in the hypothalamus and stored in the posterior pituitary that increases the permeability of the distal tubule and the collecting duct of the kidney to water, reducing the amount of urine produced but increasing the concentration of the urine

cyclic AMP (cAMP) a compound formed from ATP that is produced when peptide hormones such as ADH and adrenaline bind to membrane receptors, and acts as a second messenger in cells

LEARNING OBJECTIVES

■ Know the gross and microscopic structure of the mammalian kidney.

■ Understand how urea is produced in the liver from excess amino acids (details of the ornithine cycle are not required) and how it is removed from the bloodstream by ultrafiltration.

■ Understand how solutes are selectively reabsorbed in the proximal tubule and how the loop of Henle acts as a countercurrent multiplier to increase the reabsorption of water.

Water moves into and out of cells by osmosis. If the concentration of water and solutes inside and outside of a cell is not balanced, water may enter or leave the cells by osmosis. Water entering the cells in this way would cause the cells to become larger and burst. Water leaving the cells by osmosis would cause the cytoplasm to shrink (become smaller) and become concentrated and unable to function (see **Section 2A.3 (Book 1: IAS)**). **Osmoregulation** is the maintenance of the osmotic potential in the tissues of a living organism within narrow limits by controlling water and salt concentrations. It is part of homeostasis and it is vital for life.

LEARNING TIP

Always look at unfamiliar words and try to break them down. This can help you understand the meaning. 'Osmo' means something to do with osmosis and 'regulation' means controlling.

Animals that live on the land have to drink all the water that they need, so they must conserve water. The cells of a land-living mammal are surrounded by tissue fluid that comes from the blood capillaries. By controlling the water potential of the blood (both the water content and the solute concentration), the body can control the water potential of the tissue fluid and so protect the cells from osmotic damage.

In mammals, the main organ involved in the homeostatic control of the water balance of the body is the kidney. The liver is also involved in homeostasis, for example in the breakdown of excess amino acids and the removal of toxins. You will look at the liver first, because urea and other toxins from the liver are removed from the body by the kidneys.

THE LIVER, PROTEIN METABOLISM AND HOMEOSTASIS

The liver has around 500 different functions in the body, many of them involved in homeostasis. It plays an important role in the **deamination** of excess amino acids in protein metabolism. Your body cannot store protein or amino acids, and without the action of the liver, any excess protein you eat would be excreted and wasted. The hepatocytes (liver cells) deaminate excess amino acids. They remove the amino group and convert it first into ammonia, which is very toxic, and then to urea which is less toxic and can be excreted by the kidneys (remind yourself of the structure of amino acids in **Section 1A.5 (Book 1: IAS)**). The ammonia produced in the deamination of proteins is converted

into urea by a series of enzyme-controlled reactions known as the **ornithine cycle** (see **fig A**). The remainder of the amino acid can then be used in aerobic respiration, entering at the beginning of the Krebs cycle as you saw in **Section 7A.5**, or be converted into lipids for storage.

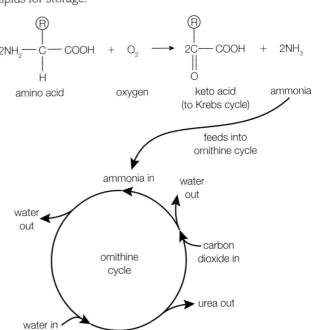

$$2NH_2 - \underset{\underset{H}{|}}{\overset{\textcircled{R}}{C}} - COOH \ + \ O_2 \ \longrightarrow \ 2\underset{\underset{O}{\|}}{\overset{\textcircled{R}}{C}} - COOH \ + \ 2NH_3$$

amino acid   oxygen   keto acid (to Krebs cycle)   ammonia

▲ **fig A** The deamination of excess amino acids and the production of urea in the ornithine cycle in the liver is important. It prevents the waste of excess protein from the diet. The toxic part of amino acid molecules is removed and the remaining parts are used in cellular respiration.

EXAM HINT

Many cells are specially adapted to their functions. The adaptations of liver cells are not obvious under a microscope. Their main specialisation is in the numbers and activity of their mitochondria.

OSMOREGULATION IN MAMMALS

Osmoregulation in mammals is largely brought about by the kidneys. The kidneys are a pair of organs capable of producing urine, which can be hypertonic to (more concentrated than) the body fluids. This makes it possible to conserve water, an ability that has allowed mammals to spread into most of the land environments of the Earth.

In humans, as in other mammals, the kidneys are a pair of dark reddish brown organs attached to the back of the abdominal cavity. They are surrounded by a thick layer of fat, which helps to protect them from mechanical damage. They control the water potential of the blood plasma that passes through them. They remove urea and substances that would affect the water balance. Blood plasma from the body passes through the kidneys where the urea and excess salts and water are removed to form urine. The urine is stored in the bladder and released from the body at intervals (see **fig B**).

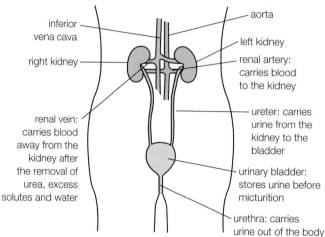

▲ **fig B** The human urinary system.

THE STRUCTURE AND FUNCTIONS OF THE KIDNEY

The mammalian kidney has two main roles in the body. One is excretion: the removal of urea from the body. The other is osmoregulation. In humans, around 120 cm³ of blood per minute pass through the kidneys, a rate that means all of the blood in your body travels through the kidneys and is filtered and balanced approximately once every hour. Around 180 dm³ of blood is filtered every day. You can see the external appearance of a kidney and the internal arrangement of tissues in **fig C**.

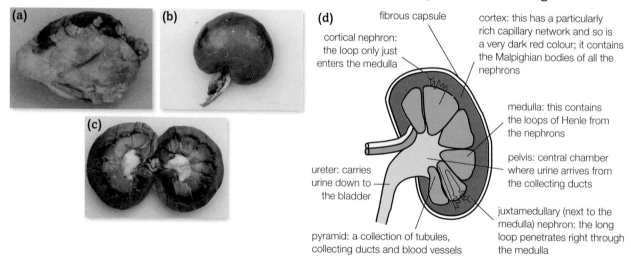

▲ **fig C** (a), (b) and (c) The gross structure of the mammalian kidney as seen with the naked eye. (d) The basic structure of the mammalian kidney. The position of the two main types of kidney tubule (see below) have been superimposed on this diagram.

The kidney performs three main functions in its osmoregulatory role. These are **ultrafiltration**, **selective reabsorption** and **tubular secretion**.

Each kidney is made up of microscopic tubules called **nephrons**, which are 2–4 cm long. There are about 1.5 million nephrons in each kidney filtering and balancing the blood. You can see the basic structure of a nephron in **fig D**. There are two main types of nephron.

- Cortical nephrons are found mainly in the renal cortex. They have a loop of Henle (a U-shaped tubule) that only just reaches into the medulla. About 85% of human nephrons are cortical.
- Juxtamedullary nephrons have long loops of Henle that penetrate right through the medulla. They are particularly efficient at producing concentrated urine.

The balance of these different types of nephron varies in different organisms.

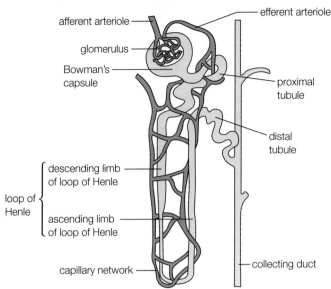

▲ **fig D** The structure of a nephron

ULTRAFILTRATION

The first stage in the osmoregulation of the blood is ultrafiltration. Ultrafiltration in the kidney tubules occurs because of a combination of very high blood pressure in the glomerular capillaries and the structure of the Bowman's capsule and glomerulus. The glomerulus and the Bowman's capsule together make up the Malpighian body.

High blood pressure develops in the capillaries of the glomerulus, because the diameter of the blood vessel coming into the glomerulus is greater than that of the blood vessel leaving. The high pressure squeezes the blood out through the pores in the capillary wall; this is like water passing along a hosepipe with holes in it. The size of the pores means that almost all the contents of the plasma can pass out of the capillary. It is only the blood cells and the largest plasma proteins that cannot pass through the pores.

The cells of the Bowman's capsule next to the capillaries act as an additional filter. The filtrate that enters the capsule contains glucose, salt, urea and many other substances in the same concentrations as they are in the blood plasma (see **table A**).

SUBSTANCE	APPROXIMATE CONCENTRATION/g dm^{-3}	
	IN PLASMA	IN FILTRATE
water	900	900
protein	80	0
inorganic ions	7.2	7.2
glucose	1	1
amino acids	0.5	0.5
urea	0.3	0.3

table A Comparison of the composition of human plasma and glomerular filtrate.

If all of the filtrate produced in the Malpighian bodies over a 24-hour period then passed out of the body, we would produce around 200 dm^3 of urine a day and we would have to drink continually to replace it. In fact, the average daily urine production is 1–2 dm^3, so ultrafiltration is only the first step in the process. Most of the filtrate is reabsorbed into the blood later.

SELECTIVE REABSORPTION

Ultrafiltration is passive and not selective. It removes urea from the blood, but it also removes a lot of water with glucose, salt and other substances that are present in the plasma. Glucose is needed for cellular respiration and is never excreted under normal circumstances. Most of the water, salt and other inorganic ions passed into the tubule during ultrafiltration are also needed by the body. After the ultrafiltrate has entered the nephron, the main function of the kidney tubule is to return most of what has been removed during ultrafiltration back into the blood.

THE PROXIMAL TUBULE

The **proximal tubule** reabsorbs over 80% of the glomerular filtrate into the blood. The cells lining this tubule are covered with microvilli; these greatly increase the surface area through which substances can be absorbed. The cells also have large numbers of mitochondria. This shows that they are involved in active processes.

Active transport in the proximal tubule results in all the glucose, amino acids, vitamins and most hormones being returned to the blood. About 85% of the sodium chloride and water is reabsorbed as well. The sodium ions are actively transported, and the chloride ions and water follow passively down concentration gradients. Once these substances are removed from the tubule cells into the intracellular spaces, they then pass by diffusion into the extensive capillary network that surrounds the tubules. The blood is constantly moving through the capillaries, maintaining a concentration gradient for diffusion. By the time the filtrate reaches the loop of Henle, it is isotonic with the tissue fluid that surrounds the tubule. The amount of reabsorption in the proximal tubule is always the same. The adjustment of the water balance to the needs of the body occurs further along the nephron in the loop of Henle.

THE LOOP OF HENLE

Each loop of Henle is found in the medulla of the kidney, in close contact with a network of capillaries. Together they create a water potential gradient between the filtrate and the medullary tissue fluid. This makes it possible for water to be reabsorbed from the distal tubule and collecting duct. It is this water potential gradient that allows mammals to produce urine that is more concentrated than their own blood.

The creation of the high concentration of sodium and chloride ions in the tissue fluid of the medulla happens as a result of the flow of fluid in opposite directions in the adjacent limbs of the loop of Henle. This is combined with the different permeabilities to water of the different sections, and a region of active transport. The combination creates a **countercurrent multiplier**. This is a biological system that uses active transport to establish and maintain concentration gradients. The ability of the kidney to concentrate the fluid (urine) in the distal tubules and collecting duct depends on this countercurrent multiplier. These processes are summarised in **fig E**.

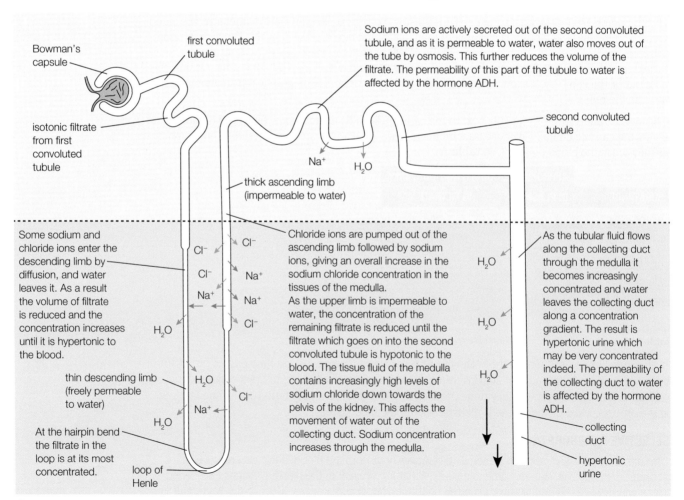

▲ **fig E** A model of the role of the loop of Henle in the reabsorption of water and the production of concentrated urine in the kidney

The changes that occur in the descending limb of the loop of Henle depend on high concentrations of sodium and chloride ions in the tissue of the medulla. These result from events in the ascending limb of the loop. The system is explained below.

- *In the descending limb* The descending limb is freely permeable to water but is not very permeable to sodium and chloride ions. No active transport occurs here. The fluid entering this limb is isotonic with the blood. As the fluid travels down the limb deeper into the medulla, the external concentration of sodium and chloride ions in the tissue fluid of the medulla and the blood in the capillaries becomes higher and higher. As a result, water moves out of the descending limb into the tissue fluid by osmosis down a concentration gradient. It then moves into the blood in the capillaries, again down a water potential gradient. By the time the fluid reaches the U-shaped bend at the bottom of the loop it is very concentrated and hypertonic to the arterial blood.

- *In the ascending limb* The first section of the ascending limb is very permeable to sodium and chloride ions but not permeable to water. No active transport occurs in this section. Sodium and chloride ions move down concentration gradients out of the very concentrated fluid in the loop of Henle into the tissue fluid of the medulla. The second, thicker section of the ascending limb is also impermeable to water, but sodium and chloride ions are actively pumped out of the tubule into the tissue fluid of the medulla and the blood of the capillary network. The tissues of the medulla, therefore, have very high sodium and chloride ion concentrations that cause the water to pass out of the descending limb by osmosis. However, the ascending limb is impermeable to water, so water cannot follow the chloride and sodium ions out down the concentration gradient. The fluid left in the ascending limb, therefore, becomes less concentrated.

THE DISTAL TUBULE

The **distal tubule** is permeable to water, but the permeability varies with the levels of antidiuretic hormone (ADH). It is here and in the collecting duct that the body balances its water needs (see **Section 7C.3**). If there is not enough salt in the body, sodium can be actively pumped out of the tubule, with chloride ions, down an electrochemical gradient. Water leaves by diffusion, if the walls of the tubule are permeable at that time.

THE COLLECTING DUCT

The permeability of the **collecting duct** is strongly affected by the hormone ADH. Water moves out of the collecting duct down a water potential gradient as it passes through the medulla, with its high levels of sodium and chloride ions. As water leaves the collecting duct, the urine becomes steadily more concentrated. The concentration of sodium ions in the surrounding fluid increases through the medulla towards the pelvis of the kidney (where the urine leaves the collecting duct), so water can be removed along the whole length of the collecting duct. This makes it possible for the kidney to produce very hypertonic urine (high osmotic pressure) when it is necessary to conserve water for the cells of the body.

THE URINE

The urine is the fluid that the kidney tubules produce. First, it is collected in the central chamber (the pelvis) of each kidney. It then passes along the ureters to the bladder and is stored there until the bladder is sufficiently stretched to stimulate urination. The urine passes out of the body along a tube called the urethra. The volume of urine produced is very variable, depending on what is taken into the body and the activity levels.

The urine contains varying amounts of water and salts, depending on the diet and the demands of the body. It also contains relatively large quantities of urea. The colour of the urine varies from almost colourless to deep yellow/brown, depending on its concentration. The colour of the urine can be affected by certain foods; for example, eating a lot of beetroot (red beets) can cause pink urine. Substances such as glucose or protein should never appear in the urine. If they do, this indicates either that there are problems elsewhere in the body (for example, the pancreas is not controlling the blood sugar levels properly) or that the kidneys themselves are not working properly.

EXAM HINT

The term *descending limb* means the part of the loop carrying fluid out of the cortex into the medulla; it does not mean fluid is moving downwards. Similarly, *ascending limb* means fluid is flowing from the medulla to the cortex; it is not necessarily flowing upwards.

LEARNING TIP

Remember that reabsorption of water and solutes in the kidney is essential; otherwise a terrestrial mammal would run out of water in a few minutes.

EXAM HINT

Do not confuse the pelvis of the kidney with the pelvic bone that forms the hips as part of the skeletal system.

DID YOU KNOW?

Kidney adaptations of desert mammals

Desert animals spend most or all of their lives short of water. In some extreme examples, such as kangaroo rats (*Dipodomys* spp.), they never drink (see fig F). A combination of behavioural, anatomical and physiological adaptations enable them to survive. Kangaroo rats spend much of their time in burrows (dug out holes) below the surface of the desert, where the temperature is cooler and more stable. This reduces the resources they use to maintain a stable body temperature (see **Section 7C.4**). They generate up to 90% of the water they need by the oxidative reactions in their cells; as you know from **Chapter 1A (Book 1: IAS)**, many metabolic processes involve condensation reactions, which produce water. In contrast, humans can obtain only 12% of the water they need from their metabolic reactions. Kangaroo rats obtain the rest of the water they need from water contained in their food.

▲ **fig F**   A number of adaptations, including modifications to their kidneys, mean kangaroo rats can live in the desert without drinking.

Kangaroo rats must still remove waste products such as urea. Therefore, they need to produce tiny amounts of very concentrated urine. In fact, they produce urine with a concentration of more than 6000 mosmol/kg H_2O. In contrast, humans can concentrate their urine to around 1400 mosmol/kg H_2O and camels to 2800 mosmol/kg H_2O. Scientists are still unsure exactly how the kidneys of kangaroo rats produce such concentrated urine, but they have a number of adaptations that seem to contribute to this ability:

- a relatively large proportion of juxtamedullary nephrons
- relatively long loops of Henle; recent research suggests that the water-permeable region of the thin descending loop is longer than in other species (this enables them to produce a very high concentration of ions in the medulla, which thus makes it possible to produce very highly concentrated urine)
- higher numbers of infoldings in the cell membranes of the epithelial cells lining the tubules (these provide increased surface area for diffusion of inorganic ions and water, making steep concentration gradients possible)
- high numbers of mitochondria with densely arranged cristae for maximum cellular respiration in the epithelial cells of the nephrons (providing the energy for the active pumping of inorganic ions into or out of the tubules as required).

It appears that a combination of these factors and others enable kangaroo rats to keep the maximum amount of water and produce very small quantities of extremely concentrated urine.

CHECKPOINT

SKILLS   CRITICAL THINKING

1. ▶ Define osmoregulation and explain why it is important in mammals.

2. Explain how the homeostatic functions of the liver are linked to the osmoregulatory functions of the kidney.

3. Explain how the kidney filters the blood and produces urine that may be more concentrated than the blood.

SKILLS   CRITICAL THINKING

4. ▶ Discuss some of the anatomical and physiological adaptations you might expect in a mammal that lives in a very hot dry environment such as a desert.

SUBJECT VOCABULARY

osmoregulation the maintenance of the osmotic potential of the tissues of a living organism within a narrow range, by controlling water and salt concentrations

deamination the removal of the amino group from excess amino acids in the ornithine cycle in the liver; the amino group is converted into ammonia and then to urea, which can be excreted by the kidneys

ornithine cycle a series of enzyme-controlled reactions that convert ammonia from excess amino acids into urea in the liver

ultrafiltration the process by which fluid is forced out of the capillaries in the glomerulus of the kidney into the kidney tubule through the epithelial walls of the capillary and the capsule

selective reabsorption the process by which substances needed by the body are reabsorbed from the kidney tubules back into the blood

tubular secretion the process by which inorganic ions are secreted into or out of the kidney tubules as needed to maintain the osmotic balance of the blood

nephrons microscopic tubules that make up most of the structure of the kidney

proximal tubule the first region of the nephron after the Bowman's capsule; it is here that over 80% of the glomerular filtrate is absorbed back into the blood

countercurrent multiplier a system that produces a concentration gradient in a living organism; energy from cellular respiration is required

distal tubule the section of the nephron after the loop of Henle that leads into the collecting duct where some of the balancing of the water needs of the body occurs

collecting duct takes urine from the distal tubule to be collected in the pelvis of the kidney; it is the region of the kidney in which most of the water balancing needed for osmoregulation takes place

3 CONTROL OF THE KIDNEY AND HOMEOSTASIS

LEARNING OBJECTIVES

■ Understand what is meant by the term homeostasis and its importance in maintaining the body in a state of dynamic equilibrium during exercise, including the role of the hypothalamus in thermoregulation.

■ Understand the principle of negative feedback in maintaining systems within narrow limits.

■ Understand how the pituitary gland and osmoreceptors in the hypothalamus, combined with the action of antidiuretic hormone (ADH), bring about negative feedback control of mammalian plasma concentration and blood volume.

The kidney is involved in the balance of both water and solutes in the body. Metabolic processes continuously produce urea and the kidney plays an important part in removing urea from the body. However, levels of other important substances vary according to the situation of the individual. A long walk on a hot sunny day, a salty meal or drinking several pints of liquid all threaten the equilibrium of the body. How is kidney function controlled to bring about homeostasis?

OSMOREGULATION

The osmotic potential of the blood is maintained within narrow boundaries by balancing the water and salts taken in by eating and drinking with the water and salts lost by sweating, defaecation and in the urine. The concentration of the urine is most important in this dynamic equilibrium and this is controlled by a negative feedback system involving antidiuretic hormone (ADH).

ADH is produced by the hypothalamus and secreted into the posterior lobe of the pituitary and stored there (see **Section 7C.1**). ADH increases the permeability to water of the distal tubule and the collecting duct.

MECHANISM OF ADH ACTION

The mechanism by which ADH increases the permeability to water of the walls of the distal tubule and the collecting duct is very elegant. ADH does not cross the membrane of the tubule cells. It binds to specific receptors, which triggers reactions that result in the formation of cAMP as the second messenger (see **Section 7C.1**). The cAMP starts a series of reactions that cause vesicles within the cells lining the tubules to move to and fuse with the cell membranes. The vesicles contain water channels which are inserted into the membrane and make it permeable to water. Water then moves through the channels out of the tubules and into the surrounding blood capillaries by osmosis.

The amount of ADH released controls the number of channels that are inserted. This means the permeability of the tubules can be very closely controlled to match the water demands of the body. When ADH levels fall, levels of cAMP also drop and the water channels are taken out of the membranes and repackaged in vesicles. This makes the tubule impermeable to water again. The channels are stored in vesicles ready for reuse when they are needed again.

ADH AND NEGATIVE FEEDBACK CONTROL

If water is in short supply or you sweat a lot as a result of exercise or you eat a very salty meal, the concentration of inorganic ions in the blood rises so its water potential becomes more negative. If this continued, the osmotic balance of the tissue fluids would become disturbed, causing cell damage (see **Section 2A.3 (Book 1: IAS)**). This is prevented by a negative feedback system involving ADH (see **fig A**).

Osmoreceptors in the hypothalamus detect an increasing plasma concentration of inorganic ions. They send nerve impulses to the posterior pituitary, which consequently releases stored ADH into the blood. The ADH is accepted by receptors in the cells of the kidney tubules. ADH increases the permeability of the distal tubule and the collecting duct to water. As a result, water leaves the tubules by osmosis into the surrounding capillary network. This means blood plasma receives more water from the filtrate, and a small volume of concentrated urine is produced.

EXAM HINT

The details of membrane structure and cell specialisation are IAS topics (see **Section 2A.1–4 (Book 1: IAS)**). But they are likely to be tested in the context of cell responses in IAL topics such as this. Remember that membranes are not permeable to polar substances such as water unless there are protein channels or carrier proteins present.

EXAM HINT

Many students become confused about how increasing the permeability of the collecting duct walls helps to conserve water. Remember that the fluid has passed into the nephron to become urine to be excreted. The water is being reabsorbed back into the blood from the urine as it is being made.

If you drink large volumes, the blood plasma becomes more dilute. The same osmoreceptors of the hypothalamus detect the change. The fall in the concentration of the blood plasma inhibits the release of ADH by the pituitary gland. The walls of the distal tubule and the collecting duct remain impermeable to water and so little or no reabsorption occurs. The kidneys therefore produce large amounts of very dilute urine and the concentration of the blood is maintained (see **fig A**).

It is very easy to test the effectiveness of this system: simply drink about a litre of water over a short period of time and wait for the results!

EXTRA FEEDBACK

Changes in the blood pressure also stimulate or inhibit the release of ADH. These changes are detected by the baroreceptors in the aortic and carotid arteries, which also help control the heart rate (see **Section 7B.7**). A rise in blood pressure (often a sign of an increase in blood volume) will suppress the release of ADH and so increase the volume of water lost in the urine. This results in a reduction in the blood volume and so the blood pressure falls.

A fall in blood pressure, which may indicate a loss of blood volume, causes an increase in the release of ADH from the pituitary and the conservation of water by the kidneys. Water is returned to the blood and a small amount of concentrated urine is produced. This is part of the normal dynamic equilibrium of the body, but it also plays an important role if you lose a lot of blood for any reason.

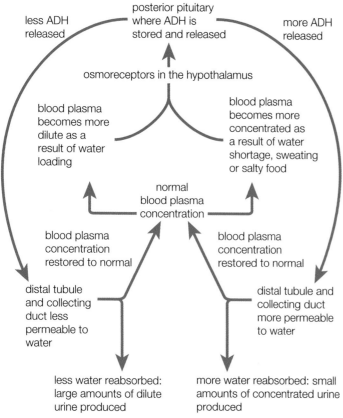

fig A A negative feedback system involving ADH maintains the concentration of the blood plasma within very narrow limits.

DID YOU KNOW?

Investigating membrane properties

ADH is not the only hormone that affects the kidney tubules and controls the amount of urine formed in response to changes in the body. There is another major homeostatic system involving the kidney that helps control the salt concentration of the body.

The hormone aldosterone, produced by the cortex of the adrenal glands, causes the active absorption of sodium ions from the filtrate in the kidney tubules into the plasma in the capillaries. Water follows by osmosis which increases the blood volume and as a result increases the blood pressure.

If sodium ions are lost from the blood, water also tends to be lost from the blood into the tissue fluid and cells. This causes a slight drop in blood pressure, which is detected by a group of cells in the kidney itself. They then produce an enzyme called rennin. Rennin acts on a protein in the blood to produce the hormone angiotensin. Angiotensin stimulates the release of aldosterone from the adrenal glands. This stimulates the active absorption of sodium ions from the filtrate to the blood. Once aldosterone is produced, the adrenal glands are also stimulated by adreno-corticotrophic hormone from the pituitary gland. This is a complex system balancing the concentration of sodium ions and blood volume.

CHECKPOINT

1. Produce a flowchart to show the role of ADH in osmoregulation:

   (a) if you drink two litres of water

   (b) if you have a fever and sweat more than normal.

2. Describe how ADH increases the permeability of the distal tubule and the collecting duct to water.

3. Explain how ADH controls the volume and concentration of the blood plasma.

SUBJECT VOCABULARY

osmoreceptors sensory receptors in the hypothalamus that detect changes in the concentration of the blood plasma

LEARNING OBJECTIVES

■ Understand what is meant by the term homeostasis and its importance in maintaining the body in a state of dynamic equilibrium during exercise, including the role of the hypothalamus in thermoregulation.

The chemical reactions in cells can only take place within a relatively narrow range of temperatures. This is mainly due to the sensitivity of the enzymes that control the reactions. For example, once temperatures rise above 40 °C, most enzymes are denatured as their protein structure is destroyed. Many organisms have evolved ways to control their internal body temperature in one way or another. **Thermoregulation** is important to their survival and is a major aspect of homeostasis.

EXAM HINT

Remember that any time changes in temperature are mentioned you should think about the effect on enzyme activity. In thermoregulation, the temperatures are not kept exactly constant but will vary a little because it takes time for the body to respond to changes.

HOW DO ORGANISMS WARM UP OR COOL DOWN?

The surface temperature of an animal or plant can change rapidly, but it is the internal core (most central) temperature that is relevant to enzyme activity. Living organisms are continually cooling down or warming up in response to their surroundings; most of the ways in which they do this are affected by the size of the animal. Small animals have a large surface area-to-volume ratio, so they transfer energy more rapidly than larger organisms.

There are several ways in which organisms warm up or cool down (see **fig A**).

- They warm up as a by-product of metabolism. Chemical inefficiency means that energy is wasted, which warms the centre of the organism.
- They cool down by the evaporation of water from the body surfaces. A certain amount of cooling always occurs in this way from the mouth and respiratory surfaces of animals which live on land and this cannot be controlled. Sweating and behavioural patterns such as wallowing (spending time in mud or water) can increase this cooling.
- Energy may be transferred to or from the environment by radiation (the transfer of energy in the form of electromagnetic waves). Infrared radiation is the most important form of energy absorbed and radiated by animals.
- Energy may be transferred to or from the environment by convection (the transfer of energy by currents of air or water). Convection currents are set up around relatively hot objects, so adaptations to prevent cooling by these currents are common in animals.

- Energy may be transferred to or from the environment by conduction (the transfer of heat by the collision of molecules). Conduction is particularly important between organisms and the ground or water, as air does not conduct well. Neither does fat, so adipose (fatty) tissue is a very valuable insulator, preventing energy exchanges with the environment.

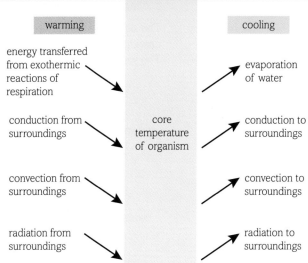

▲ **fig A** Thermoregulation in mammals: the warming and cooling experienced by an organism. It is the balance of these that determines the core temperature. Organisms use a variety of means to shift the balance and allow them to become warmer or cooler as needed.

THERMOREGULATION IN MAMMALS

Mammals, including humans, live in a wide variety of habitats. They need to regulate their temperatures, to avoid damage to their cells and to enable them to have an active way of life. Mammals are **endotherms**. An endotherm relies on its own metabolic processes to provide at least some warming and usually has a body temperature higher than the ambient temperature.

Mammals regulate their body temperature in a number of different ways. They are adapted to conserve their body temperature when necessary and also take advantage of warmth from the environment when possible. This means that there are few environments in which they cannot survive. In order to maintain their body temperature against adverse environmental conditions, the metabolic rate of endotherms has to be high. This means they have to eat a lot of food to supply their metabolic needs. They also need adaptations which allow them to thermoregulate during exercise, which generates large amounts of metabolic heat.

THERMOREGULATION IN HUMANS

Thermoregulation in humans is a good example of homeostasis, as we regulate our body temperature within a very narrow range. The main source of warming is from our metabolism, but humans survive in almost every area of the world and have effective ways of both transferring energy to the environment and conserving energy when needed.

People use a wide variety of behavioural mechanisms to keep warm or cool down as needed. The major difference between humans and other mammals with respect to temperature regulation is that we can manipulate our environment to help us to survive. People build warm houses, light fires and wear clothes to help keep them warm in cold conditions. We make cool buildings, use air-conditioning and build swimming pools to keep ourselves comfortable in high temperatures. In spite of these behavioural adaptations, we still maintain all our homeostatic adaptations for temperature regulation. They play the major role in the maintenance of our core body temperature. All we can do with our technology is reduce the extremes we experience.

THE SKIN

The major homeostatic organ involved in thermoregulation in most endotherms is the skin. As water evaporates from the surfaces of the mouth and nose, so inevitable cooling occurs because of their moist surfaces. The skin, however, has an enormous surface area that can be modified either for cooling or keeping warm (see **fig B**).

The skin is the largest single organ. It covers the entire surface of the body in a waterproof layer, providing protection against both mechanical damage and damage from ultraviolet radiation of the Sun. The surface area of your skin is about $2\,m^2$ and your skin makes up about 16% of your body weight!

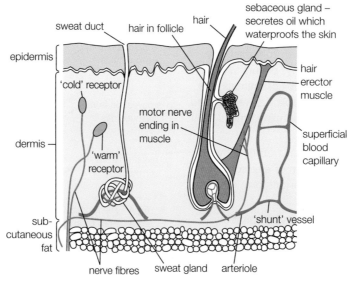

▲ **fig B** The human skin. The same major structures are present in the skin of all other mammals too. Many of the structures are involved in thermoregulation.

KEEPING COOL

The skin plays a major homeostatic role in maintaining the core body temperature within a dynamic equilibrium during exercise or in high external temperatures. It helps prevent overheating in a number of ways. A rich supply of capillaries is near to the surface of the skin. Cooling by radiation, convection and conduction to the environment occurs from the blood flowing through the skin. This cooling is controlled via the **arteriovenous shunt**. During exercise or if the external temperature rises, the shunt is closed and **vasodilation** occurs. This allows more blood to flow through the capillaries at the surface of the skin. As a result the skin appears red, and more energy is transferred to the environment by radiation. This is described clearly in **fig D (page 229)**.

If you are exercising or are in a hot environment, the erector pili muscles, which are attached to hair follicles, are relaxed and the body hairs lie flat against the body, minimising any insulating air layer that is trapped next to the skin. This has little or no impact on cooling in human beings, because we have so little body hair, but it is important in other mammals.

The rate of sweat production in the sweat glands increases with increases in the core temperature. As more sweat is released onto the skin's surface, cooling occurs as the water evaporates. We usually produce and lose almost $1\,dm^3$ of sweat in a day, but this can rise to $12\,dm^3$ under very hot conditions. This means it is very important to keep drinking plenty of water when you are hot, so that the tissues remain hydrated and sweat can be produced for thermoregulation.

Subcutaneous fat acts as insulation, reducing cooling. People who are very physically active, such as elite athletes, tend to have very little subcutaneous fat, because they use up all the energy from their food. This reduces the insulation and increases the amount of energy that can be lost by conduction from the surface of the skin. Very overweight people overheat easily, which is a disadvantage in relation to exercise.

KEEPING WARM

Homeostatic mechanisms also act if the core temperature starts to fall. Some of the energy conservation measures are the exact opposite of the cooling responses of the skin. So the arteriovenous shunt in the blood supply to the skin opens, reducing the blood flow through the capillaries. This is called **vasoconstriction**, and it reduces energy lost from the surface of the skin (see **fig D**).

People sweat less and so cooling by evaporation is reduced too. The erector pili muscles are contracted, pulling the hairs upright. In humans this is visible as 'goose-pimples' (small bumps) and has little effect on temperature regulation, but in hairy mammals and feathered birds it traps an insulating layer of air that helps to reduce cooling (see **fig C**).

▲ **fig C**   The contraction of the erector pili muscles makes this small bird very fluffy and traps a layer of insulating air. This helps to maintain the body temperature in cold external conditions.

The metabolic rate of the body also speeds up, warming the body. This occurs particularly in the liver and the muscles. Shivering, which is the involuntary contraction of the skeletal muscles, also helps with metabolic warming. The energy released raises the body temperature. Animals living in cold areas often develop thick layers of subcutaneous fat that act as an effective insulator against cooling.

IN A WARM ENVIRONMENT	IN A COLD ENVIRONMENT
• vasodilation occurs	• vasoconstriction occurs
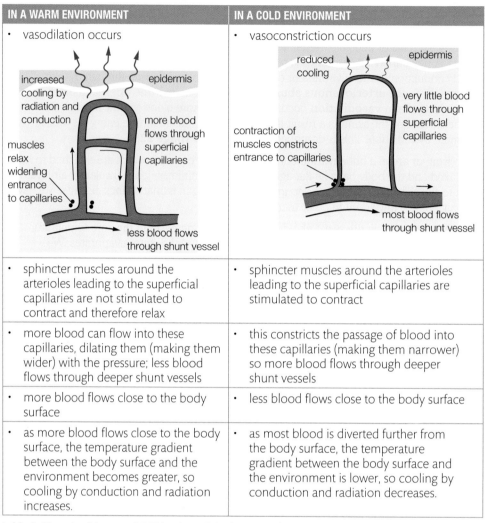	
• sphincter muscles around the arterioles leading to the superficial capillaries are not stimulated to contract and therefore relax	• sphincter muscles around the arterioles leading to the superficial capillaries are stimulated to contract
• more blood can flow into these capillaries, dilating them (making them wider) with the pressure; less blood flows through deeper shunt vessels	• this constricts the passage of blood into these capillaries (making them narrower) so more blood flows through deeper shunt vessels
• more blood flows close to the body surface	• less blood flows close to the body surface
• as more blood flows close to the body surface, the temperature gradient between the body surface and the environment becomes greater, so cooling by conduction and radiation increases.	• as most blood is diverted further from the body surface, the temperature gradient between the body surface and the environment is lower, so cooling by conduction and radiation decreases.

table A The role of the superficial blood vessels in thermoregulation

CONTROL OF THE CORE (BLOOD) TEMPERATURE

In a homeostatic feedback system, there need to be receptors which are sensitive to changes in the system. In the case of thermoregulation, there are two types of receptors. Receptors in the brain directly monitor the temperature of the blood, while receptors in the skin detect changes in the external temperature. This allows for great sensitivity, both to changes in the core temperature and to potential changes as well. The temperature receptors in the brain are in the hypothalamus and they act as the thermostat of the body to keep it at the right temperature. As a result of sensitive feedback mechanisms, the temperature of the human body is usually controlled within a 1 °C range.

If the temperature of the blood flowing through the hypothalamus increases, the **thermoregulatory centre** is activated. It sends out impulses along autonomic motor nerves to effectors that increase the blood flow through the skin and increase sweating. The erector pili muscles are relaxed so that the hairs lie flat, and any shivering stops. The metabolic rate may be reduced to lower the amount of warming in the body. The response is the same, whether the increase in core temperature comes from warming due to internal factors (such as during exercise or a fever) or external factors (such as a hot day).

If the temperature of the blood flowing through the hypothalamus drops, the thermoregulatory centre reacts by sending nerve impulses through the autonomic nervous system to the skin. These cause a reduction in the blood flow through the capillaries in the skin, along with a reduction in the production of sweat and contraction of the erector pili muscles to raise the hairs. Impulses in autonomic motor neurones from the thermoregulatory centre also stimulate involuntary contractions of muscles (known as shivering) and raise the amount of metabolic warming. Thus, the core temperature is usually maintained within very narrow limits (see **fig D**).

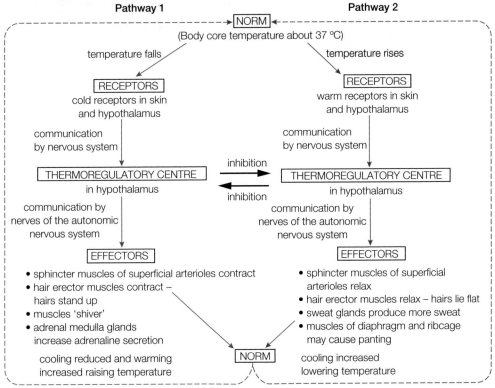

▲ **fig D** Thermoregulation in a mammal. The dilation or constriction of the blood vessels, the level of sweating and the erection of the body hair are all controlled by negative feedback systems. Each has an impact on the core temperature.

CHECKPOINT

1. Explain why thermoregulation is such an important aspect of homeostasis.

2. Produce a table to show how a mammal regulates its body temperature when:

   (a) the external temperature drops

   (b) the external temperature rises.

3. Some mammals give birth to immature, bald babies that are kept in nests or dens (an enclosed space) lined with moss, fur or feathers. Other mammals produce relatively mature offspring that are often very fluffy.

   (a) What is the main thermoregulatory problem for young endotherms?

   ▶ (b) Explain how the adaptations described in this chapter help them to survive.

SKILLS ▷ REASONING

SUBJECT VOCABULARY

thermoregulation a homeostatic mechanism that enables organisms to control their internal body temperature within set limits

endotherms animals that warm their bodies through metabolic processes at least in part and usually have a body temperature higher than the ambient temperature

arteriovenous shunt a system which closes to allow blood to flow through the major capillary networks near the surface of the skin, or opens to allow blood along a 'shortcut' between the arterioles and venules, so it does not flow through the capillaries near the surface of the skin

vasodilation the process by which the blood vessels become wider by relaxation of their muscle walls, which increases blood flow

vasoconstriction the process by which the blood vessels become narrower by contraction of their muscle walls, which reduces blood flow

thermoregulatory centre region of the hypothalamus in the brain that acts as the thermostat of the body keeping it at a set temperature; it contains temperature receptors which cause nerve impulses to be sent to different parts of the body if the temperature of the blood flowing through the hypothalamus increases or decreases

HYPERTHERMIA: HOW TO SURVIVE IN THE HEAT

SKILLS INTERPRETATION, DECISION MAKING, ADAPTIVE LEARNING, CREATIVITY, PERSONAL AND SOCIAL RESPONSIBILITY

People, and other 'warm-blooded' animals, battle all the time to maintain a constant body temperature regardless of the temperature of their surroundings. When we live in or visit extreme temperatures, this can be difficult.

INFORMATIVE JOURNALISM

Heat kills!

People can die in extreme weather. Don't be a victim of soaring temperatures!

▲ **fig A** When temperatures soar over 40 °C regularly in the summer heat, local people and tourists alike have to be very careful.

What is hyperthermia?

Hyperthermia happens when the normal core temperature of the human body goes up from 37 °C to 40 °C or over, but even before that your body will begin to malfunction. If your body temperature begins to climb above the normal 37 °C, what can you expect?

Heat stress, also known as heat exhaustion, causes headaches, nausea, vomiting, muscle cramps, weakness and fatigue, and either a lot of sweat or very little sweat. If you ignore the early warning signs of heat stress you may develop heat stroke which is potentially life-threatening. The body temperature goes above 40 °C, enzymes start to denature and cell membranes are damaged. Affected people develop neurological symptoms, becoming confused and even unconscious. If the core temperature is not brought down, the internal organs begin to fail and eventually people will die.

What causes hyperthermia?

There are two main causes. When the external temperatures get very high, people are at risk. Very young children, very old people and people who are seriously ill are most affected in these conditions. The other main cause of hyperthermia is when people are very active, especially in a very hot environment. This form of hyperthermia often affects young, healthy people who don't think about the effect of heat on their health.

Treating hyperthermia

It is very important to cool affected people from the outside and bring their core temperature down. This can be done by removing tight or unnecessary clothing, spraying them with cool water, blowing cool air on them, wrapping them in wet sheets or applying ice packs to the neck and other areas of the body. In hospital, cool liquids may be placed inside the body and in extreme cases the blood may be removed and passed through a heart bypass machine to cool it down.

SCIENCE COMMUNICATION

In 2018, many countries around the world experienced very hot summers. More people than ever before have been affected by hyperthermia as a result. This article describes some of the basic features of hyperthermia.

1 Make a bullet pointed list of the main symptoms of hyperthermia.

2 What causes hyperthermia?

3 Explain, with reasons, whether you think an article like this gives you the information you need to avoid hyperthermia yourself or to help you prevent or treat hyperthermia in others.

BIOLOGY IN DETAIL

Now let's examine the biology. You already know about the effect of temperature on reaction rates and enzyme structures, the main metabolic reactions of the body, hormonal controls of the body and thermoregulation. This will help you answer the questions below.

4 Reflect on the symptoms of hyperthermia. Explain what is happening biologically, and why the body reacts as it does.

5 Hypothermia is when the core temperature of the body falls from the normal 37 °C to below 35 °C. The skin turns greyish-blue, with a puffy face and blue lips, and the person is cold to the touch all over the body. The sufferer becomes drowsy and confused, with slurred speech and slow reflexes. The breathing and heart rate slow down and the blood pressure drops. If the temperature continues to fall they will become unconscious and eventually die. If people are found in time, they recover completely if they are warmed up gently. However, rapid warming leads to a severe drop in blood pressure and death.

(a) Give biological explanations for the symptoms of hypothermia described here.

(b) You should never give someone suffering from hypothermia a very hot drink, although a warm drink is helpful. Explain this statement.

ACTIVITY

- People travel and visit different countries around the world for many different reasons. These include visiting family and friends, for business, for sporting events, for a holiday or for a pilgrimage, such as the Hajj when millions of people from all over the world travel to Mecca.

- People from relatively cool areas of the world are at much greater risk of heat stroke when they visit countries where the environmental temperature is very high. This is partly because their bodies are not adapted to high environmental temperatures, and partly because they do not realise how dangerous any exercise (even just visiting tourist attractions) can be in the heat.

- Work in groups. You are going to develop some resources to help tourists visiting a very hot country. You need to explain what heat stroke is, how you get heat stroke, the symptoms, how to treat heat stroke and how to avoid heat stroke in the first place.

- **Either** use a smart phone to make a short movie which visitors can download

- **Or** produce a leaflet which can be handed out at the airport when visitors arrive.

- You could start by looking at *https://www.nih.gov/news-events/news-releases/hyperthermia-too-hot-your-health-1*

- Use online resources as well as scientific magazines and books. Evaluate your sources before using them and reference all sources used.

1 (a) What name is given to hormone-secreting glands in mammals? [1]

    A endocrine glands

    B exocrine glands

    C hypocrine glands

    D endothalamus glands

  (b) How are hormones transported to their target cells in mammals? [1]

    A through plasmodesmata

    B in the red blood cells

    C in the plasma

    D along neurones

  (c) Peptide hormones, such as insulin and adrenaline, affect the activity of their target cells by binding to receptors in the cell surface membrane. This initiates cascade effects in the cells, resulting in the activation of enzymes.

    Steroid hormones, however, diffuse through the cell surface membrane. The diagram below shows the mechanism of action of a steroid hormone.

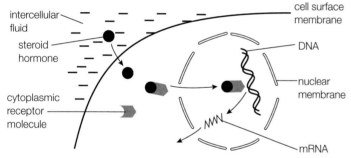

    (i) Name **one** steroid hormone that could act in this way. [1]

    (ii) Explain why steroid hormones can diffuse through the cell surface membrane while insulin and adrenaline cannot. [2]

    (iii) Using the information in the diagram, and your own knowledge, suggest how the presence of only a very small amount of steroid hormone in the intercellular fluid can result in a large change in the activity of the cell. [3]

    (iv) Adrenaline has an effect on its target cells within seconds of its release from the adrenal glands, and its effects last only for a short time. Explain why steroid hormones take much longer to act on their target cells, and their effects last longer. [2]

    (Total for Question 1 = 10 marks)

2 (a) A diet cola tastes sweet because it contains an artificial sweetener, aspartame, rather than glucose syrup or sucrose. Aspartame is very sweet so very little needs to be present and the diet cola has a very high water potential. If you consume a large bottle of this drink your body will respond in a certain way. Which of these best describes the way your body responds? [1]

    A More ADH is produced, increasing the volume of urine.

    B Less ADH is produced, decreasing the volume of urine.

    C Less ADH is produced, increasing the volume of urine.

    D More ADH is produced, decreasing the volume of urine.

  (b) Describe the roles of the loop of Henle in the production of concentrated urine in a mammal. [3]

  (c) Describe in detail the role of antidiuretic hormone. [7]

    (Total for Question 2 = 11 marks)

3 (a) At high environmental temperatures, the rate of sweating in humans increases. Explain how sweating is involved in the regulation of body temperature. [2]

  (b) In an investigation, a healthy volunteer measured his body temperature. After 5 minutes, he got into a bath of water at a temperature of 18 °C. He stayed in the bath for 10 minutes, then got out and sat on a chair. During the investigation, he recorded his body temperature at regular time intervals The results of this investigation are shown in the table below.

Time/min	Activity	Body temperature/°C
0	started investigation	37.0
5	got into bath	36.9
10	lying in bath	36.7
15	got out of bath	36.5
20	sitting on a chair	36.8
25	sitting on a chair	37.0

    (i) Describe the changes in body temperature that occurred during this investigation. [3]

    (ii) Explain the changes in body temperature that occurred between 5 and 10 minutes. [3]

    (iii) Suggest why the temperature loss was not more rapid. [3]

    (Total for Question 3 = 11 marks)

4 The diagram below represents a nephron (kidney tubule).

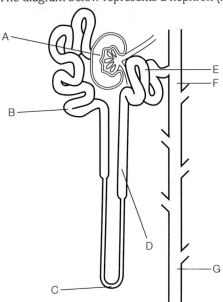

(a) (i) Which letter correctly identifies the glomerulus?   [1]

A

B

C

D

(ii) All the glucose in the fluid is reabsorbed back into the bloodstream as the fluid in the nephron passes from region **A** to region **B**. Which two cell adaptations would enable the cells of the nephron to absorb all the glucose?   [1]

A cell membrane and many ribosomes

B nucleus and many mitochondria

C microvilli and many mitochondria

D many ribosomes and many mitochondria

(b) The graph below shows the concentration of solutes in the fluid in the nephron in each of the labelled regions shown in the diagram. The graph shows the concentration of solutes when there is a high level of ADH (antidiuretic hormone) in the blood and when there is a low level of ADH in the blood.

(i) Calculate the percentage decrease in the concentration of solutes between regions A and G when there is a low level of ADH in the blood. Show your working.   [3]

(ii) The concentration of solutes in the fluid changes as it passes from region A to region G. Compare and contrast the changes that occur when the level of ADH in the blood is high with changes that occur when the level of ADH is low.   [4]

(iii) Use the information in the graph to explain how a rise in the level of ADH results in the production of a more concentrated urine.   [2]

(Total for Question 4 = 11 marks)

5 (a) Define the term *homeostasis*.   [2]

(b) Explain why negative feedback is essential in a homeostatic mechanism.   [4]

(c) The graph below shows the rate of flow of heat through the skin of a person during intense exercise. The exercise started at time 0s.

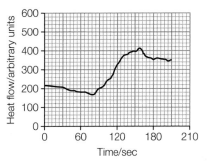

(i) Suggest why the heat flow decreased during the first 90 seconds of exercise.   [3]

(ii) Explain why heat flow through the skin increases after a period of intense exercise.   [3]

(Total for Question 5 = 12 marks)

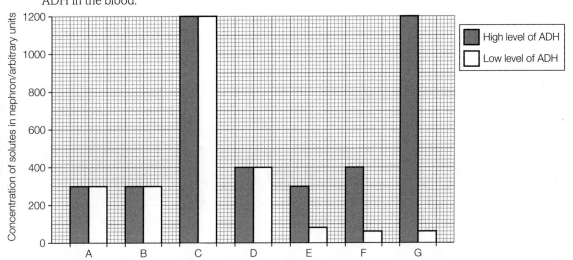

TOPIC 8 COORDINATION, RESPONSE AND GENE TECHNOLOGY

8A THE NERVOUS SYSTEM AND NEURONES

Think about the importance of your senses. Recall the sounds you hear when you first wake up in the morning, the sight of the people you care about and the beauty of the natural world. Think how important the sense of touch is to you, and why you need to be able to distinguish between hot and cold. All of your senses depend on the functioning of your nervous system, supplying you with information about the world (and the ability to react to that information) in milliseconds.

In this chapter, you will learn about the mammalian nervous system. You will consider the structure of individual neurones and consider how this structure is related to their functions. You will study the movements of ions into and out of axons that result in the passage of an action potential. You will learn about synapses, the junctions between neurones, that have a complex structure that is tightly and clearly related to their functions. You will consider how scientists have found out how this amazing system works and look at the effect of various drugs on transmission across the synapse.

The sensory organs are vital in the functioning of the nervous system and you will consider in some detail the way in which the eye works.

MATHS SKILLS FOR THIS CHAPTER

* **Construct and interpret frequency tables and diagrams, bar charts and histograms** (*e.g. interpreting bar charts on efficacy of lidocaine*)

* **Translate information between graphical, numerical and algebraic forms** (*e.g. interpretation of action potentials*)

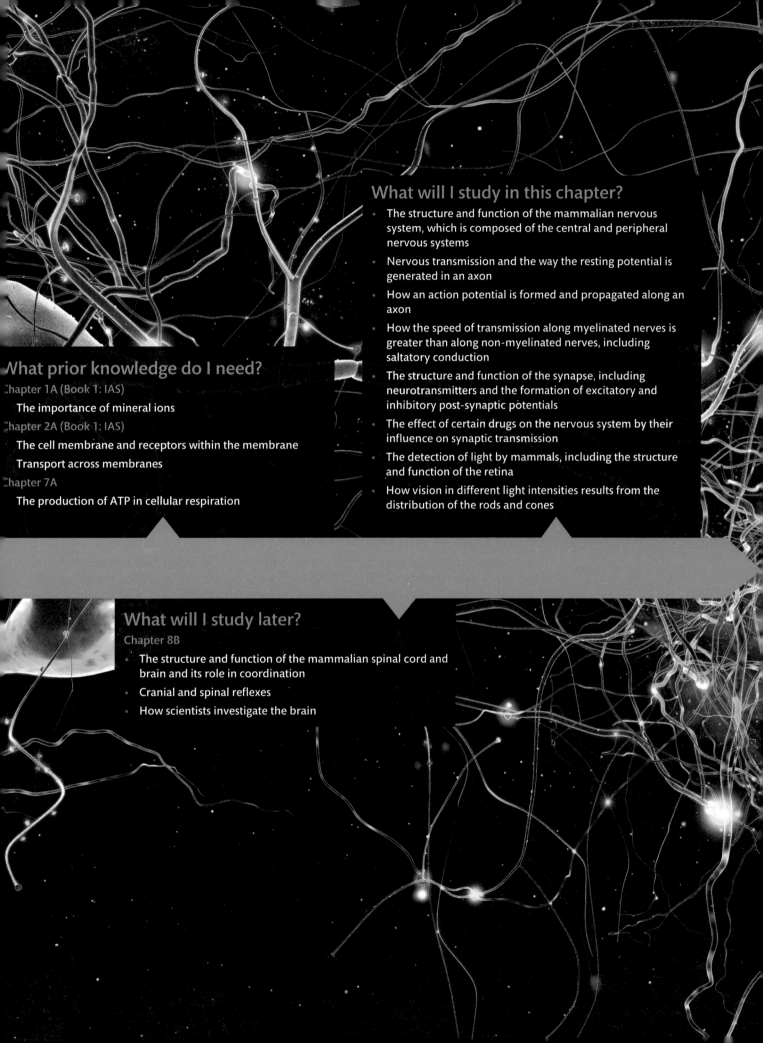

What will I study in this chapter?

- The structure and function of the mammalian nervous system, which is composed of the central and peripheral nervous systems
- Nervous transmission and the way the resting potential is generated in an axon
- How an action potential is formed and propagated along an axon
- How the speed of transmission along myelinated nerves is greater than along non-myelinated nerves, including saltatory conduction
- The structure and function of the synapse, including neurotransmitters and the formation of excitatory and inhibitory post-synaptic potentials
- The effect of certain drugs on the nervous system by their influence on synaptic transmission
- The detection of light by mammals, including the structure and function of the retina
- How vision in different light intensities results from the distribution of the rods and cones

What prior knowledge do I need?

Chapter 1A (Book 1: IAS)

The importance of mineral ions

Chapter 2A (Book 1: IAS)

The cell membrane and receptors within the membrane

Transport across membranes

Chapter 7A

The production of ATP in cellular respiration

What will I study later?

Chapter 8B

- The structure and function of the mammalian spinal cord and brain and its role in coordination
- Cranial and spinal reflexes
- How scientists investigate the brain

LEARNING OBJECTIVES

■ Know the structure and function of sensory, relay and motor neurones, including Schwann cells and myelination.
■ Understand how the nervous system of organisms can cause effectors to respond to a stimulus.
■ Understand how a nerve impulse (action potential) is conducted along an axon, including changes in membrane permeability to sodium and potassium ions.

In **Topic 7**, you looked at the need for homeostasis in mammals. Control systems are needed to maintain a dynamic equilibrium in the body. You learned that both the nervous system and the hormonal system are involved in maintaining homeostasis. These systems control the heart rate and breathing rate during exercise. They also maintain the osmotic balance of the body fluids and the core body temperature within narrow limits. Nervous coordination is rapid and very specifically targeted. In this section, you will look in detail at how the nervous system works.

THE BASIC STRUCTURE OF THE NERVOUS SYSTEM

A nervous system is made up of interconnected **neurones** (nerve cells) specialised for the rapid transmission of impulses throughout the organism. Neurones carry impulses from **receptor cells**, giving information about both the internal and the external environment. They also carry impulses to specialised **effector cells**, which are muscles or glands. These cause the appropriate response.

The simplest nervous systems are made up of receptor cells, neurones and the nerve endings associated with the effectors. Many nervous systems are much more complex. Groups of receptors have evolved to work together in **sense organs** such as the eye and the ear, and simple nerve nets have been replaced by complex nerve pathways.

As animals increase in size and complexity, they develop more specialised groups of nerve cells, which form a **central nervous system (CNS)**. In vertebrates, the central nervous system consists of the brain and spinal cord. It processes incoming information from sensory neurones and sends out impulses through motor neurones. **Sensory neurones** carry only impulses from receptor cells about the internal or external environment into the CNS. **Motor neurones** carry impulses from the CNS to the effector organs.

Neurones are individual cells and each one has a long nerve fibre that carries the nerve impulse. **Nerves** are bundles of nerve fibres. Some nerves carry only fibres from motor neurones and are called motor nerves, some carry only sensory fibres and are known as sensory nerves, and others carry a mixture of motor and sensory fibres. These are called mixed nerves.

The parts of the nervous system that are not within the central nervous system are known as the **peripheral nervous system**. You will learn more about this in **Section 8B.2**.

THE STRUCTURE AND FUNCTION OF NEURONES

Neurones are the basic unit of a nervous system and millions of neurones work together as an integrated whole in mammals, including humans. Neurones are cells which are specialised in the transmission of electrical signals called **nerve impulses**. They have a cell body that contains the cell nucleus, mitochondria, other organelles, and the rough endoplasmic reticulum (ER) and ribosomes needed for the synthesis of the neurotransmitter molecules. Neurones have many thin extensions from the cell body. These are called **dendrites** and they connect to neighbouring nerve cells (see **fig A**). The most distinctive feature of all nerve cells is the nerve fibre, which is extremely long and thin and carries the nerve impulse. Fibres that carry impulses away from the nerve cell body (in motor neurones) are called **axons**. Fibres that transmit impulses towards the cell body (in sensory neurones) are known

EXAM HINT

When describing nervous responses and the nervous system, always be clear whether you are talking about neurones or nerves.

as **dendrons**. Short relay or connector neurones are found in the CNS and they connect motor and sensory neurones. They are also known as bipolar neurones, because two fibres leave the same cell body. The different types of neurone are shown in **fig A**.

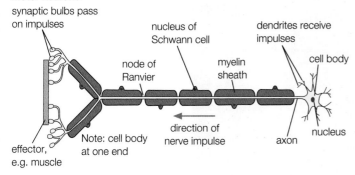

motor neurone

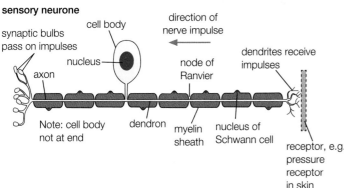

sensory neurone

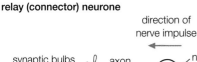

relay (connector) neurone

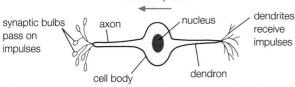

▲ **fig A** All neurones have the same basic structure of a cell body, dendrites and axons or dendrons. The detailed arrangements of these structures vary depending on the function of the neurone.

EXAM HINT

As part of your IAS work, you studied features of cells and how they are specialised (see **Chapter 3A (Book 1: IAS)**). Neurones are very specialised: their shape, the proportions of different organelles and their surface membrane are all specialised features. This could be tested as a link back to your IAS knowledge.

DEFINING A NERVE IMPULSE

The current model of a nerve impulse is of a minute electrical event produced by charge differences between the outside and inside of the neurone membrane. It is based on ion movements through specialised protein pores and by an active pumping mechanism.

MYELINATED NERVE FIBRES

Most vertebrate neurones are associated with another very specialised type of cell, the **Schwann cell** (see **fig B**). The Schwann cell membrane wraps itself around the nerve fibre many times, forming a fatty layer known as the **myelin sheath**. Gaps exist between the Schwann cells; they are known as the **nodes of Ranvier**. The myelin sheath is important for two reasons: it protects the nerves from damage and speeds up the transmission of the nerve impulse (see **Sections 8A.2** and **8A.3**).

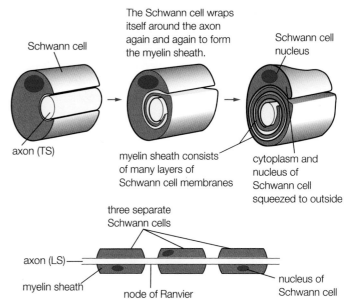

▲ **fig B** The myelin sheath forms a protective, insulating layer around vertebrate nerve fibres. (TS = transverse section, LS = longitudinal section.)

SPEEDY NERVE IMPULSES

The role of the nerve cells is to carry electrical impulses rapidly from one area of the body to another. The speed at which the impulses can be carried depends on two things.

- The diameter of the nerve fibre. Generally, the thicker the fibre is, the more rapidly impulses travel along it.

- The presence or absence of a myelin sheath. Myelinated nerve fibres can carry impulses much faster than unmyelinated ones.

Invertebrates do not have myelin sheaths on their nerve fibres, and many of their axons and dendrons are thin, measuring less than 0.1 mm in diameter. Therefore, the nerve impulses in many invertebrate animals travel quite slowly, at around $0.5\,ms^{-1}$. But there are times when even a relatively slow-moving invertebrate needs to react quickly to avoid danger. Many invertebrate groups have evolved a number of giant axons, which are nerve fibres with diameters of around 1 mm. These allow impulses to travel at approximately $100\,ms^{-1}$. This is fast enough for the animals to escape successfully if they are attacked. These giant axons have been used for much of the research into how axons work. They are relatively easy to use because they are so big, and they are found in invertebrates, which raises fewer ethical issues than working with vertebrates. **Fig C** shows you an example of giant axons in a squid.

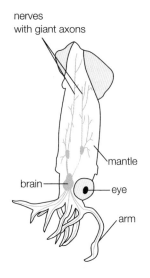
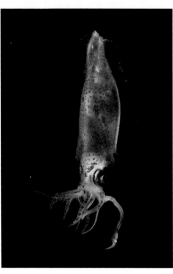

▲ **fig C** Squids are invertebrates with giant axons in their bodies and in their tentacles. This allows them to escape their predators by moving very rapidly.

Vertebrates have both unmyelinated and myelinated nerves. The voluntary motor neurones that transmit impulses to voluntary muscles to control movement are all myelinated. However, the autonomic neurones that control involuntary muscles, like those of the digestive system, have some unmyelinated fibres. The effect of the myelin sheath is to speed up the transmission of a nerve impulse without the need for giant axons, which require quite a lot of space because they are so big. A more adaptable network of relatively small myelinated nerve fibres can carry impulses extremely rapidly, at speeds of up to $120\,\text{ms}^{-1}$.

INVESTIGATING NERVE IMPULSES

To look at the events of a nerve impulse it is easiest to consider a 'typical' nerve fibre, ignoring size, myelination or type for the moment. Nerve impulses are electrical events, so a very effective way to investigate them is to record and measure the tiny electrical changes. Scientists did early work using a pair of recording electrodes placed on the outside of a nerve, which was then given a controlled stimulus. They recorded the resulting impulses with the electrodes and displayed them on a screen.

External electrodes record the responses of entire nerves. As you have learned, each nerve is made up of large numbers of different nerve fibres. These fibres have varying diameters and levels of sensitivity. Therefore, it can be difficult to interpret the results of these recordings correctly and the technique has been of limited value.

It was necessary to make recordings from within individual nerve fibres to make real progress. Motor axons are often found in large motor nerves that run directly to muscles. This makes them relatively easy to access for experiments, and you can immediately see the effect of stimulating them with the twitch of a muscle. Sensory nerve fibres often run from a sense organ in the head directly to the brain or from individual sensory receptors in the skin to the spinal cord, making them difficult to reach. Much of what you will learn about nerve impulses refers to axons, because these are the nerve fibres of motor neurones, but you can see the same events in all nerve fibres.

Most axons are extremely small, around $20\,\mu\text{m}$ in diameter, and this makes recording from inside the axon very difficult. Over 70 years ago Alan Hodgkin and Andrew Huxley began work on the giant unmyelinated axons of the squid (see **fig C**). They are around 0.5–1.0 mm in diameter and so are relatively easy to work with. These giant axons allow for very rapid nerve transmission to particular muscles, in situations when the squid needs to move quickly.

Hodgkin and Huxley used very fine glass microelectrodes inserted into the giant axon. Another electrode recorded the electrical potential from the outside of the axon. This allowed them to accurately record the changes that occur during the passage of an individual nerve impulse for the first time (see **fig D**). Scientists have refined this technique so that we can now use internal electrodes with almost any nerve fibre. We no longer have to base all our information on invertebrates.

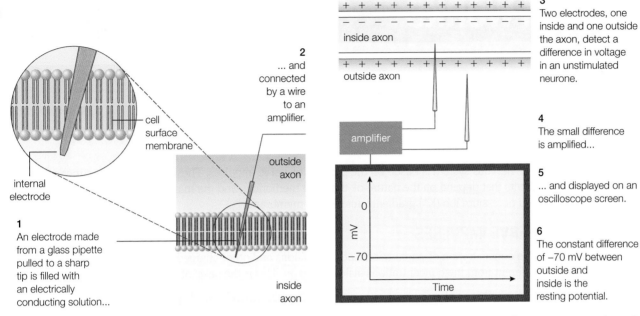

3 Two electrodes, one inside and one outside the axon, detect a difference in voltage in an unstimulated neurone.

2 ... and connected by a wire to an amplifier.

4 The small difference is amplified...

5 ... and displayed on an oscilloscope screen.

6 The constant difference of −70 mV between outside and inside is the resting potential.

1 An electrode made from a glass pipette pulled to a sharp tip is filled with an electrically conducting solution...

▲ **fig D** This apparatus, with an internal and an external electrode, has been used to research how neurones work. Here you can see the resting potential of a neurone being measured. The resting potential is the potential difference across the membrane of the nerve fibre in millivolts.

CHECKPOINT

1. Compare and contrast a nerve and a nerve fibre.

2. Describe how the structure of a motor neurone is related to its function.

3. ▶ The speed of transmission of a nerve impulse is in part related to the diameter of the nerve fibre. Vertebrate nerve fibres always have relatively small diameters, but the speed of the transmission of the nerve impulses of vertebrates is much faster than that in invertebrates. Explain these observations.

4. Squid giant axons are widely used in experiments on nerve transmission. Explain why.

SKILLS ▶ REASONING

SUBJECT VOCABULARY

neurones cells specialised for the rapid transmission of impulses throughout an organism
receptor cells specialised neurones that respond to changes in the environment
effector cells specialised cells that bring about a response if stimulated by a neurone
sense organs groups of receptors working together to detect changes in the environment
central nervous system (CNS) a specialised concentration of nerve cells which process incoming information and which send out impulses through motor neurones, which carry impulses to the effector organs
sensory neurones neurones that carry impulses from receptors about the internal or external environment into the central nervous system
motor neurones neurones that carry impulses from the central nervous system to the effector organs
nerves bundles (groups) of nerve fibres which may be all axons from motor neurones, all dendrons from sensory neurones or a mixture of both in a mixed nerve
peripheral nervous system the parts of the nervous system that spread through the body and are not involved in the central nervous system
nerve impulses the electrical signals transmitted through the neurones of the nervous system
dendrites the thin, finger-like extensions from the cell body of a neurone that connect with neighbouring neurones
axon the long nerve fibre of a motor neurone, which carries the nerve impulse
dendron the long nerve fibre of a sensory neurone, which carries the nerve impulse
Schwann cell a specialised type of cell associated with myelinated neurones which forms the myelin sheath
myelin sheath a fatty insulating layer around some vertebrate neurones produced by the Schwann cell
nodes of Ranvier gaps between the Schwann cells that enable saltatory conduction

LEARNING OBJECTIVES

■ Understand how a nerve impulse (action potential) is conducted along an axon, including changes in membrane permeability to sodium and potassium ions.

The nervous system is based on the passage of nerve impulses. These are minute electrochemical events that depend on the nature of the axon membrane, and the maintenance of sodium ion (Na$^+$) and potassium ion (K$^+$) gradients across that membrane.

NERVE IMPULSES

The concentration of sodium ions, potassium ions and other charged particles outside the axon is different from the concentrations inside the axon. This is the basis of the nerve impulse.

The membrane of an axon, like any other cell surface membrane, is partially permeable. It is the difference in permeability of this membrane to positively charged sodium and potassium ions that gives it special conducting properties. The axon membrane is relatively impermeable to sodium ions, but quite freely permeable to potassium ions.

THE RESTING POTENTIAL

If an axon is not conducting a nerve impulse, we say it is resting. The concentration of ions outside the nerve fibre is greater than the concentration in the cytoplasm of the axon. This gradient is created by a very active sodium/potassium ion pump. This is often referred to as the sodium pump or sodium/potassium pump. This pump has an enzyme called Na$^+$/K$^+$ ATPase that uses ATP (active transport) to move sodium ions out of the axon and potassium ions into the axon (see **Section 2A.4 (Book 1: IAS)**). It pumps sodium ions out of the axon, lowering the concentration of sodium ions inside the axon. The axon membrane is relatively impermeable to sodium ions, so they cannot diffuse back in again. At the same time, potassium ions are actively moved into the axon by the pump. However, because the axon membrane is permeable to potassium ions, they passively diffuse out again along the concentration gradient through open potassium ion channels. As a result, the inside of the cell is left slightly negatively charged relative to the outside. We say the membrane is **polarised**.

When you look at the events in neurones, the figures are always relative, comparing the inside of the nerve fibre to the outside solution. There is a potential difference across the membrane of around −70 mV. This is called the **resting potential** (see **fig A**).

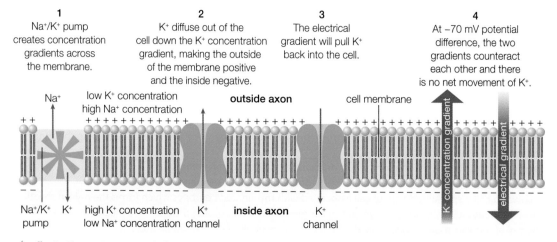

1
Na$^+$/K$^+$ pump creates concentration gradients across the membrane.

2
K$^+$ diffuse out of the cell down the K$^+$ concentration gradient, making the outside of the membrane positive and the inside negative.

3
The electrical gradient will pull K$^+$ back into the cell.

4
At −70 mV potential difference, the two gradients counteract each other and there is no net movement of K$^+$.

Na$^+$

low K$^+$ concentration
high Na$^+$ concentration

outside axon

cell membrane

K$^+$ concentration gradient

electrical gradient

++

+ + + + + + + + + +

+ + + + + + + + +

+ + + + + + + + + + +

+ + + + + + +

Na$^+$/K$^+$ pump

K$^+$

high K$^+$ concentration
low Na$^+$ concentration

K$^+$ channel

inside axon

K$^+$ channel

▲ **fig A** The resting potential of the axon is maintained by the sodium pump, the relative permeability of the membrane and the movement of potassium ions along concentration and electrochemical gradients.

THE ACTION POTENTIAL

When an impulse travels along an axon, there is a change in the permeability of the cell membrane to sodium ions. This change occurs in response to a stimulus (e.g. light, sound, touch, taste or smell) in a sensory neurone; or the arrival of a **neurotransmitter** in a motor neurone. In the experimental situation, the stimulus is usually a very small, precisely controlled electrical impulse.

When a neurone is stimulated, the axon membrane shows a sudden and dramatic increase in its permeability to sodium ions. Specific sodium ion channels or **sodium gates** open up, allowing sodium ions to diffuse rapidly down their concentration and electrochemical gradients. As a result, the potential difference across the membrane is briefly reversed, with the cell becoming positive on the inside with respect to the outside. This **depolarisation** lasts about 1 millisecond. The potential difference across the membrane at this point is about +40 mV. This is known as the **action potential**. Remember that these events happen in any nerve fibre, not just axons.

At the end of this very short depolarisation, the sodium ion channels close again and the excess sodium ions are rapidly pumped out by the active sodium pump. Also, the permeability of the membrane to potassium ions is temporarily increased. This happens because voltage-dependent potassium ion channels open as a result of the repolarisation. Consequently, potassium ions diffuse out of the axon down both a concentration gradient and an electrochemical gradient, attracted by the negative charge on the outside of the membrane. The inside of the axon becomes negative relative to the outside once again. It takes a few milliseconds before the resting potential is restored and the axon is ready to carry another impulse (see **fig B**).

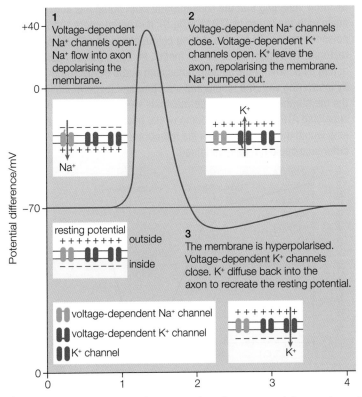

▲ **fig B** The ionic changes during excitation of an axon result in an action potential.

The events of the action potential can be recorded clearly using the internal/external electrode combination you have already seen. The oscilloscope trace is often referred to as a 'spike', because of its shape. It shows the change in the potential difference across the membrane with the inward movement of sodium ions followed by a return to the resting potential as the permeability of the membrane changes again. The **threshold** for any nerve fibre is the point when sufficient sodium ion channels open for the rush of sodium ions into the axon to exceed the outflow of potassium ions. Once the threshold has been reached, the action potential occurs. The size of this action potential is always the same. It is an all-or-nothing response (see **fig C**).

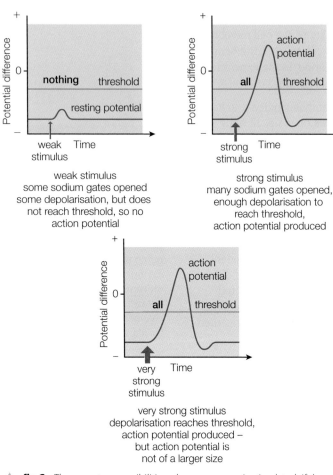

fig C There are two possibilities when a neurone is stimulated: if the stimulus is too small, the action potential does not happen; if the stimulus is large enough, the action potential will happen. If the stimulus increases, the action potential stays the same.

The recovery time of an axon is called the **refractory period**. This is the time it takes for an area of the axon membrane to recover after an action potential. The length of the refractory period is the time it takes for ionic movements to repolarise the membrane and restore the resting potential. This depends on both the sodium/potassium pump and the membrane permeability to potassium ions. For the first millisecond or so after the action potential, it is impossible to restimulate the fibre, the sodium ion channels are completely blocked and the resting potential has not been restored. This is known as the **absolute refractory period**. After this, there is a period of several milliseconds when the axon may be restimulated, but it will only respond to a much stronger stimulus than before because the threshold has been raised. This is known as the **relative refractory period**. During this time, the voltage-dependent potassium ion channels are still open. It is not until they are closed that the normal resting potential can be fully restored.

The refractory period is important in the functioning of the nervous system as a whole. It limits the rate at which impulses may flow along a fibre to 500–1000 per second. It also ensures that impulses flow in only one direction along nerves. Until the resting potential is restored, the part of the nerve fibre that the impulse has just left cannot conduct another impulse. This means the impulse can only continue travelling in the same direction.

DID YOU KNOW?

Building up the evidence

Some of the most convincing evidence for this model of the nerve impulse comes from work using poisons. A metabolic poison such as dinitrophenol (DNP) prevents the production of ATP. It also prevents the axon from functioning properly.

When an axon is treated with a metabolic poison, the sodium/potassium pump stops working as ATP is used up and the resting potential is lost at the same rate as the decrease in concentration of ATP. This implies that the ATP is being used to power the sodium/potassium pump; when there is no more ATP, the pump no longer works. If the poison is washed away, the metabolism returns to normal and ATP production begins again. The resting potential is restored, meaning that the sodium/potassium pump has started again with the return of ATP (see **fig D**). If experimenters supply a poisoned axon with ATP, the resting potential will be at least partly restored. This again confirms our model, implying that the poison is acting by depriving the sodium/potassium pump of ATP rather than by interfering with the membrane structure and its permeability. If the membrane structure and permeability were responsible, even supplying ATP directly to the pump would have no effect, because ions would move freely across the membrane and a potential difference could not be maintained.

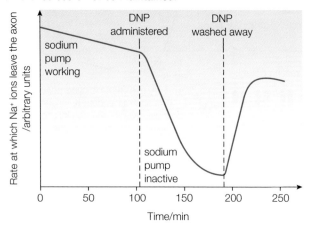

fig D This graph, based on work by Hodgkin and Keynes in 1955, illustrates clearly the effect of dinitrophenol (DNP) on the removal of sodium ions from the giant axon of a cuttlefish.

CHECKPOINT

1. Describe the resting potential of a neurone and explain how it is maintained.
2. (a) Describe an action potential and explain the importance of the refractory period.

   (b) Explain how an action potential can most accurately be measured.
3. ▶ Explain how the graph in **fig D** provides evidence for the role of the sodium/potassium ATPase pump in maintaining the resting potential.

SKILLS ▷ INTERPRETATION

SUBJECT VOCABULARY

polarised the condition of a neurone when the movement of positively charged potassium ions out of the cell down the concentration gradient is opposed by the actively produced electrochemical gradient, leaving the inside of the cell slightly negative relative to the outside

resting potential the potential difference across the membrane of around −70 mV when the neurone is not transmitting an impulse

neurotransmitter a chemical which transmits an impulse across a synapse

sodium gates specific sodium ion channels in the nerve fibre membrane that open up, allowing sodium ions to diffuse rapidly down their concentration and electrochemical gradients

depolarisation the condition of the neurone when the potential difference across the membrane is briefly reversed during an action potential, with the cell becoming positive on the inside with respect to the outside for about 1 millisecond

action potential when the potential difference across the membrane is briefly reversed to about +40 mV on the inside with respect to the outside for about 1 millisecond

threshold the point when sufficient sodium ion channels open for the rush of sodium ions into the axon to be greater than the outflow of potassium ions, resulting in an action potential

refractory period the time it takes for ionic movements to repolarise an area of the membrane and restore the resting potential after an action potential

absolute refractory period the first millisecond or so after the action potential during which it is impossible to re-stimulate the fibre, the sodium ion channels are completely blocked and the resting potential has not been restored

relative refractory period a period of several milliseconds after an action potential and the absolute refractory period during which an axon may be re-stimulated, but only by a much stronger stimulus than before

LEARNING OBJECTIVES

■ Understand how the nervous system of organisms can cause effectors to respond to a stimulus.
■ Understand the role of myelination in saltatory conduction.
■ Know the structure and function of synapses in nerve impulse transmission, including the role of neurotransmitters and acetylcholine.

Once an action potential is set up in response to a stimulus, it will travel the entire length of that nerve fibre, which may be many centimetres or even metres long. The movement of the nerve impulse along the fibre is the result of local currents set up by the ion movements at the action potential itself. These ion movements occur both in front of and behind the action potential. They depolarise the membrane in front of the action potential sufficiently to cause the sodium ion channels to open. The sodium ion channels behind the action potential cannot open because of the refractory period of the membrane behind the spike. In this way, the impulse is continually conducted in the required direction (see **fig A**).

EXAM HINT

This is a rare example of positive feedback. A small depolarisation caused by the local currents ahead of the action potential causes the sodium ion channels to open which increases the depolarisation.

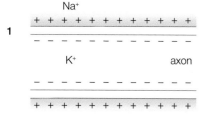

At resting potential there is positive charge on the outside of the membrane and negative charge on the inside, with high sodium ion concentration outside and high potassium ion concentration inside.

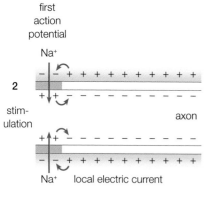

When stimulated, voltage-dependent sodium ion channels open, and sodium ions flow into the axon, depolarising the membrane. Localised electric currents are generated in the membrane.

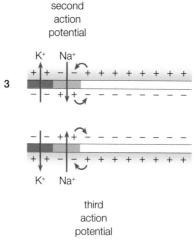

The potential difference in the membrane adjacent to the first action potential changes. A second action potential is initiated. At the site of the first action potential the voltage-dependent sodium ion channels close and voltage-dependent potassium ion channels open. Potassium ions leave the axon, repolarising the membrane. The membrane becomes hyperpolarised.

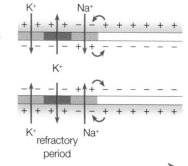

A third action potential is initiated by the second. In this way local electric currents cause the nerve impulse to move along the axon. At the site of the first action potential, potassium ions diffuse back into the axon, restoring the resting potential.

▲ **fig A** The action potential is conducted along a nerve fibre only in the direction in which it needs to go by a combination of tiny local currents and the inhibiting effect of the refractory period.

SALTATORY CONDUCTION

In myelinated neurones (see **Section 8A.1**), the situation is more complex. Ions can only pass in and out of the axon freely at the nodes of Ranvier, which are about 1 mm apart. This means that action potentials can only occur at the nodes, and so they appear to jump from one node to the next (see **fig B**). This speeds up transmission as the ionic movements associated with the action potential occur much less frequently, taking less time. It is known as **saltatory conduction**, from the Latin verb *saltare*, which means 'to jump'.

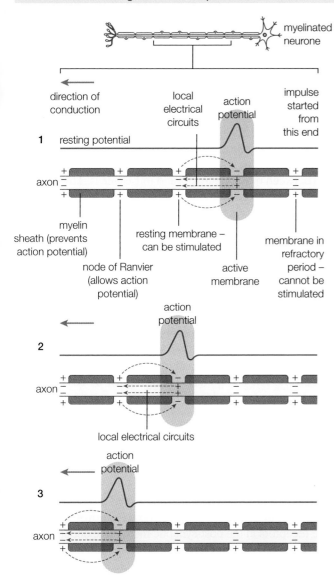

▲ **fig B** Saltatory conduction. By 'jumping' from node to node along a myelinated nerve fibre, the nerve impulses in vertebrate neurones can travel very rapidly along very narrow nerve fibres. This makes the development of complex but compact nervous systems possible.

SYNAPSES

Neurones must be able to communicate with each other. Receptors must pass their information into the sensory nerves, which must then relay the information to the central nervous system. Information needs to pass freely around the central nervous system and the impulses sent along the motor nerves must be communicated to the effector organs so that action can be taken. Wherever two neurones meet they are linked by a **synapse** (see **fig C**). Every cell in the central nervous system is covered with **synaptic knobs** from other cells: several hundred in some cases. Neurones never actually touch their target cell: a synapse involves a gap between two cells. Nerve impulses must somehow cross this gap.

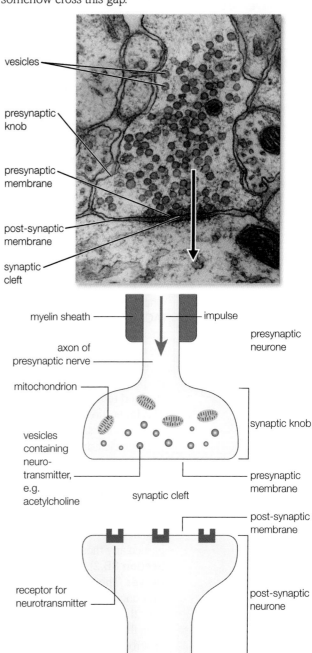

▲ **fig C** The structure of a synapse, based on information revealed by electron microscopy.

The functioning of synapses depends on the movement of calcium ions. The arrival of an impulse at the synaptic knob increases the permeability of the **presynaptic membrane** to calcium ions because calcium ion channels open up. Calcium ions then move into the synaptic knob down their concentration gradient. The effect of calcium ions flowing in is to cause the **synaptic vesicles**, which contain a transmitter substance or neurotransmitter, to move to the presynaptic membrane. Each vesicle contains about 3000 molecules of transmitter. Some of the vesicles fuse with the

EXAM HINT

Many candidates make a simple error in their description when they say that the vesicle of neurotransmitter diffuses across the gap. The vesicle is not released; it is the individual molecules that diffuse. This is an example of exocytosis.

EXAM HINT

Write out this sort of sequence of events as a numbered list. This will help you to ensure you have all the steps in the correct order.

presynaptic membrane and release the transmitter substance into the **synaptic cleft**. These molecules diffuse across the gap and become attached to specific protein receptor sites on the sodium channels of the post-synaptic membrane. This causes sodium ion channels in the membrane to open, resulting in sodium ions flowing into the nerve fibre, causing a change in the potential difference across the membrane and an **excitatory post-synaptic potential (EPSP)** to be set up.

If there are enough EPSPs, the positive charge in the post-synaptic cell becomes greater than the threshold level and an action potential is triggered which then travels on along the post-synaptic neurone.

In some cases, the neurotransmitter has the opposite effect. Different ion channels open in the membrane, allowing the movement of negative ions inwards, which makes the inside more negative than the normal resting potential. The result is an **inhibitory post-synaptic potential (IPSP)**; this makes it less likely that an action potential will occur in the post-synaptic fibre. These IPSPs play a part in the way we hear patterns of sound for example, and they are thought to be important in the way birds learn songs.

Once the transmitter has had an effect, it is destroyed by enzymes in the synaptic cleft so that the receptors on the post-synaptic membrane are emptied and can react to a subsequent impulse. Something similar to a synapse occurs at the site of a motor neurone meeting an effector. For example, neuromuscular junctions are found between motor neurones and muscle cells. The release of the transmitter stimulates a contraction in the muscle cell.

WHAT ARE THE TRANSMITTER SUBSTANCES?

One common neurotransmitter, found at the majority of synapses in humans, is **acetylcholine (ACh)**. The chemical is made in the synaptic knob using ATP produced in the many mitochondria present. Nerves that use acetylcholine as their transmitter are known as **cholinergic nerves**. Once the acetylcholine has travelled across the synaptic gap and triggered a post-synaptic potential, it is very rapidly broken down by the enzyme **acetylcholinesterase**. This enzyme is embedded in the post-synaptic membrane next to the acetylcholine receptors. Once the neurotransmitter has attached to the receptor and initiated a response, it is rapidly hydrolysed into acetate and choline. This (a) ensures that the acetylcholine no longer affects the post-synaptic membrane and (b) releases the components acetate and choline to be recycled. The components rapidly diffuse across the synaptic cleft down a concentration gradient and are taken back into the synaptic knob through the presynaptic membrane. In the presynaptic knob they are re-synthesised into more acetylcholine. Acetylcholine is the neurotransmitter in all motor neurones, the parasympathetic nervous system (part of the autonomic nervous system, see **Section 8B.2**), and cholinergic synapses in the CNS. It usually results in excitation at the post-synaptic membrane.

Not all vertebrate nerves have acetylcholine as their synaptic transmitter substance. Some neurones, particularly those of the sympathetic nervous system (part of the autonomic nervous system, see **Section 8B.2**), produce **noradrenaline** in their synaptic vesicles and are known as **adrenergic nerves**. The binding of noradrenaline to the receptors in the post-synaptic membrane depends on the concentration of the neurotransmitter in the synaptic cleft. After the release of noradrenaline from the presynaptic knob has stopped, levels in the synaptic cleft fall. Noradrenaline is then released from the post-synaptic receptors back into the synaptic cleft. Up to 90% of the noradrenaline is then absorbed by the presynaptic knob, and much of it is repacked and reused the moment another action potential comes along. Noradrenaline is the neurotransmitter in the sympathetic nervous system and in adrenergic synapses in the brain.

SKILLS CRITICAL THINKING

CHECKPOINT

1. ▶ Enzymes play several important roles in a synapse. Describe them.

2. Produce flow diagrams to summarise the sequence of events at:
    (a) a cholinergic synapse
    (b) an adrenergic synapse.

3. (a) State the difference between an EPSP and an IPSP.
    (b) Give reasons why they are important in the functioning of the nervous system.

SUBJECT VOCABULARY

saltatory conduction the process by which action potentials are transmitted from one node of Ranvier to the next in a myelinated nerve

synapse the junction between two neurones that nerve impulses cross via neurotransmitters

synaptic knobs the bulges (bumps) at the end of the presynaptic neurones in which neurotransmitters are made

presynaptic membrane the membrane on the side of the synapse which receives the first impulse and from which neurotransmitters are released

synaptic vesicles membrane-bound sacs (tiny bags) in the presynaptic knob which each contain about 3000 molecules of neurotransmitter; they move to fuse with the presynaptic membrane after an impulse arrives in the presynaptic knob

synaptic cleft the gap between the pre- and post-synaptic membranes in a synapse

excitatory post-synaptic potential (EPSP) the potential difference across the post-synaptic membrane caused by an influx of sodium ions into the nerve fibre as the result of the arrival of a molecule of neurotransmitter on the receptors of the post-synaptic membrane; this makes the inside more positive than the normal resting potential, increasing the chance of a new action potential

inhibitory post-synaptic potential (IPSP) the potential difference across the post-synaptic membrane caused by an influx of negative ions as the result of the arrival of a molecule of neurotransmitter on the receptors of the post-synaptic membrane; this makes the inside more negative than the normal resting potential, decreasing the chance of a new action potential

acetylcholine (ACh) a neurotransmitter found in the parasympathetic nervous system, the synapses of motor neurones, and cholinergic synapses in the brain

cholinergic nerves nerves using acetylcholine as their synaptic neurotransmitter

acetylcholinesterase an enzyme found within the post-synaptic membrane of cholinergic nerves that breaks down acetylcholine in the synapses after it has triggered a post-synaptic potential

noradrenaline a neurotransmitter found in the sympathetic nervous system and adrenergic synapses of the brain

adrenergic nerves nerves using noradrenaline as their synaptic neurotransmitter

LEARNING OBJECTIVES

■ Understand how the effects of drugs can be caused by their influence on nerve impulse transmission, illustrated by nicotine, lidocaine and cobra venom alpha toxin, the use of L-DOPA in the treatment of Parkinson's disease, and the action of MDMA (ecstasy).

The more we understand about synapses, the better we will understand many of the diseases that affect the peripheral and central nervous systems. Understanding synapses also helps us to decipher how many drugs in common use have their effect (see **fig A**).

THE EFFECTS OF DRUGS ON THE NERVOUS SYSTEM

Fig A shows the main ways in which drugs affect synapses.

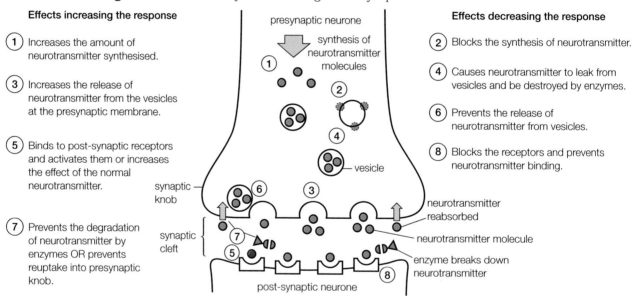

Effects increasing the response

(1) Increases the amount of neurotransmitter synthesised.

(3) Increases the release of neurotransmitter from the vesicles at the presynaptic membrane.

(5) Binds to post-synaptic receptors and activates them or increases the effect of the normal neurotransmitter.

(7) Prevents the degradation of neurotransmitter by enzymes OR prevents reuptake into presynaptic knob.

Effects decreasing the response

(2) Blocks the synthesis of neurotransmitter.

(4) Causes neurotransmitter to leak from vesicles and be destroyed by enzymes.

(6) Prevents the release of neurotransmitter from vesicles.

(8) Blocks the receptors and prevents neurotransmitter binding.

presynaptic neurone — synthesis of neurotransmitter molecules — vesicle — synaptic knob — synaptic cleft — neurotransmitter reabsorbed — neurotransmitter molecule — enzyme breaks down neurotransmitter — post-synaptic neurone

▲ **fig A** The main ways in which drugs affect synapses

This information enables us to understand how many different substances affect our bodies. Here are three specific examples.

- **Nicotine** is the addictive drug found in tobacco. It mimics the effect of acetylcholine and binds to specific acetylcholine receptors in post-synaptic membranes known as nicotinic receptors. It triggers an action potential in the post-synaptic neurone, but then the receptor remains unresponsive to more stimulation for some time. Nicotine causes raised heart rate and blood pressure. It also triggers the release of another type of neurotransmitter in the brain called dopamine. This is associated with pleasure sensations. At low doses, nicotine has a stimulating effect, but at high doses it blocks the acetylcholine receptors and can kill. It is a highly addictive drug. One cigarette contains 1–6 mg nicotine. Even at a low dose, nicotine has a big effect on your acetylcholine synapses, and dopamine release is also large. People trying to stop smoking cigarettes, with their lethal load of carcinogens and other toxins, often need to replace the nicotine to help them do so.

- **Lidocaine** is a drug used as a local anaesthetic. It is commonly used by dentists before drilling or removing a tooth. If you have ever had an injection at the dentist you will probably have been injected with lidocaine. Lidocaine molecules block voltage-gated sodium channels, preventing the production of an action potential in sensory nerves and so it prevents you from feeling pain (see **fig B**). It is also used to prevent some heart arrhythmias (irregular heartbeat). In this situation, it works in the same way (i.e. blocking sodium channels) therefore raising the depolarisation threshold. This reduces or prevents early or extra action potentials from the pacemaker region that can cause arrhythmias.

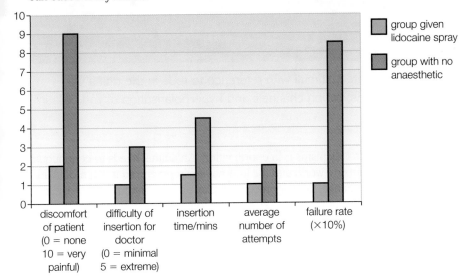

▲ **fig B** Inserting a tube into the nose of patients in hospital is not easy and can be very painful. Research done in 2010 showed that using a topical spray of lidocaine and giving plenty of time for it to work, made the process much less painful for the patient, and faster and much more successful for the doctor.

- **Cobra venom (α-cobratoxin)** is a substance made by several species of cobra. It is toxic and often fatal in snake bites. It binds reversibly to acetylcholine receptors in post-synaptic membranes and neuromuscular junctions. Thus, it prevents the transmission of impulses across synapses, including the neuromuscular junctions between motor neurones and muscles. Consequently, muscles are not stimulated to contract and gradually the person affected becomes paralysed. When the toxin reaches the muscles involved in breathing it causes death. However, in very low concentrations, it can relax the muscles of the trachea and bronchi in severe asthma attacks, and so save lives.

CHECKPOINT

SKILLS   INTERPRETATION

1. ▶ Strychnine inactivates cholinesterase at the post-synaptic membrane. Botulinus toxin affects the presynaptic membrane, preventing the release of acetylcholine. Explain how electron micrographs and these toxins could be used as evidence for the current model of synaptic transmission.

2. Make a table to summarise the main ways in which drugs may affect synapses.

3. Use the data in **fig B**, taken from a sample of over 200 patients randomly assigned lidocaine or a placebo before having a nasogastric tube inserted, to answer the following questions.

   (a) The first two scores refer to the pain recorded by the patient during the procedure and the difficulty recorded by the doctor in getting the nasogastric tube in place. Determine the percentage reduction in pain for the patients and difficulty for the doctors.

   (b) State the mean time taken to insert the tube with and without lidocaine.

   (c) Give the percentage of procedures that failed completely with and without lidocaine.

   (d) Describe how lidocaine works and state why it has such a marked effect on the success of inserting a nasogastric tube.

SUBJECT VOCABULARY

nicotine a drug found in cigarettes that mimics the effect of acetylcholine and binds to specific acetylcholine receptors in post-synaptic membranes known as nicotinic receptors

lidocaine a drug used as a local anaesthetic that works by blocking the voltage-gated sodium channels in post-synaptic membranes in sensory neurones, preventing the production of an action potential

cobra venom (α-cobratoxin) a substance made by several species of cobra that binds reversibly to acetylcholine receptors in post-synaptic membranes in motor neurones, preventing the production of a post-synaptic action potential

5 SENSORY SYSTEMS AND THE DETECTION OF LIGHT

LEARNING OBJECTIVES

■ Understand how the nervous systems of organisms can detect stimuli with reference to rods in the retina of mammals, the roles of rhodopsin, opsin, retinal, sodium ions, cation channels and hyperpolarisation of rod cells in forming action potentials in the optic neurones.

The main role of nervous systems in organisms is in coordination. An animal needs to continuously receive information from both the outside world and the internal environment, so it can modify its behaviour and survive as situations change. Sensory receptors play an important role in providing an animal with information about both its internal and external environment.

SENSORY RECEPTORS

EXAM HINT

Remember that a stimulus is a change in the environment which is detected by a receptor. The receptor translates the energy in the stimulus to electrical impulses in the neurones.

Simple sensory receptors are neurones with a dendrite that is sensitive to one particular stimulus. After the dendrite receives a stimulus, chemical events occur that result in an action potential in the nerve fibre of the neurone. This type of cell is known as a primary receptor. A secondary receptor is slightly more complicated. It consists of one or more completely specialised cells (not neurones) that are sensitive to a particular type of stimulus. These cells synapse with a normal sensory neurone, which carries the impulse to the central nervous system. The retinal cells in the retina of the eye are an example of these secondary receptors.

There are various different types of receptors and a variety of ways of categorising them.

As animals become increasingly complex, so do their sensory systems. Many sensory receptors are always found as isolated entities, but in higher animals sensory receptors are often found in systems known as sense organs.

The different types of receptor include the following.

- Proprioceptors are sensitive to the relative positions of the skeleton and degrees of muscle contraction. They are very important for maintaining posture.
- Chemoreceptors are sensitive to chemical stimuli such as smell, taste and the pH levels of the blood.
- Mechanoreceptors are sensitive to mechanical stimuli such as pressure, tension, movement and gravity.
- Photoreceptors are sensitive to electromagnetic stimuli. For humans, this is usually visible light, but many insects are sensitive to UV light.
- Thermoreceptors are sensitive to temperature changes and differences in temperature.

HOW DO SENSORY RECEPTORS WORK?

Receptor cells, similar to nerve fibres, have a resting potential that depends on maintaining a negative charge of the cell interior in relation to the outside. This potential difference is maintained by membrane sodium pumps. When a receptor cell receives a stimulus, sodium ions move rapidly into the cell, along concentration and electrochemical gradients, and this sets up a **generator potential**. A small stimulus results in a small generator potential and a large stimulus results in a large generator potential. Generator potentials do not obey the all-or-nothing law. If the generator potential produced is large enough to reach the threshold of the sensory neurone, an action potential will occur in that neurone. If it does not reach the threshold then there will be no action potential so the action potential does obey the all-or-nothing law.

The following process is common to most sensory receptors:

Stimulus → local change in permeability → generator potential → action potential

In sense organs such as the eye, several receptor cells will often synapse with a single sensory neurone. If the generator potential from an individual receptor cell is insufficient to trigger an action potential, the generator potentials from several receptor cells may add together and trigger an action potential. This is known as **convergence** and it is a useful adaptation for increasing the sensitivity of a sensory system to low-level stimuli. It is an important feature of the light-sensitive cells of the retina of the eye.

A weak stimulus results in a low frequency of action potentials along the sensory neurone. A strong stimulus results in a rapid stream of action potentials being fired along the sensory neurone. As a result, the axon obeys the all-or-nothing law in terms of each individual action potential, but a graded response is still possible which gives information about the strength of the stimulus. In the eye, this means that we are not only aware of the difference between light and dark but also of all the varying degrees of light and shade (see **fig A**).

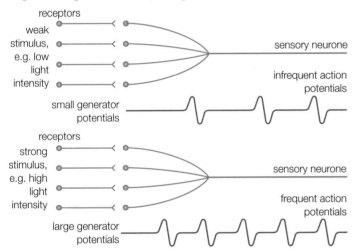

▲ **fig A** The interactions between sensory receptors and sensory neurones give high levels of sensitivity to different strength stimuli, as seen in the eye and other sense organs.

THE HUMAN EYE

Sensitivity to light is found in some of the simplest organisms. In humans, this has developed far beyond a simple sensitivity to the absence or presence of light. The human eye is a sophisticated optical system giving us clear and focused vision in a wide variety of circumstances. Our eyes are sensitive to electromagnetic radiation with a wavelength between 400 and 700 nm. The ranges of other species are often different from this. The structure of the human eye can be seen in **fig B**.

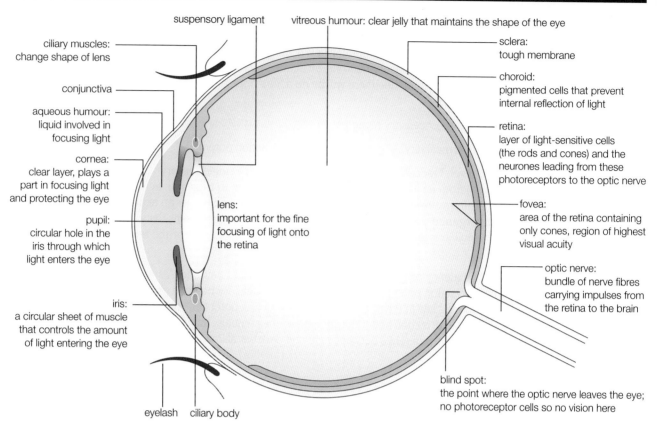

suspensory ligament

vitreous humour: clear jelly that maintains the shape of the eye

ciliary muscles:
change shape of lens

sclera:
tough membrane

conjunctiva

choroid:
pigmented cells that prevent
internal reflection of light

aqueous humour:
liquid involved in
focusing light

retina:
layer of light-sensitive cells
(the rods and cones) and the
neurones leading from these
photoreceptors to the optic nerve

cornea:
clear layer, plays a
part in focusing light
and protecting the eye

lens:
important for the fine
focusing of light onto
the retina

fovea:
area of the retina containing
only cones, region of highest
visual acuity

pupil:
circular hole in the
iris through which
light enters the eye

optic nerve:
bundle of nerve fibres
carrying impulses from
the retina to the brain

iris:
a circular sheet of muscle
that controls the amount
of light entering the eye

eyelash   ciliary body

blind spot:
the point where the optic nerve leaves the eye;
no photoreceptor cells so no vision here

▲ **fig B** The main structures of the human eye.

THE ROLE OF THE RETINA

The structures at the front of the eye include the cornea, iris, pupil and lens. All of these plus the aqueous and vitreous humour are involved in controlling the amount of light that enters the eye (see **Section 8B.2**) and focusing that light on the retina. But focusing the light onto the retina is only the first step. The retina must then perceive the light and provide the information the **brain** needs to interpret the image. The human retina contains around 100 million light-sensitive cells (photoreceptors), and the neurones with which they synapse. There are two main types of photoreceptors in the retina, known as the **rods** and the **cones**. Both types are secondary receptors signalling changes in the external environment.

You have around 90 million rods in each eye. They are spread evenly across the retina except at the **fovea** where there are none. The rods provide black and white vision only and they are very sensitive to light. They are used mainly for seeing in low light intensities or at night. The rods are not very tightly packed together, and several of them synapse with the same sensory neurone as shown in **fig C**. Consequently, they do not give a very clear picture because several rods need to be stimulated at the same time to cause an action potential in the sensory neurone. However, this also means they are extremely sensitive to low light levels and to movements in the visual field. Because several sensory receptors converge on one sensory neurone, each one only needs a small stimulus to produce several small generator potentials which can add together to trigger an action potential to the CNS by summation.

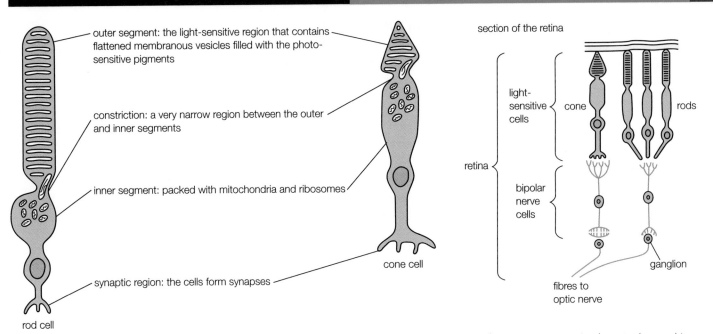

▲ **fig C**   The two types of receptor cell found in the human retina, along with their different arrangement of synapses, give us a visual system that combines great sensitivity to low levels of light with high visual acuity and clarity of colour vision.

Cones, on the other hand, are found tightly packed together in the fovea. There are only around 6 million of them. They are used mainly for vision in bright light and they also provide colour vision. Because they are so tightly packed in the fovea, in addition to each cone usually having its own sensory neurone (see **fig C**), in bright light cones provide great **visual acuity** (ability to see very clearly in sharp focus). It is only when light is focused directly on the fovea that an image is clearly in focus.

The arrangement of the retina is surprising because it appears to be back-to-front. The 'outer segments' are next to the choroid, and the neurones are at the interior edge of the eyeball. The light has to pass through the synapses and the inner segments before reaching the outer segments containing the visual pigments. The reason for this unexpected arrangement lies in the origin of the retina and the way in which the eye is formed during embryonic development. The apparent back-to-front structure is also why there is a blind spot: all the nerve fibres are gathered together to pass through all the layers of the eye to go to the brain (see **fig D**). This means there is a spot where there cannot be any sensory cells.

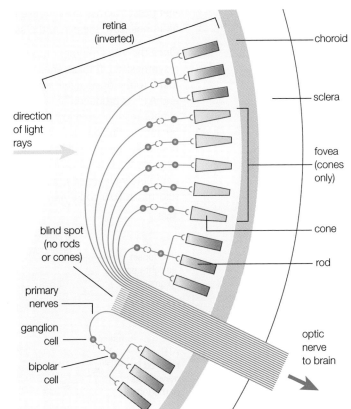

▲ **fig D**   The structure of the retina. (The rods and cones are shown as being much larger than they really are, to make the arrangement of sensory cells and neurones clear.)

HOW DO THE RODS AND CONES WORK?

The rods and cones work in a similar way founded on the reactions of a visual pigment with light. In the rods this visual pigment is **rhodopsin (visual purple)**, which is formed from two components, opsin and retinal. Opsin is a lipoprotein and retinal is a light-absorbing derivative of vitamin A. Retinal exists as two different isomers, cis-retinal and trans-retinal. In the dark, it is all in the cis-form. When a photon of light (the smallest unit of light) hits a molecule of rhodopsin, it converts the cis-retinal into trans-retinal. This changes the shape of the retinal which puts a strain on the bonding between the opsin and retinal. As a result, the rhodopsin breaks into opsin and retinal. This breaking of the molecule is referred to as **bleaching**.

The membranes of most neurones are relatively impermeable to sodium ions, but rod cell membranes are normally very permeable to them. Sodium ions move into the rod cell through sodium ion channels, and the sodium/potassium pump moves them out again. When rhodopsin is bleached, it triggers a cascade reaction (a series of reactions each caused by the reaction before) leading to the sodium ion channels closing, so the rod cell membrane becomes much less permeable to sodium ions and fewer sodium ions diffuse into the cell. The sodium pump continues to work at the same rate, pumping sodium ions out of the rod cell, so the interior becomes more negative than usual. This hyperpolarisation is known as the generator potential in the rod. The size of the generator potential depends on the amount of light hitting the rod, and therefore the amount of rhodopsin bleaching that follows. If it is large enough to reach the threshold, or if several rods are stimulated at once, neurotransmitter substances are released into the synapse with the bipolar cell, which synapses on both the photoreceptor and the nerve fibre. An action potential is then triggered in the bipolar cell that passes across the synapse to cause an action potential in the sensory neurone. All the sensory neurones leave the eye at the same point to form the optic nerve leading to the brain.

Once the visual pigment has been bleached, the rod cannot be stimulated again until the rhodopsin is resynthesised. It takes energy in the form of ATP to convert the trans-retinal back to cis-retinal and join it to the opsin again to form rhodopsin. Rods have many mitochondria in their inner segments and ATP is produced here (see **fig C**). In normal daylight the rods are almost entirely bleached and cannot respond to a low level or intensity of light and the eye is said to be light-adapted. After about 30 minutes in complete darkness the rhodopsin will have almost completely reformed from opsin and retinal, so the eye is fully sensitive to low intensity light and is said to be dark-adapted. This explains why you become almost blind when you walk from a sunny street into a house. The bright light has completely bleached the rhodopsin you need to see in the darker interior light. As the rhodopsin reforms in your rods, your vision returns as your eyes become dark-adapted again.

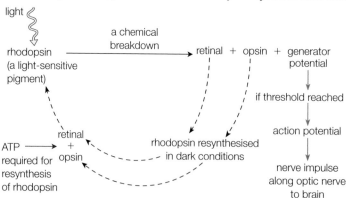

▲ **fig E**   The reactions of rhodopsin in light.

THE CONES AND COLOUR VISION

The cones work in a very similar way to rods, except that their visual pigment is known as **iodopsin**. There appear to be three types of iodopsin, each sensitive to one of the primary colours of light. Iodopsin needs to be hit with more light energy than rhodopsin in order to break down, and so it is not sensitive to low light intensities. The cones provide colour vision because the brain interprets the numbers of different types of cones stimulated as different colours. This complements the low light vision and sensitivity to movement provided by the rods.

> **LEARNING TIP**
>
> Remember that a rod cell operates in a different way from most neurones. In a rod, the sodium ion channels are open, which usually prevents the inside becoming too negative. Light causes the channels to close allowing a negative potential to develop inside the cell and reach the threshold.

CHECKPOINT

1. (a) Describe how a sensory receptor works.

    (b) Explain how convergence increases the sensitivity of a system to low level stimuli.

2. (a) Name the light-sensitive pigment in the rod cells and describe the effect that light has on the pigment.

    (b) Explain how this change in the visual pigment is translated into a visual image in the brain.

3. Explain how the different neural connections of the rods and the cones, which you can see in **fig C**, account for the different visual acuity they give.

4. Account for the following:

    (a) rods transmit information in dimmer light than cones

    (b) rods are more sensitive to movement than cones

    (c) cones give a clearer image than rods.

SKILLS   REASONING

SUBJECT VOCABULARY

generator potential a graded response to a stimulus across the synapse of a sensory receptor

convergence when several sensory receptors all synapse with one sensory neurone so the neurotransmitters add together to trigger an action potential in the sensory neurone

brain the area of the CNS in which information can be processed and from which instructions can be issued as required to give fully coordinated responses to a range of situations

rods photoreceptors found in the retina which contain the visual pigment rhodopsin; they respond to low light intensities, give black and white vision and are very sensitive to movement

cones photoreceptors found in the fovea of the retina which contain the visual pigment iodopsin; they respond to bright light, give great clarity of vision and colour vision

fovea area of the retina packed with cones which provides colour vision and great visual acuity

visual acuity the ability to see very clearly in sharp focus

rhodopsin (visual purple) the visual pigment in the rods

bleaching the photochemical breakdown of visual pigments (e.g. rhodopsin to opsin and retinal)

iodopsin the visual pigment in the cones

LEARNING OBJECTIVES

■ Understand what is meant by the term habituation.

Habituation is part of the learning process in almost all animals, including human beings. It is very important in the development of young animals, as they learn not to react to the neutral elements in the world around them. Habituation may be relatively short term or it may become long term so that a response is lost permanently. Memory is a crucial element of learning and scientists think that habituation is involved in the brain development which happens as memories are formed, when permanent changes occur in the synapses.

STUDYING HABITUATION

Much of the work studying habituation has used invertebrate organisms such as the nematode worm *Caenorhabditis elegans* and the giant sea slug *Aplysia* (see **fig A**).

▲ **fig A** The sea slug *Aplysia* has about 20 000 neurones, some of which are giant axons. It is a very useful experimental animal and can be used to investigate habituation.

Aplysia breathes using a gill in a cavity on the upper side of the body; water passes through and is forced out through a siphon tube at one end. If you touch the siphon tube, the whole gill is withdrawn into the body cavity as a defence mechanism. However, as the sea slug lives in the sea, the movement of the water constantly stimulates the siphon. The animal learns by habituation not to withdraw its gill every time a wave hits it.

We can investigate the mechanism of this habituation experimentally. *Aplysia* grown in the laboratory are not habituated to waves or water movement. Eric Kandel was one of many researchers to use this animal to study the nervous system and he performed some classic experiments. He stimulated the siphon of the sea slug with a jet of water which made it withdraw its gill. The stimulation was applied repeatedly. The response of the animal

became less and less until eventually water squirted at the siphon had no effect and the gill was not withdrawn at all. The animal had habituated, that is to say it had learned to ignore the stimulus. This habituation was retained over time (see **fig B (a)** and **(b)**).

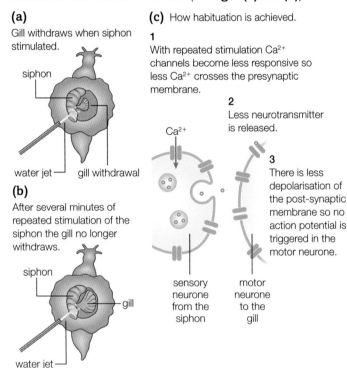

(a)
Gill withdraws when siphon stimulated.

siphon

water jet — gill withdrawal

(b)
After several minutes of repeated stimulation of the siphon the gill no longer withdraws.

siphon

gill

water jet —

(c) How habituation is achieved.

1
With repeated stimulation Ca²⁺ channels become less responsive so less Ca²⁺ crosses the presynaptic membrane.

2
Less neurotransmitter is released.

Ca²⁺

3
There is less depolarisation of the post-synaptic membrane so no action potential is triggered in the motor neurone.

sensory neurone from the siphon

motor neurone to the gill

▲ **fig B** Repeated stimulation of the siphon of *Aplysia* results in habituation (a) and (b). This is a simple form of learning whereby the sea slug stops responding to the stimulus as a result of changes in the response of the synapse, as shown in (c).

Kandel then identified the neurones involved in the gill reflex and investigated what happened to them when the siphon was repeatedly stimulated. He discovered changes in the synapses (see **Section 8A.3**). The calcium channels in the presynaptic membrane become less responsive with repeated stimulations. With fewer calcium channels open, fewer calcium ions cross into the presynaptic knob. Consequently, fewer vesicles move to the presynaptic membrane, fuse and discharge their neurotransmitter. When there is less neurotransmitter available to bind to the post-synaptic membrane, the post-synaptic excitatory potential is not high enough to trigger an action potential so there is no response (see **fig B (c)**).

DID YOU KNOW?

Investigating habituation

It is possible to investigate habituation to a stimulus in the school laboratory. Various organisms can be used to demonstrate this simple form of learning, including worms, snails and people. It is important to maintain high standards of ethics and welfare when carrying out any investigation using living animals. For example, you can investigate the way the responses of invertebrates such as the Giant African land snail habituate to gentle touch.

▲ **fig C** Giant African land snails can be used to investigate habituation in a school laboratory.

EXAM HINT

You should try to learn lots of examples of habituation because exam questions could feature organisms you have never heard of. Reading will make you familiar with more examples and you are then more likely to recognise what the exam question is asking.

HABITUATION IN HUMANS

Habituation is an important form of learning in humans in the short term and in the long term. For example, if you visit someone who lives near an airport, the first few times a plane takes off or lands you will be surprised by or at least aware of the noise. After a few hours, you will no longer notice the planes because you have become habituated. If you don't visit again for some time, your awareness of the planes taking off and landing will return but you will habituate more rapidly on your second visit. Habituation is very important in the development of newborn babies who have a number of startle reflexes when disturbed suddenly. These gradually become habituated as the baby grows and matures; eventually, they are lost completely. An example is the Moro reflex (see **fig D**). In this response, a baby arches its back and throws

its arms up and out in a grasping movement if it is startled or its head is lowered suddenly. This reflex response reduces and is lost in a normal baby by about 8 weeks of age. This allows the baby to interact more calmly with the world around it. Modern research suggests that one of the problems for people affected by schizophrenia may be that they do not habituate as readily as other people.

The study of the brain and behaviour continues and we still have enormous amounts to discover about how our own brains work. You will learn more about brains in **Chapter 8B**.

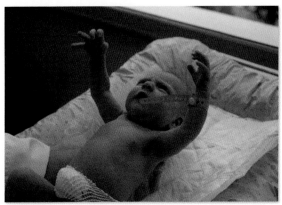

▲ **fig D** Habituation plays an important part in the development of a baby's brain. You can imagine the problems it would cause if adults still showed a startle reaction like this every time there was a loud noise.

CHECKPOINT

1. Sea slugs and other similar non-vertebrates can transmit nervous impulses rapidly although they do not have myelinated neurones. Explain why this makes them such useful experimental animals.

2. If someone makes a sudden movement towards your eyes, you will blink. If they do this repeatedly, you stop blinking. This is habituation. Explain how changes in the synapses bring about this change in behaviour.

SKILLS ▶ CREATIVITY

3. ▶ Explain the importance of habituation in development and learning.

SUBJECT VOCABULARY

habituation diminishing of an innate response to a frequently repeated stimulus

1 (a) Which of these neurones will transmit an impulse
fastest? [1]

   A myelinated axon 30 μm diameter

   B myelinated axon 2 μm diameter

   C unmyelinated axon 30 μm diameter

   D unmyelinated axon 20 μm diameter

(b) The graph below shows the potential difference across an
axon membrane during an action potential.

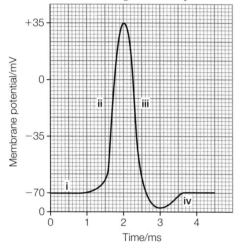

(i) Which row correctly identifies stages i–iv of the
action potential? [1]

	i	ii	iii	iv
A	resting potential	hyperpolarisation	depolarisation	repolarisation
B	resting potential	depolarisation	repolarisation	hyperpolarisation
C	resting potential	repolarisation	hyperpolarisation	depolarisation
D	hyper-polarisation	depolarisation	repolarisation	resting potential

(ii) Which row correctly identifies the movement of ions at
each stage of the action potential? [1]

	i	ii	iii	iv
A	Na^+ out K^+ in	K^+ in	Na^+ out	Na^+ out K^+ in
B	Na^+ out K^+ in	Na^+ out	K^+ in	K^+ out
C	Na^+ out K^+ in	Na^+ in	K^+ out	Na^+ out K^+ in
D	Na^+ in K^+ out	Na^+ in	K^+ out	Na^+ out K^+ in

(c) Myelinated and non-myelinated neurones carry impulses
at various speeds. The speed of a nerve impulse along an
axon is known as the conduction velocity. The graph below
shows the conduction velocities of myelinated and non-
myelinated neurones of different axon diameters.

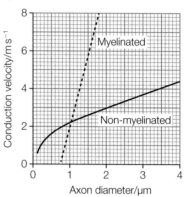

Compare and contrast the conduction velocities of
myelinated and non-myelinated neurones. [3]

(d) Explain how the presence of myelin affects the conduction
velocity of nerve impulses along an axon. [4]

(e) An investigation was carried out to determine the
conduction velocity of a nerve impulse along an axon
of a myelinated neurone. A stimulus was applied at a
point on the axon that was **25 mm** from the synapse. It
took 3.4 milliseconds (ms) for an action potential to
arrive at the synapse.

(i) Calculate the conduction velocity in **metres per
second**. Show your working. [2]

(ii) Using the information in the graph, estimate the
diameter of this axon. [1]

(Total for Question 1 = 13 marks)

2 (a) Which of these is a drug that interferes with the action of
acetylcholine? [1]

   A cobra venom

   B lidocaine

   C MDMA

   D rhodopsin

(b) When an action potential arrives at a synaptic knob,
acetylcholine is released. Describe how acetylcholine is
released into the synaptic cleft. [3]

(c) The graph below shows a recording of one complete action potential produced after the binding of acetylcholine to receptors on a post-synaptic membrane.

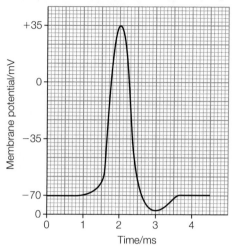

(i) Using the graph, state the time at which the sodium channels open to allow an increased flow of sodium ions into the neurone. [1]

(ii) Using the graph, state the time at which the hyperpolarisation is at its greatest. [1]

(iii) Calculate the number of action potentials that could occur in **one second** if the stimulus is maintained. Show your working. [2]

(d) When a transmitter substance called gamma-aminobutyric acid (GABA) is released at a synapse, it causes chloride ion (Cl^-) channels and potassium ion (K^+) channels to open in the post-synaptic membrane. This results in chloride ions moving into the post-synaptic neurone and potassium ions moving out.

Explain why an action potential is less likely to develop when GABA is released at the same time as acetylcholine. [2]

(Total for Question 2 = 10 marks)

3 (a) Draw a labelled diagram of a synapse. [4]

(b) Nicotine acts as a stimulant. The effects of nicotine from cigarette smoke can be felt in the body within 15 to 30 seconds of being inhaled.

(i) Describe how nicotine from cigarette smoke can enter the body and reach the synapse. [3]

(ii) Explain how nicotine can act as a stimulant. [3]

(c) Some snakes produce a toxin that has a similar structure to acetylcholine. When a person has been bitten by a snake, this toxin blocks nerve pathways. Explain how this toxin could stop post-synaptic neurones from being stimulated. [2]

(Total for Question 3 = 12 marks)

4 (a) The presynaptic end of a neurone contains all the normal organelles found in many cells.

(i) Explain the function of the mitochondria in the neurone. [2]

(ii) Suggest why mitochondria may be found in high numbers. [1]

(iii) Describe the role of the synaptic vesicle. [2]

(b) What is the name of the neurotransmitter in a cholinergic synapse?

A acetate

B acetylcholine

C GABA

D acetyl cholinesterase [1]

(c) Some headphones come with 'noise-cancelling technology'. These work by emitting a constant frequency behind the music being played. Suggest how this could work to cancel other background sounds and explain how this works. [5]

(Total for Question 4 = 11 marks)

5 Rhodopsin is the light-sensitive pigment contained in rod cells. The diagram below shows a rod cell from the retina of a mammal.

(a) Use the letter **R** to label on the diagram where rhodopsin is found in the rod cell. [1]

(b) A person enters a dimly lit room after being in bright sunlight. Explain why objects in the room only gradually become more visible. [3]

(Total for Question 5 = 4 marks)

COORDINATION, RESPONSE AND GENE TECHNOLOGY

8B COORDINATION IN ANIMALS AND PLANTS

If someone loses a limb because of cancer or an accident or is born without a limb, they may be provided with a prosthetic replacement. New prosthetic limbs are being developed that can be controlled by electrical signals in the nerves or muscles remaining in the stump. It may even be possible to control a prosthetic limb by thought. Work is already progressing using computer technology with amputees learning to control a limb on a computer avatar, before progressing to using a similar system integral to their prosthetic limb. Scientists are now moving towards linking sensors in the limbs to the remaining sensory nerves. Eventually, people may be able to both control and sense their prosthesis almost like their own limb.

Plants, like animals, respond to their environment. Complex chemical interactions (in response to the hours of darkness and light) control when many plants flower. These responses are exploited commercially: huge greenhouses are filled with plants exposed to artificial days and nights, so they flower when we are most likely to buy them.

In this chapter, you will learn about the mammalian brain, including the basic anatomy of the spinal cord and the brain and how this is related to their functions. You will investigate the role of reflexes compared to conscious thought. You will consider the different ways in which scientists find out about the working of our brains. You will also discover the importance within the body of the voluntary and autonomic nervous systems and the way they interact to bring about nervous coordination, and compare nervous coordination with hormonal coordination.

You will also discover how hormones control the growth of plants and consider the role of phytochromes in the control of flowering.

MATHS SKILLS FOR THIS CHAPTER

- Construct and interpret frequency tables and diagrams, bar charts and histograms (*e.g. interpreting data on day length responses in plants*)
- Translate information between graphical, numerical and algebraic forms (*e.g. interpretation of graphs of light absorption of different phytochromes*)

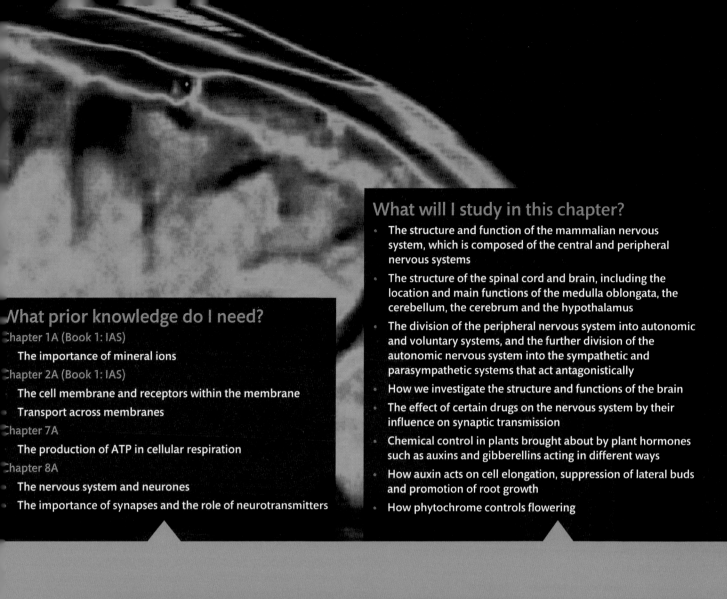

What will I study in this chapter?

- The structure and function of the mammalian nervous system, which is composed of the central and peripheral nervous systems
- The structure of the spinal cord and brain, including the location and main functions of the medulla oblongata, the cerebellum, the cerebrum and the hypothalamus
- The division of the peripheral nervous system into autonomic and voluntary systems, and the further division of the autonomic nervous system into the sympathetic and parasympathetic systems that act antagonistically
- How we investigate the structure and functions of the brain
- The effect of certain drugs on the nervous system by their influence on synaptic transmission
- Chemical control in plants brought about by plant hormones such as auxins and gibberellins acting in different ways
- How auxin acts on cell elongation, suppression of lateral buds and promotion of root growth
- How phytochrome controls flowering

What prior knowledge do I need?

Chapter 1A (Book 1: IAS)

   The importance of mineral ions

Chapter 2A (Book 1: IAS)

   The cell membrane and receptors within the membrane

- Transport across membranes

Chapter 7A

   The production of ATP in cellular respiration

Chapter 8A

- The nervous system and neurones
- The importance of synapses and the role of neurotransmitters

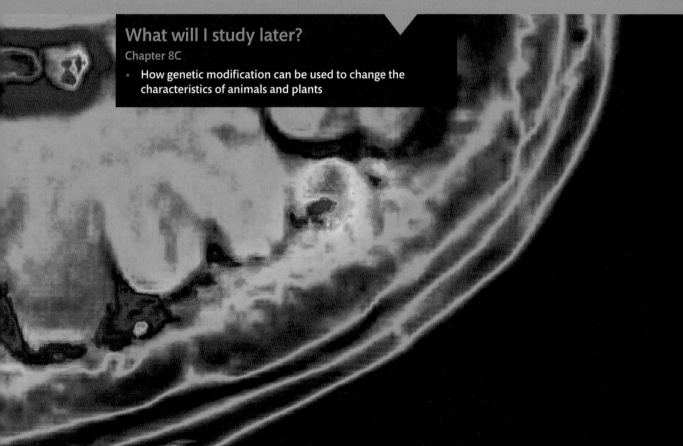

What will I study later?

Chapter 8C

- How genetic modification can be used to change the characteristics of animals and plants

LEARNING OBJECTIVES

- ◼ Know that the mammalian nervous system consists of the central and peripheral nervous systems.
- ◼ Know the location and main functions of the cerebral hemispheres, hypothalamus, pituitary gland, cerebellum and medulla oblongata of the human brain.
- ◼ Know the structure and function of a spinal reflex, including grey matter and white matter of the spinal cord.
- ◼ Understand how the pupil dilates and contracts.

A nervous system of receptors, nerves and effectors allows an animal to respond to basic stimuli from the environment. However, large complex animals need more than this. Mammals, including humans, have a central nervous system (CNS), which includes:

- the **brain**, in which information can be processed and from which instructions can be issued as required to give fully coordinated responses to a range of situations
- the **spinal cord**, which carries the nerve fibres into and out of the brain and also coordinates many unconscious reflex actions.

THE FORMATION OF THE BRAIN

In vertebrates (including humans) the brain forms as a swelling (a larger area) in the hollow neural tube at the anterior (front) end of a vertebrate embryo that folds back on itself. The basic brain pattern has three distinct areas: the forebrain, midbrain and hindbrain (see **fig A**).

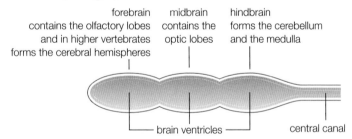

forebrain
contains the olfactory lobes
and in higher vertebrates
forms the cerebral hemispheres

midbrain
contains the
optic lobes

hindbrain
forms the cerebellum
and the medulla

brain ventricles — central canal

▲ **fig A** The basic pattern of the vertebrate brain

In some vertebrates, the brain remains simple with areas specific to particular functions such as sight or smell. In mammals, including humans, the brain becomes an extremely complex structure. The original arrangement of the brain into three areas is very difficult to see, because a part called the **cerebrum** (made up of the two **cerebral hemispheres**) is folded back over the entire brain.

The brain is made up of a combination of **grey matter** (neurone cell bodies) and **white matter** (nerve fibres). Some areas of the human brain have very specific functions concerned with the major senses and control of basic bodily functions. There are also many regions of the brain for which we still do not clearly understand the precise functions and interrelationships with other areas of the brain. Scientists have estimated that there are around

100 000 million neurones working together in the human brain and that each neurone synapses with up to 10 000 other neurones. The brain contains centres or nuclei made up of cell bodies that make intercommunication between millions of cells possible. The great nerve tracts from the spinal cord cross over as they enter and leave the brain, so that the left-hand side of the brain receives information from and controls the right-hand side of the body, and vice versa.

SOME OF THE MAJOR AREAS OF THE BRAIN

The two cerebral hemispheres are the site of many of the higher functions of the brain. They are the biggest and most highly developed area of the human brain: about 65–67% of the mass of brain tissue. Our abilities to see, think, learn and feel emotions are focused here. The cerebral hemispheres also control our motor functions (all our conscious movements). The outer layer of the cerebral hemispheres is known as the cerebral cortex. This layer is only 2–4 mm thick, but it is made up almost entirely of grey matter. The cerebral cortex is also deeply folded to give a huge surface area. The left and right cerebral hemispheres are connected by a band of axons (white matter) known as the corpus callosum. The hemispheres are subdivided into a number of lobes (rounded parts) that are associated with particular functions. For example, the frontal lobe is associated with the higher brain functions such as emotional responses, planning ahead, reasoning and decision making.

There are a number of other areas of the brain that we know are linked to specific aspects of the way the body works. Many of these are involved in the unconscious responses which maintain the processes of life.

OTHER REGIONS OF THE BRAIN

Inside the brain are other structures that have important functions.

- The **hypothalamus** coordinates the autonomic (unconscious) nervous system (see **Section 8B.2**). It plays a major part in thermoregulation (the regulation of the core temperature of the body) and osmoregulation (the regulation of the osmotic potential of the body fluids) (see **Chapter 7C**). It monitors the chemistry of the blood and controls the hormone secretions of the pituitary gland. It also controls many basic drives, including thirst, hunger, aggression and reproductive behaviour.

- The **cerebellum** coordinates smooth movements. It uses information from the muscles and the ears to control balance and maintain posture.
- The **medulla oblongata (medulla)** is the most primitive part of the brain. It contains reflex centres that control functions such as the breathing rate, heart rate, blood pressure, coughing, sneezing, swallowing, saliva production and peristalsis. It is this region that may maintain the basic life responses, even when the higher areas of the brain have been destroyed. You learned about some of these in **Chapters 7B** and **7C**.

Fig B gives you a more detailed illustration of the various areas of the brain.

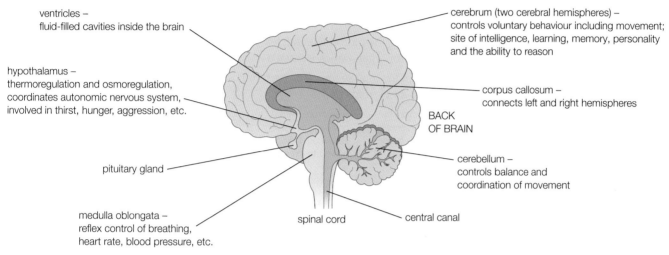

ventricles –
fluid-filled cavities inside the brain

cerebrum (two cerebral hemispheres) –
controls voluntary behaviour including movement;
site of intelligence, learning, memory, personality
and the ability to reason

hypothalamus –
thermoregulation and osmoregulation,
coordinates autonomic nervous system,
involved in thirst, hunger, aggression, etc.

corpus callosum –
connects left and right hemispheres

BACK
OF BRAIN

pituitary gland

cerebellum –
controls balance and
coordination of movement

medulla oblongata –
reflex control of breathing,
heart rate, blood pressure, etc.

spinal cord

central canal

▲ **fig B**   The brain is a very complex organ. Our knowledge of it remains limited, although we are discovering more all the time.

THE STRUCTURE AND FUNCTIONS OF THE SPINAL CORD

To coordinate the functions of the body, the brain needs a way of communicating with the body and this is one of the major functions of the spinal cord.

The spinal cord is a tube made up of a core of grey matter surrounded by white matter, which runs out from the base of the brain (the medulla oblongata) through the vertebrae. It is approximately 43–45 cm long. Impulses from sensory receptors travel along sensory nerve fibres into the spinal cord through the dorsal roots, and then travel in sensory fibres up the spinal cord to the brain. Instructions from the brain travel as impulses down motor fibres in the spinal cord and out in motor neurones through the ventral roots to the effector organs (see **fig C**).

The spinal cord is also an important coordination centre. Many simple organisms have little CNS and certainly no complex brain. Their actions take place without conscious thought, as a result of **reflex responses**. Many of the actions of more complex animals are also the result of unconscious reflex actions. Well-known examples of human reflexes include:

- moving a hand or foot rapidly away from something hot or sharp
- swallowing as food moves to the back of the throat
- blinking if an object approaches the eyes
- contracting and dilating of the pupils in response to changing light levels.

SPINAL AND CRANIAL REFLEXES

The simplest type of nerve pathway in the body, known as a reflex arc, controls these unconditioned reflexes. In vertebrates, including mammals, this involves a minimum of a receptor, a motor neurone and a sensory neurone.

Part of the pathway occurs in the CNS, often in the spinal cord. The reflex arc may simply involve a sensory and motor neurone. But there is often a small third relay neurone situated in the CNS. The function of the reflex arc is to cause an appropriate response to a particular stimulus as rapidly as possible without the time delay that occurs if the conscious centres become involved. There are two main types of reflex: the spinal reflex (e.g. hand withdrawing from heat or sharp object) and the cranial reflex (e.g. blinking, pupil reflexes). Both involve parts of the CNS. However, sensory neurones also transfer information to the conscious areas of the brain, so you know what has happened.

SPINAL REFLEXES

Fig C shows you the pathway involved in a spinal reflex.

- A stimulus is received by a sensory receptor; for example, heat receptors in the skin or pain receptors.

- An impulse travels up the sensory neurone through the dorsal root ganglion into the grey matter of the spinal cord. It synapses with a relay neurone which then synapses with a motor neurone within the grey matter.

- The impulse passes along the motor neurone, leaving the spinal cord through the ventral root. It then travels to an effector organ, which is often a muscle.

- The motor end plate in the muscle transfers the stimulus to the muscle which then contracts, moving the body part away from danger.

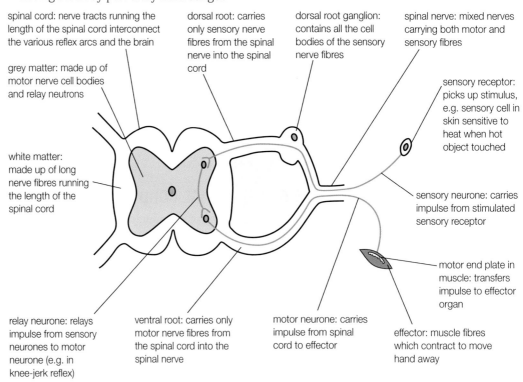

spinal cord: nerve tracts running the length of the spinal cord interconnect the various reflex arcs and the brain

dorsal root: carries only sensory nerve fibres from the spinal nerve into the spinal cord

dorsal root ganglion: contains all the cell bodies of the sensory nerve fibres

spinal nerve: mixed nerves carrying both motor and sensory fibres

grey matter: made up of motor nerve cell bodies and relay neutrons

sensory receptor: picks up stimulus, e.g. sensory cell in skin sensitive to heat when hot object touched

white matter: made up of long nerve fibres running the length of the spinal cord

sensory neurone: carries impulse from stimulated sensory receptor

motor end plate in muscle: transfers impulse to effector organ

relay neurone: relays impulse from sensory neurones to motor neurone (e.g. in knee-jerk reflex)

ventral root: carries only motor nerve fibres from the spinal cord into the spinal nerve

motor neurone: carries impulse from spinal cord to effector

effector: muscle fibres which contract to move hand away

▲ **fig C**   The spinal cord and the structures involved in spinal reflexes.

CRANIAL REFLEXES

An example of a cranial reflex is the control of the amount of light that enters the eye (see **Section 8A.5**). The iris is a muscular diaphragm with a hole (the pupil) in the middle. Pigments in the iris absorb light, making sure that the only way light can enter the eye is through the pupil. The amount of light that gets into your eyes is controlled by the size of the pupil. The iris has both circular and radial muscles that work antagonistically. In bright light, the circular muscle is contracted and the radial muscles are relaxed, so the pupil is reduced to a narrow aperture (small hole). This reduces the amount of light that enters the eye to avoid damage to the delicate rods and cones by overstimulating them. In low light levels, the circular muscles relax and the radial muscles contract, opening the pupil wide so as much light as possible falls on the rods to maximise your ability to see.

The muscles of the iris control the size of the pupil through the iris reflex. This involves the cranial nerves of the autonomic nervous system and the brain in a cranial reflex. The muscles act as effectors and they respond in a reflex action to the levels of light which enter the eye through the pupil and fall on the retina. In this way, the system constantly adapts to changes in light level (see **fig D**).

EXAM HINT

If you are asked to describe a reflex arc then use a simple diagram and annotate it fully.

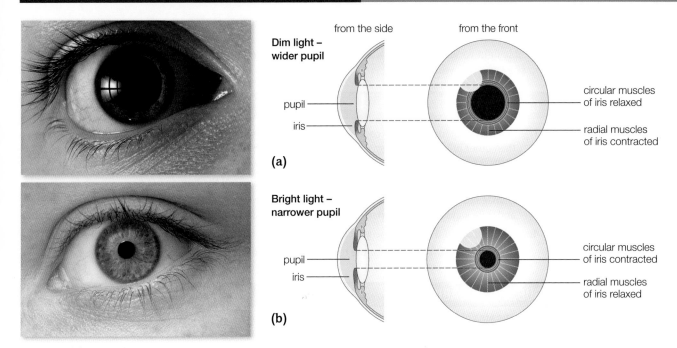

▲ fig D The size of the pupil of your eyes can change dramatically in different light conditions. Pale blue eyes have been used in these illustrations. The pupil reflexes are exactly the same whatever the eye colour but you can see them more easily in blue eyes.

The basic principle of the pupil reflex involves the following stages.

- Light may enter one or both eyes at the same time and the effect on the pupils is the same.
- Light falling on the sensory cells of the retina causes impulses to travel along neurones in the optic nerve to the brain. The brighter the light, the bigger the frequency of action potentials.
- The impulse is detected in a control centre in the midbrain.
- The impulse then travels along two neurones to further control centres.
- In the control centres the nerve impulses synapse with branches of the parasympathetic cranial nerve (the oculomotor) which transmits impulses to the iris.
- These impulses stimulate the effectors (the muscles of the iris).
- The circular muscles contract and the radial muscles relax so the pupil constricts.

If the frequency of action potentials from the retina falls (when light levels drop), impulses travel from the control centres along sympathetic nerves to the iris, causing the circular muscles to relax and the radial muscles to contract and so the pupil becomes wider. This negative feedback system controls the amount of light entering the eye (see **fig E**). The reflex response of either eye controls the dilation or constriction of the pupils of both eyes, so a bright light which is shone into one eye only will cause the pupils of both eyes to constrict.

> **EXAM HINT**
>
> Write out this type of sequence as a list of numbered points. Recalling the number of points will help you to remember all the points in your exam.

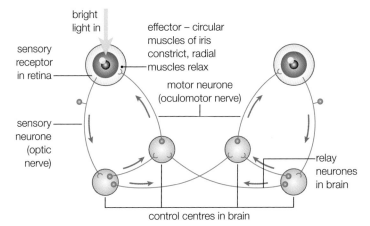

▲ fig E The cranial reflexes that control the dilation and contraction of the pupils of your eyes are much more complex than the simpler spinal reflex arc. Again you can see the changes in the size of the pupil more easily in blue eyes.

The pupils respond to emotional cues as well as light. So, for example, the release of the stress hormone adrenaline causes the pupils to become wider, which makes sure you can use all the available light to see as well as possible. Pupils also dilate if you see someone you like and constrict if you see someone you don't like.

SKILLS   CREATIVITY

CHECKPOINT

1. ▶ The degree of folding in the cerebral hemispheres varies among species, with more folding in humans than other primates, for example. Dolphins have much more folding than people but the layer of tissue is thinner. Explain these differences.

2. Make a table to summarise the functions of the main areas of the human brain.

3. Damage to both the cerebrum and the cerebellum can affect the way a person moves. Explain why, and state ways in which damage to either area might show itself.

4. (a) Compare and contrast the structure of the brain and the spinal cord.

   (b) Produce a flow diagram for the response if a person treads on a sharp stone in bare feet.

SUBJECT VOCABULARY

brain the area of the CNS in which information can be processed and from which instructions can be issued as required to give fully coordinated responses to a range of situations

spinal cord the area of the CNS that carries the nerve fibres into and out of the brain and also coordinates many unconscious reflex actions

cerebrum the area of the brain responsible for conscious thought, personality, control of movement and much more

cerebral hemispheres the two parts of the cerebrum, joined by the corpus callosum

grey matter the cell bodies of neurones in the CNS

white matter the nerve fibres of neurones in the CNS

hypothalamus a small area of brain directly above the pituitary gland that controls the activities of the pituitary gland and coordinates the autonomic (unconscious) nervous system

cerebellum the area of the brain that coordinates smooth movements; it uses information from the muscles and the ears to control balance and maintain posture

medulla oblongata (medulla) the most primitive part of the brain that controls reflex centres controlling functions such as the breathing rate, heart rate, blood pressure, coughing, sneezing, swallowing, saliva production and peristalsis

reflex responses rapid responses that occur without conscious thought

2 THE PERIPHERAL NERVOUS SYSTEM

LEARNING OBJECTIVES

■ Know that the mammalian nervous system consists of the central and peripheral nervous systems.

■ Understand how coordination in animals is brought about through nervous and hormonal control.

For nervous coordination to work there are two important stages:

- first, changes in the internal or external environment are detected by sensory receptors and are carried to the CNS
- second, the instructions from the CNS must be carried to the effector organs.

This is the role of the peripheral nervous system.

THE PERIPHERAL NERVOUS SYSTEM

The peripheral nerves are divided into two systems, which you have already met. The sensory nerves carry impulses from the receptors about changes in both the external and internal environment into the CNS. The motor nerves carry impulses out from the CNS to the effectors in the body. All the sensory nerves of the peripheral system function in much the same way. The motor nerves are of two main types.

- The **voluntary nervous system** involves motor neurones that are under voluntary or conscious control involving the cerebrum. Voluntary motor neurones function as a result of conscious thought. When you consider an action, such as picking up a drink or switching on the computer, the instructions that need to be issued to the muscles will be carried along voluntary nerve fibres.

- The **autonomic nervous system** (the involuntary nervous system) involves motor neurones that the conscious areas of the brain do not control. These motor neurones control bodily functions that are normally involuntary. Examples include control of the heart and breathing rate, the movements and secretions of the gut, sweating, the dilation or constriction of the iris of the eye in response to changing light levels and the dilation or constriction of the blood vessels in response to changing demands for blood.

The autonomic nervous system is sub-divided into the **sympathetic nervous system** and the **parasympathetic nervous system**.

Most of the body organs have connections with both the parasympathetic and the sympathetic nervous system. The differences between them are both anatomical and functional.

STRUCTURAL DIFFERENCES BETWEEN THE PARASYMPATHETIC AND SYMPATHETIC SYSTEMS

Both the parasympathetic and the sympathetic autonomic nervous systems have myelinated preganglionic fibres that leave the CNS and synapse in a **ganglion** (a collection of cell bodies outside the central nervous system) with unmyelinated postganglionic fibres.

In the sympathetic system, the ganglia are very close to the CNS, so the preganglionic fibres are short and the postganglionic fibres are long.

In the parasympathetic system the situation is reversed. The ganglia are near to or in the effector organ, so the preganglionic fibres are very long and the postganglionic fibres are very short. These structural differences can be seen clearly in **fig A**.

EXAM HINT

If you are asked to compare or contrast the sympathetic and parasympathetic nervous systems use a table. The table should have three columns: the first column is the feature being compared while the second and third columns describe each of the two systems.

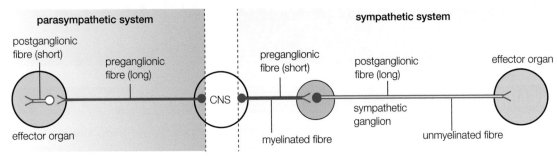

▲ **fig A** Structural differences between the parasympathetic and sympathetic autonomic nervous systems.

FUNCTIONAL DIFFERENCES BETWEEN THE PARASYMPATHETIC AND SYMPATHETIC AUTONOMIC NERVOUS SYSTEMS

There are two basic differences between these two autonomic nervous systems.

- The sympathetic nervous system produces *noradrenaline* at the synapses and usually produces a rapid response in the target organ system. This sympathetic autonomic nervous system is sometimes referred to as the 'fight-or-flight' system. When you are active, or under physical or psychological stress, the sympathetic nervous system will dominate, stimulating the organs of the body to cope with the stress you experience.

- The parasympathetic nervous system often has a slower, calming or inhibitory effect on organ systems and it produces the neurotransmitter *acetylcholine* at the synapses. The parasympathetic system maintains normal functioning of the body and restores calm after a stressful situation. It is sometimes referred to as the 'rest and digest' or 'feed and breed' system, in contrast to the action of the sympathetic system.

The sympathetic and parasympathetic nervous systems act antagonistically, rather like the accelerator and brake of a car. For example, the sympathetic system speeds up the breathing rate and the heart rate, whereas the parasympathetic slows them down (see **table A**). However, as with the accelerator and brake, it is often not a case of all or nothing. The way each of these complementary systems affects the other allows for fine control. This allows the body to match its responses exactly to the demands it meets.

	SYMPATHETIC SYSTEM	PARASYMPATHETIC SYSTEM
EYES	dilates pupil	constricts pupil
SALIVARY GLANDS	inhibits flow of saliva	stimulates flow of saliva
LACRIMAL GLANDS	–	stimulates flow of tears
LUNGS	dilates bronchi	constricts bronchi
HEART	accelerates heartbeat	slows heartbeat
LIVER	stimulates conversion of glycogen to glucose	stimulates release of bile
STOMACH	inhibits peristalsis and secretion	stimulates peristalsis
KIDNEYS	stimulates secretion of adrenaline and noradrenaline	–
INTESTINES	inhibits peristalsis and anal sphincter contraction	stimulates peristalsis and contracts anal sphincter
BLADDER	inhibits bladder contraction	contracts bladder

table A The opposing effects of the sympathetic and parasympathetic nerves on body systems.

Several bodily functions which we might consider to be under voluntary control (for example, opening the bowel and bladder sphincters) appear in **table A** as under the control of the autonomic system. However, the nervous system is very complex and many areas of the body are supplied with voluntary nerves as well as involuntary nerves. Most of us have control over our bladder and bowels and can control our breathing rate if we want to. It is relatively easy even to control the heart rate to some degree, as well as other normally involuntary activities. Scientists still do not understand all of the ways in which the voluntary and autonomic nervous systems and the CNS interact to control both our body functions and our behaviour.

COMPARING MECHANISMS OF COORDINATION

In **Chapter 7C**, you looked at hormones and their role in coordination and control in animals. Chemical coordination is important in many different situations in a mammal, from aspects of growth and sexual maturity to stress responses and the control of blood sugar levels. However, animals also have nervous systems. How do they compare?

Chemical control is often relatively slow but it can be very long lasting. Hormones travel around the body of an animal in the plasma of the blood as it moves around the circulatory system. They move into the target cells by diffusion and attach to receptors on cell membranes. Chemical control is often linked to changes which involve growth of an organism. It allows for long-term responses to environmental changes. However, it can also be used for rapid day-to-day responses such as the control of blood sugar levels and is well suited to delicate control mechanisms such as negative feedback systems.

Nervous control is usually very rapid, making it an ideal form of internal communication and control for organisms that move their whole bodies about. If you need to respond quickly to environmental cues, nerve impulses give you the speed you need.

CHECKPOINT

1. Describe the advantages of having both autonomic and voluntary motor nerves.

2. (a) State what the parasympathetic and sympathetic nervous systems are.

   (b) State why they are sometimes described as antagonistic and give a reason why this is now regarded as a rather limited description.

3. ▶ Most animals use both chemical and electrical means of coordination. Compare the two methods of coordination.

SKILLS ▷ CRITICAL THINKING

SUBJECT VOCABULARY

voluntary nervous system involves motor neurones that are under voluntary or conscious control involving the cerebrum

autonomic (involuntary) nervous system the involuntary nervous system; autonomic motor neurones control bodily functions that the conscious area of the brain does not normally control

sympathetic nervous system involves autonomic motor neurones which produce noradrenaline as their neurotransmitter and often have a rapid response, activating an organ system; these neurones have very short myelinated preganglionic fibres that leave the CNS and synapse in a ganglion very close to the CNS; postganglionic fibres are long and unmyelinated

parasympathetic nervous system involves autonomic motor neurones which produce acetylcholine as their neurotransmitter and often have a relatively slow, inhibitory effect on an organ system; these neurones have very long myelinated preganglionic fibres that leave the CNS and synapse in a ganglion very close to the effector organ; postganglionic fibres are very short and unmyelinated

ganglion (plural: ganglia) a collection of nerve cell bodies outside of the central nervous system

LEARNING OBJECTIVES

■ Understand how magnetic resonance imaging (MRI), functional magnetic resonance imaging (fMRI), positron emission tomography (PET) and computed tomography (CT) are used in medical diagnosis and the investigation of brain structure and function.

There are several hundred million nerve cells working together in the human brain. The great nerve tracts from the spinal cord cross over as they enter and leave the brain, so that the left-hand side of the brain receives information from and controls the right-hand side of the body, and vice versa. The cerebral cortex is only about 3 mm thick, but it controls most of the functions that make us what we are. The brain contains centres (called nuclei) formed of cell bodies which may have hundreds of synapses, making intercommunication between thousands and indeed millions of cells possible.

The bones of the skull hide and protect the brain. This makes it very difficult for scientists to discover how it works. It also makes it hard to find out what goes wrong in the case of diseases that affect the brain. You will look at some of the ways in which scientists use technology to investigate the brain and see how this affects the way we understand and treat diseases.

STUDYING HUMAN BRAINS

We cannot carry out experiments directly on human brains to find out how they work. The nearest kind of research to this is when brain surgery patients have allowed their brains to be artificially stimulated. Brain surgery is often completed under local anaesthetic because there are no sensory nerve endings in the brain. The conscious patients describe the sensations relating to the stimulation, and this has shown that certain areas of the brain are associated with very particular functions.

Most of our information about the functions of the human brain comes from situations in which parts of the brain are damaged or missing at birth or as a result of illness or injury. There are some famous examples of such situations.

In 1848, Phineas Gage was a foreman on a US railway construction company. He was likeable, reliable, hardworking and very responsible. In an explosives accident, an iron bar passed through his head. It didn't kill him, but it destroyed much of the front part of the left-hand side of his brain. Gage could still walk, talk and continue what seemed to be a normal life, but his personality changed dramatically. He became impatient, irresponsible, rude and unpleasant. He lost his job and died 12 years later. Researchers have used computer graphics along with images of the injury and the skull to show that the bar destroyed much of the connection between the left-side frontal lobes and the midbrain (see **fig A**). As a result, Phineas Gage lost the ability to control his emotional behaviour.

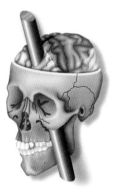

▲ **fig A**   The metal bar lodged in the skull of Phineas Gage (which he sold to scientists while he was still alive) damaged his frontal lobe. This enabled researchers to localize the frontal lobe as the area of the brain which houses our emotions.

Diseases that affect the brain can cause terrible problems for the people affected, but they can also help scientists understand how the healthy brain works. In some cases, they can also indicate possible treatments. Dr Oliver Sacks, a clinical neurologist, has described a number of unusual cases, such as a man who mistook his wife for a hat. This man had a disease affecting the visual areas of the brain. He could see and describe things but he had lost the ability to make the normal connections between what he saw and what the object was. Another patient had a massive stroke (a bleed into the brain) which affected the deeper and back portions of her cerebral hemispheres. Her personality and ability to talk were not affected but she completely lost the concept of the left-hand side, both of her own body and of the world around her. She would eat only the right-hand portion of her food and applied make-up only to the right-hand side of her face. These patients show clearly the location of function and awareness in the different sides of the brain.

MEDICAL IMAGING TECHNOLOGY

The development of medical imaging technology has had a huge effect on our ability to investigate the development of the brain and to link structures to functions. It has also made it possible to understand the progress of many brain diseases and how they affect the people who have them.

CT SCANS

X-rays can be used to take very effective images of hard tissue such as bones. They are much less useful for producing images of soft tissues such as the brain. The development of **computed tomography (CT) scans** has allowed scientists and doctors to see inside the brain. A CT scan involves thousands of tiny beams of X-rays which are passed through an area of the body such as the head. Each beam is reduced in strength by the density of the tissue it passes through. The X-rays which pass through the tissue are detected and measured. A computer system puts together all the data to produce a cross-sectional image of a thin slice through the body. Sometimes special dyes are injected into the blood or tissues which stop X-rays from passing, so they show up more clearly in the scan.

A CT scan can identify major structures in the brain and detect problems such as brain tumours, bleeding in the brain or swellings of the arteries in the brain (aneurysms). It does not detect very fine structural details. The images are like photographs; they are images frozen in time. CT scans cannot be used to show how areas of the brain are used or change during different activities. However, we can use evidence from CT scan images linked to changes in behaviour to understand the importance of certain areas of the brain in particular functions (see **fig B**).

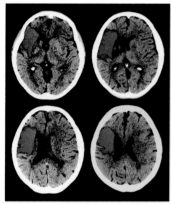

▲ **fig B**   These CT scans show the extent of a massive stroke in the brain of a patient. The bleeding in the left hemisphere affected the language areas of the brain. This patient could no longer speak but could still sing perfectly well.

MRI AND fMRI SCANS

Magnetic resonance imaging (MRI) scans produce images that show much finer detail than CT scans (see **fig C**). They are produced using magnetic fields and radio waves to image the soft tissues. This removes even the small risk of damage from the X-rays used in CT scans. Hydrogen atoms are the most commonly imaged element, partly because so much of the body is made of water (every molecule of water contains two hydrogen atoms), and partly because hydrogen atoms produce a particularly strong MRI signal. A computer system analyses the signals that are produced and produces an image.

Different tissues respond differently to the magnetic field from the radio waves. The way they respond depends on many factors, including the amount of water in the structure. As a result, we can recognise distinct regions of the brain in the image. The thin sections of the body that are usually examined produce two-dimensional (2D) MRI scans. However, the computer can put these slices together to produce a three-dimensional (3D) image.

MRI produces very detailed images which give doctors a great deal of information about the living brain. Doctors use it to

diagnose brain injuries, strokes, tumours and infections of the brain or the spine. By showing up areas of damage in the brain very clearly, it has enabled doctors and scientists to make links between the structures in the brain and patterns of behaviour seen in their patients. However, MRI scans give a historical image of the brain, like CT scans, but they don't show it while it is working.

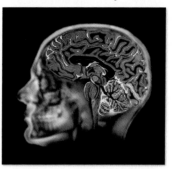

▲ **fig C**   MRI scans provide a very detailed image of the structure of the brain, as you can see in this image.

A more recent development in technology, **functional magnetic resonance imaging (fMRI) scans**, make it possible to watch the different areas of the brain in action while people perform different tasks. The basis of fMRI scanning is monitoring the absorption of oxygen in different brain areas. Deoxyhaemoglobin absorbs the radio-wave signal and later re-emits it, whereas oxyhaemoglobin does not. The blood flow to an area of the brain which is active increases and more oxyhaemoglobin is delivered to supply the active cells with the oxygen they need for aerobic respiration. Less of the signal is absorbed as a result, so an active area of the brain absorbs (and so emits) less energy than a less active area. We can observe this in real time because different areas of the brain 'light up' on the images as they become active (see **fig D**).

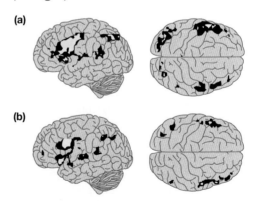

▲ **fig D**   fMRI scans show the difference in brain activation during a memory test between two groups of subjects, all apparently normal. (a) People in this group carry an allele which is a known risk factor for Alzheimer's disease. The brain activity is spread over a bigger area of the brain in the people with the high-risk allele. Scientists think that this may show that their brains are working extra hard to compensate for minor problems which are already starting in their memory. (b) People in this group do not carry the high-risk allele.

The number of fMRI images which can be recorded each second is constantly increasing. Therefore, as technology improves, scientists are able to observe more precisely the changes in brain activity while different tasks are performed. It is possible to generate an extremely spatially accurate image of the brain using fMRI. At the moment, we use it mainly to investigate normal brain structure and function. For example, we can see clear differences in the activity of the brains of people with dyslexia when they are reading compared with the brains of non-dyslexic readers. However, increasingly researchers are looking at ways of using fMRI to diagnose diseases such as the early signs of stroke damage and the beginning of Alzheimer's.

However, there are some disadvantages with fMRI. Like MRI, it is a noisy procedure which some people find very stressful. Patients must keep their head completely still because any movement reduces the accuracy of the image. This mainly limits what we can test although increasingly scientists are finding ways to allow movement in other body areas. For example, in one investigation scientists studied the brains of dancers as they moved their feet as if dancing to music. The areas which lit up were strongly linked to the speech areas of the brain.

Some neurophysiologists have questioned the validity of fMRI scans. They argue that the blood flow to different areas of the brain when a subject is looking at different stimuli is a case of correlation, not causation. Scientists must do more research to confirm whether or not it is a causal link. For now, the majority of scientists in this field are convinced that the areas which are highlighted by this imaging technique really are the active regions of the brain involved in an activity.

PET SCANS

Positive emission tomography (PET) scans give scientists and doctors another way of forming detailed, 3D images of the inside of the body, including the brain. PET scans reveal abnormal areas in the body and can also show how well different areas are working. PET scans can be combined with CT scans and MRI scans to produce very detailed images to help with diagnosis. The great advantage of PET scans is that they can show how parts of your brain are actually working. They can be used to plan surgery by giving surgeons a 3D image of the areas of the brain that are affected.

To produce a PET scan, the patient is injected with a **radiotracer** (radioactive isotope) which is similar to glucose. This means the body treats it in a similar way and it is carried to all the cells. The scanner works by detecting the radiation which the radiotracer gives off and the computer system analyses where it accumulates and where it does not. So, for example, cancer cells absorb much more of the radiotracer than normal cells, so a PET scan shows cancerous cells clearly. Areas of the brain that are less active than they should be (for example, areas that have died as a result of diseases such as Alzheimer's, which causes dementia) absorb less of the radiotracer than expected. You can see the difference between a PET scan of a healthy brain and one affected by dementia in **fig E**.

(a)   (b)

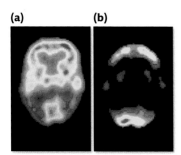

▲ **fig E** These PET scans show clearly the differences between (a) a healthy brain and (b) the brain of someone who is severely affected by Alzheimer's.

CHECKPOINT

1. Explain why different forms of imaging are so important in helping us understand how brains function and what happens when brains are affected by disease.

2. Summarise the main imaging techniques used to investigate brain structure and function. Explain how they work, the advantages and limitations of each and give examples of their use in both understanding brain structure and function and in diagnosing disease.

SUBJECT VOCABULARY

computed tomography (CT) scans scans using thousands of tiny beams of X-rays which are passed through an area of the body such as the head to produce an image of the brain

magnetic resonance imaging (MRI) scans scans produced using magnetic fields and radio waves to image the soft tissues; they produce images showing much finer detail than CT scans

functional magnetic resonance imaging (fMRI) scans scans which monitor the uptake of oxygen in different brain areas, making it possible to watch the different areas of the brain in action while people conduct different tasks

positive emission tomography (PET) scans scans produced by detecting the radiation given off by a radiotracer injected into a patient; computer analysis shows areas in which the radiotracer builds up, so detailed three-dimensional images of the inside of the body, including the brain, are formed

radiotracer any radioactive isotope introduced into the body to study metabolic processes

4 THE CHEMICAL BALANCE OF THE BRAIN

LEARNING OBJECTIVES

■ Understand how the effects of drugs can be caused by their influence on nerve impulse transmission, illustrated by nicotine, lidocaine and cobra venom alpha toxin, the use of L-DOPA in the treatment of Parkinson's disease, and the action of MDMA (ecstasy).

■ Understand how imbalances in certain naturally occurring brain chemicals can contribute to ill health, including dopamine in Parkinson's disease and serotonin in depression, and to the development of new drugs.

Communication between the neurones around your body and in your brain depends on a delicate balance of naturally occurring chemicals. The brain uses a number of different neurotransmitters, some of which only occur in brain synapses (e.g. dopamine, serotonin). An imbalance in these transmitters can result in both mental and physical symptoms.

Treating diseases which are caused by such imbalances means transferring drugs across the **blood–brain barrier**. This barrier is formed by the endothelial cells that line the capillaries of the brain and these are very tightly joined together. This makes it difficult for bacteria to cross into the brain and cause infections, which is a good thing. However, the blood–brain barrier also makes it difficult for therapeutic drugs to enter the brain.

Drugs that do affect the brain are usually active at the synapses and there are a number of stages in synaptic transmission at which they can be targeted (see **fig A**).

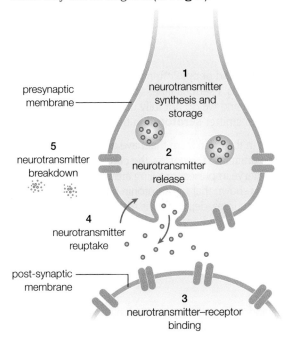

▲ **fig A** Brain function depends a lot on synapses. The five stages of synaptic transmission that you can see here can all be affected by drugs which may cure a problem, or cause one.

PARKINSON'S DISEASE

Parkinson's disease involves the loss of nerve cells in an area of the midbrain known as the **substantia nigra**. In a healthy brain, these cells produce the neurotransmitter **dopamine**, and the axons from them spread through the frontal cortex, the brain stem and the spinal cord. So they are closely involved in the control and coordination of movement. In Parkinson's disease, these dopamine-producing cells die and motor control is gradually lost. For a long time, the brain compensates for the loss. Most people don't show symptoms of the disease until around 80% of their dopamine-producing cells have disappeared.

Parkinson's disease usually develops in people over 50 years old, although it can appear in younger people. The causes are still not clear. In young-onset Parkinson's disease, which is very rare, there appears to be a relatively strong genetic link. However, in the most common form of Parkinson's disease, the genetic link is much weaker and other factors are also involved. Environmental factors may play a part and there may be a link with a number of toxins, herbicides and pesticides. Parkinson's disease affects around 7–10 million people in the world.

The ways individuals react to the falling levels of dopamine in Parkinson's disease vary. Common symptoms include the following.

• Tremor (shaking) which usually begins in just one hand. This is the first symptom for around 70% of affected people.

• Slowness of movements. It is hard to start to make a movement, and movements take longer to perform.

• Stiffness (rigidity) of the muscles. This can make it difficult to turn over in bed or stand up after sitting in a chair.

Other problems that appear as the disease progresses may include poor balance, difficulty in walking, problems with sleeping, depression and even difficulties with speech and breathing.

TREATING THE SYMPTOMS

At the moment, there is no cure for Parkinson's disease. However, there are a number of drugs which can be very effective at easing or delaying the symptoms. Most of them aim either to replace the natural dopamine in the brain or to allow the body to make the best use of the dopamine it still produces. Treatment drugs include the following.

• **Levodopa (L-dopa)** is a precursor to dopamine; it can cross the blood–brain barrier whereas dopamine cannot. L-dopa has been used to treat Parkinson's since the 1960s. Supplying the brain with L-dopa allows the remaining cells to make as much dopamine as possible. This can greatly reduce stiffness and slowness of movements. Eventually, the drug becomes less effective as brain cells continue to die.

- **Dopamine agonists** are chemicals that bind to dopamine receptors in brain synapses and thereby mimic the effect of dopamine. They are often used at the beginning of the disease when they are most effective, so L-dopa can be used later.
- **Monoamine oxidase B (MAOB) inhibitors** inhibit the enzyme monoamine oxidase B (MAOB) which breaks down dopamine in brain synapses. Thus, MAOB inhibitors reduce the destruction of the dopamine made by the cells.

DID YOU KNOW?

Developing new treatments for Parkinson's disease
Research into possible new drug treatments for Parkinson's disease continues. However, there are two more radical areas of research.

Gene therapy
Scientists are using knowledge obtained from the results of the Human Genome Project to help them develop gene therapies. Although Parkinson's disease is not a genetic disease, scientists are investigating the possibility of inserting healthy genes into the affected cells. There are two main approaches: adding genes to prevent the dopamine-producing cells of the midbrain from dying, and adding genes to enhance the levels of dopamine production in the remaining cells. Both approaches have produced some encouraging results. However, there are major problems in delivering the healthy genes to the cells of the midbrain, and safety is essential, so using gene therapy like this is still some years away (you will find out more about gene therapy in **Chapter 8C**).

Stem cell therapy
This research aims to provide a cure for Parkinson's disease rather than a therapy to relieve the symptoms. Scientists hope to use stem cells to replace the failing dopamine-producing cells in the brains of people with Parkinson's disease. Current research using mouse models is very promising, but the ethical issues of using stem cells remain. With stem cell therapy, there is also the risk of uncontrolled growth and, therefore, cancer may result. As scientists learn how to stimulate stem cells to become dopamine-producing cells in the laboratory, it will also provide them with a valuable tool for developing new and possibly more effective drug therapies.

DEPRESSION

Everyone gets a low mood or feels sad from time to time, and people often say they are feeling 'a bit depressed'. However, this is not clinical depression, which is a serious illness that affects many people during their lives. The causes of depression are complex and we do not fully understand them. However, one cause may be problems with the neurotransmitter **serotonin** in the brain.

Serotonin is the synaptic transmitter in a group of cells in the brain stem that have an extensive influence. They have axons that spread throughout the brain into the cortex, the cerebellum and the spinal cord. Low levels of serotonin result in fewer nerve impulses travelling around the brain. This prevents the brain working effectively. Research has shown that the serotonin pathways are often abnormal in people suffering from depression. Sometimes, depression is triggered by external factors such as work or relationship stress, or the death of a friend or family member. In some cases, depression seems to be simply the result of chemical changes in the brain. Serotonin is not the only neurotransmitter implicated in depression. Dopamine and noradrenalin may also play a role in the condition for some people.

TREATING DEPRESSION

Treatment for depression includes the use of drugs and 'talking therapies' (which can help a patient accept adverse life events). Many of the drugs used for depression are linked to the serotonin synapse systems and other neurotransmitters. Some of the best-known antidepressant drugs are the **SSRIs (selective serotonin reuptake inhibitors)**. These drugs inhibit the reuptake proteins in the presynaptic membrane. As a result, more serotonin remains in the synaptic cleft, more impulses travel along the post-synaptic axon and this reduces symptoms by producing a more positive mood and improving the ability to sleep.

Other treatments also involve neurotransmitters. **Tricyclic antidepressants (TCAs)** work by increasing the levels of serotonin and noradrenalin in the brain, and monoamine oxidase inhibitors (like the MAOB inhibitors used in treating Parkinson's disease) inhibit the enzymes which usually cause the breakdown of neurotransmitters in the synapses of the brain.

ILLEGAL DRUGS AND THE BRAIN

Drugs such as L-dopa and SSRIs can have a clear and measurable effect on the way the brain works. These drugs are therapeutic and legal. Other drugs are enjoyable and legal, such as the caffeine from coffee, tea, cola or energy drinks. Caffeine crosses the blood–brain barrier and affects the brain in a number of ways which include slowing down the rate of dopamine reabsorption at dopamine synapses. But there are some drugs which used specifically because they have an impact on the way the brain works. These drugs are often illegal. One which is widely used is **ecstasy (MDMA: 3,4-methylenedioxy-N-methylamphetamine)**.

Ecstasy has a marked effect on the brain. It acts as a stimulant, increasing the heart rate (similar to amphetamine), and also as a psychotropic drug which alters the way a person 'sees' the world. The short-term effects of the drug are to change mood and make people feel sociable, full of energy, warm and empathetic. The drug causes these changes by affecting the serotonin synapses of the brain.

Serotonin also has a major effect on the way the brain works. As you have just learned, depression seems to be the result of too little serotonin in some way. Ecstasy blocks the serotonin reuptake transport system so that the synapses are completely flooded with serotonin, which cannot be returned to the presynaptic knob. There is also some evidence that the drug makes the transport system work in reverse, so that all of the serotonin in the presynaptic knob is moved out into the synapse which affects the post-synaptic membrane and floods the brain with impulses. Ecstasy may also affect the dopamine systems, with the high levels of serotonin stimulating the release of more dopamine, adding to the 'pleasure sensation', but there is some debate among scientists about whether this is really the case.

Using ecstasy causes physical changes, such as increased heart rate, and can cause problems in the body's thermoregulatory system. There may be no desire to drink, which can lead to hyperthermia (overheating) with raised blood pressure and irregular heartbeat. In a small number of cases, this can lead to death. Ecstasy can also affect the hypothalamus so that it secretes more antidiuretic hormone, which is normally secreted when the body needs to conserve water. This effectively stops the kidneys from producing urine and can lead to problems if someone keeps drinking water in an attempt to stay hydrated and cool down.

They retain so much water that osmosis destroys their cells.

EXAM HINT

Ensure you are ready to link the effects of illegal drugs to homeostasis in the body. There are also clear indications that the use of ecstasy can cause long-term damage to the brain, particularly the serotonin neurones, although research on this is still in progress.

CHECKPOINT

SKILLS ▶ INTERPRETATION

1. ▶ Explain how a drug affecting each of the five stages of neurotransmission would have an effect on how nerve impulses are transmitted and what this might mean in the body.

2. Describe the difference between dopamine and serotonin synapses and their roles in the body.

3. Compare the action of the drugs L-dopa and ecstasy on the brain and their impact on human health.

SUBJECT VOCABULARY

blood–brain barrier a barrier formed by the endothelial cells that line the capillaries of the brain which are very tightly joined together, making it difficult for bacteria to cross into the brain but also making it difficult for therapeutic drugs to enter the brain

substantia nigra the area of the midbrain involved in the control and coordination of movement; it is affected by Parkinson's disease

dopamine the neurotransmitter produced by nerve cells in the substantia nigra, which is closely involved in the control and coordination of movement

levodopa (L-dopa) a precursor of dopamine which can cross the blood–brain barrier and has been in use since the 1960s to treat Parkinson's disease

dopamine agonists chemicals that bind to dopamine receptors in brain synapses and mimic the effect of dopamine

monoamine oxidase B (MAOB) inhibitors drugs which inhibit the enzyme monoamine oxidase B (MAOB), which breaks down dopamine in brain synapses; thus, MAOB inhibitors reduce the destruction of the dopamine made by the cells

serotonin the neurotransmitter in a group of cells in the brain stem which have axons that spread throughout the brain into the cortex, the cerebellum and the spinal cord, and have an extensive influence over much of the brain

SSRIs (selective serotonin reuptake inhibitors) antidepressant drugs that inhibit the reuptake proteins in the presynaptic membrane so more serotonin remains in the synaptic cleft and more impulses travel along the post-synaptic axon, reducing the symptoms of depression by producing a more positive mood and improving the ability to sleep

tricyclic antidepressants (TCAs) antidepressant drugs that work by increasing the levels of serotonin and noradrenalin in the brain

ecstasy (MDMA: 3,4-methylenedioxy-N-methylamphetamine) an illegal drug that acts as a stimulant and has a psychotropic effect by blocking the serotonin reuptake transport system

As you have learned, specific chemicals released by cells in response to a stimulus can act as messages. In animals, these chemicals may be hormones, or neurotransmitters in synapses in the nervous system. Plants also rely on chemical messages for communication between different parts of the plant. This makes it possible for them to respond to factors such as light and gravity (see **fig A**). The chemicals move from cell to cell, and also through the plant transport system. They act as plant hormones.

WHICH STIMULI AFFECT PLANTS?

Although most plant movements are very slow, plants respond to a variety of stimuli, most of which are cues from the external environmental that have a direct impact on the well-being of the plant.

(a)  (b)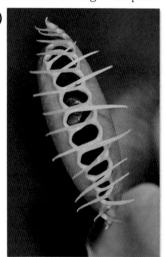

▲ **fig A** Chemical control allows plants to respond to light and gravity. Some plants also respond to touch. The slow responses of the bean plant (a) and the fast responses of a Venus fly trap (b) are both examples of sensitivity to touch.

Plants respond to the presence or absence of light. They also respond to the direction from which light comes, the intensity of the light and the length of daily exposure to light. Light affects how much plants grow, the direction in which they grow and when they reproduce. Plants are also sensitive to gravity, water and temperature, and in some cases to touch and chemicals. Different parts of the same plant may react differently to the same stimulus. For example, shoots grow towards light but roots grow away from it.

As well as responding to external stimuli, plants also respond to internal chemical signals. Most of the responses of plants are concerned with maximising the opportunities for photosynthesis and reproduction.

PLANT RESPONSES

Chemical control in plants is given by a number of different chemicals produced in response to specific stimuli. Many of these chemicals act as plant growth regulators, but they do more than simply control the growth of plants. The chemicals which control growth and development in plants are called plant hormones. They are produced in one area of the plant, are transported around the body of the plant and have their effect on cells in another area. Some of their effects involve growth, but many do not. Animals respond to nervous and chemical messages in a variety of ways that include the release of further chemicals, the contraction of muscle cells, and growth. Plants also respond to chemical messages in a number of different ways. In some cases, growth is stimulated and in others it is inhibited. For example, sometimes one side of a plant grows more than the other in response to a particular stimulus, resulting in the bending of the shoots or roots.

These directional growth responses to specific environmental cues are known as **tropisms**. Other plant hormones affect the differentiation and development of plant cells and processes, such as the ripening of fruit and leaf fall.

HOW PLANTS GROW

Growth is a permanent increase in the size of an organism or of some part of it. It is brought about by cell division (see **Section 3B.2 (Book 1: IAS)**) and the assimilation of new material within the cells that result from the division, followed by cell expansion (see **fig B**). The main areas of cell division in plants are known as the meristems and they occur just behind the tip of a root or shoot. The regions of cell division and cell elongation are particularly sensitive to plant growth substances. These chemical messages act in a number of ways. Some affect cell division, increasing the number of divisions that occur. Others make it easier for the cellulose walls to be stretched, and this makes it easier for the cells to expand and grow.

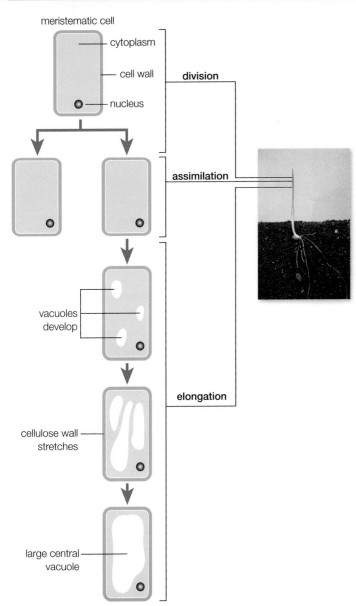

▲ **fig B**   Some plant hormones affect the area of cell division. Others affect the region of cell elongation.

AUXINS

Auxins, for example indoleacetic acid (IAA), are powerful growth stimulants that are effective in very low concentrations. Auxins are produced in young shoots and always move down the plant from the shoots to the roots. This movement involves some active transport and calcium ions. Auxins are involved in apical dominance, by suppressing the growth of lateral shoots so that the main stem grows fastest. In low concentrations, they promote root growth. The more auxin that is transported down the stem, the more roots grow. If the tips of the stems are removed, removing the source of auxins, the stimulation of root growth is removed

and root growth slows and stops. Auxins are also involved in the tropic responses of plant shoots to unilateral light (light from one side). The response of a plant to auxins often depends on both the concentration of the hormone and the region of the plant.

> ### DID YOU KNOW?
>
> **Auxins in the garden**
> Gardeners and horticulturists make use of auxins in two important ways:
> - *Taking cuttings (a method of obtaining a new plant):* applying rooting powder containing different auxins to the cut end of stems when cuttings are taken encourages the development of new roots and so helps the cutting to form a new independent plant.
> - *Weedkillers:* synthetic auxins mimic the effects of natural plant hormones, but cannot be synthesised and regulated by the plants. They are used as weedkillers. Broadleaved dicot plants absorb them much more effectively than narrow-leaved monocots, so they are selectively absorbed by the broadleaved weeds in lawns and cereal crops. The hormone analogues are also taken down into the roots, so they affect the whole plant. They interfere with the growth of the plant, so that different parts grow at different rates and the internal metabolism is so disturbed that the plant dies. These hormone weedkillers are selective to dicot plants and are also relatively harmless to pets and garden birds.

HOW DO AUXINS WORK IN A PLANT?

Auxins affect the ability of the plant cell walls to stretch. For example, IAA is made in the tip of the shoot and diffuses back towards the zone of elongation. The molecules of IAA bind to specific receptor sites on the cell surface membranes, activating the active pumping of hydrogen ions into the cell wall spaces. This changes the hydrogen ion concentration, providing the optimum pH of around 5 for enzymes that break bonds between neighbouring cellulose microfibrils (see **Section 4A.1 (Book 1: IAS)**). This allows the microfibrils to slide past each other very easily, so the walls stay very plastic and flexible. The cells absorb water by osmosis and, as a result of turgor pressure, the very flexible cell walls stretch allowing the cells to elongate and expand. Eventually, as the cells mature, the IAA is destroyed by enzymes, the pH of the cell walls rises, the enzymes are inhibited and bonds form between the cellulose microfibrils. Consequently, the cell wall becomes more rigid again and the cell can no longer expand (see **fig C**).

> ### EXAM HINT
>
> Remember that your knowledge from your **IAS (Book 1)** studies can be tested in unfamiliar contexts. This has clear links to both plant cell structure and the structure of cellulose which you studied at IAS.

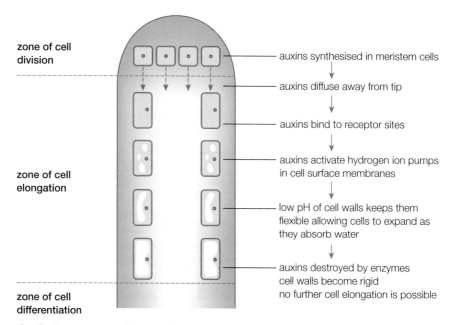

zone of cell division

zone of cell elongation

zone of cell differentiation

- auxins synthesised in meristem cells
- auxins diffuse away from tip
- auxins bind to receptor sites
- auxins activate hydrogen ion pumps in cell surface membranes
- low pH of cell walls keeps them flexible allowing cells to expand as they absorb water
- auxins destroyed by enzymes cell walls become rigid no further cell elongation is possible

▲ **fig C**   A summary of the role of auxins, such as IAA, in the growth of a plant shoot.

This basic model of the way plants grow was based on shoots which were kept entirely in the dark or in full light. However, in real life, plants are usually in a situation where the light on one side is stronger than the other. Research shows that the side of a shoot exposed to light contains less auxin than the side that is not. Light seems to cause the auxin to move laterally across the shoot, producing a greater concentration on the unlit side. This movement means the shoot tip acts as a photoreceptor. More of the hormone diffuses down to the region of cell elongation on the darker side. This stimulates cell elongation, and therefore growth, on the darker side. Consequently, the shoot bends towards the light. Once the shoot is growing directly towards the light, there is no longer a light side and a dark side. The asymmetric transport of auxin finishes and the shoot grows straight towards the light. The original theory was that light destroyed the auxin. However, more recent evidence suggests that the levels of auxin in shoots are much the same, whether they have been kept in the dark or under light from one side. Although scientific work on tropisms began over a century ago, by Charles Darwin and his son Francis, there is still much to learn.

GIBBERELLINS AND SEED GERMINATION

Auxins do not act alone to control and coordinate plant responses. There are a number of other classes of plant hormones that have different effects on plant cells. They work individually and also interact to give extremely sensitive responses to a constantly changing environment. **Gibberellins** are another important group of plant hormones. These compounds act in several ways, including as growth regulators. They affect the internodes of stems, stimulating elongation of the growing cells. They also promote the growth of fruit. They are involved in breaking dormancy (inactive period) in seeds and in germination, because they stimulate the formation of enzymes in seeds (see **fig D**). For example, they stimulate the production of amylase, which breaks down starch stores in cereal plants. This makes glucose available for respiration in the embryo plant as it develops during germination. Gibberellins also stimulate bolting, a period of sudden rapid growth and flowering, in biennial plants.

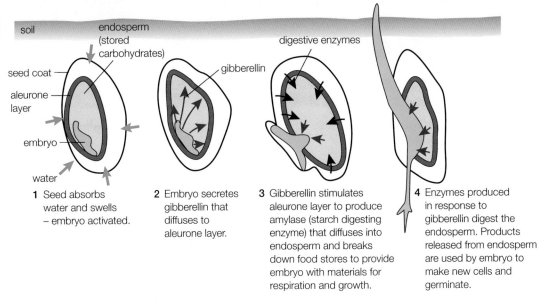

soil
endosperm (stored carbohydrates)
seed coat
aleurone layer
embryo
water
digestive enzymes
gibberellin

1 Seed absorbs water and swells – embryo activated.

2 Embryo secretes gibberellin that diffuses to aleurone layer.

3 Gibberellin stimulates aleurone layer to produce amylase (starch digesting enzyme) that diffuses into endosperm and breaks down food stores to provide embryo with materials for respiration and growth.

4 Enzymes produced in response to gibberellin digest the endosperm. Products released from endosperm are used by embryo to make new cells and germinate.

▲ **fig D** The role of gibberellin in seed germination.

EXAM HINT

As part of your study of this topic, you will conduct **Core Practical 18: Investigate the production of amylase in germinating cereals using a starch agar assay**. Make sure you understand this practical well because your understanding of the experimental method may be assessed in your examination. Be clear about the effect of amylase on starch, and the use of iodine indicators to show the digestion of starch by amylase. You can look back to **Section 1A.3 (Book 1: IAS)** to remind yourself of the iodine test for starch.

DID YOU KNOW?

A gaseous hormone
Ethene (ethylene) is the only plant hormone in gas form. It promotes fruit ripening, and the fall of ripe fruits and leaves from plants. If you put a ripe banana in a bag of green bananas, the unripe bananas will ripen quickly in response to the ethene produced by the ripe fruit.

EXAM HINT

Many plant hormones have a range of effects. These can be tested through questions that provide you with data from experiments and ask you to interpret or analyse the data. You may also be asked to evaluate experiments such as this core practical.

SYNERGY AND ANTAGONISM

Most plant hormones do not work in isolation, they interact with other substances. This adaptation means that plants can have a very fine control over their responses. The growth regulators interact in one of two ways. If they work together, complementing each other and giving a greater response than the simple addition of their two responses, the interaction is known as synergy. For example, auxins and gibberellins work synergistically in the growth of stems. If the hormones have opposite effects they are known as antagonistic.

CHECKPOINT

SKILLS ▷ CRITICAL THINKING

1. ▷ List as many of the different environmental stimuli that elicit a response in plants as you can. For each of them, try to explain why it is important for the plant to respond to that stimulus.

2. Growth in animals usually stops at a certain stage. The meristems of plants, where growth occurs, remain active throughout the life of the plant. Explain why this difference is important in the way the organisms respond to stimuli.

SUBJECT VOCABULARY

tropisms plant growth responses to environmental cues; they are also known as tropic responses
auxins plant hormones that act as powerful growth stimulants (e.g. indoleacetic acid, IAA) and are involved in apical dominance, stem and root growth, and tropic responses to unilateral light
gibberellins plant hormones that act as growth regulators, particularly in the internodes of stems, by stimulating elongation of the growing cells; they also promote the growth of fruit and are involved in breaking dormancy in seeds and in germination

Plants need light. Without light, the metabolism of a plant is severely disrupted and no chlorophyll is formed. Chlorophyll is needed for photosynthesis to occur. If a plant is deprived of light for a long time it will die. Day length (or night length) is the environmental cue that controls changes such as bud development, flowering, fruit ripening and leaf fall. Plants have evolved very sensitive mechanisms for detecting light, and many aspects of plant development are controlled by the levels and type of light available.

SENSORY SYSTEMS IN PLANTS

Scientists have known for a long time that the seeds of many plants will germinate only if they are exposed, even very briefly, to light. Researchers in the US Department of Agriculture showed that **red light** (wavelength 620–700 nm) is the most effective at stimulating germination in lettuce seeds, and **far red light** (wavelength 700–800 nm) inhibits germination.

If you expose seeds to a flash of red light they will germinate. If you expose them to a flash of red light followed by a flash of far red light, they will not germinate. With any series of flashes of light, it is the colour of the final flash that determines whether or not the seeds will germinate. Scientists hypothesised the existence of a plant pigment that reacts with different types of light, and then affects the responses of the plant, so acting as part of the system that controls photomorphogenesis. In 1960, this pigment was isolated from plants and called **phytochrome**.

PHYTOCHROMES

Plants make a number of different phytochromes, but they all respond to light in the same way. A phytochrome is a blue–green pigment that exists in two interconvertible forms: P_r (or P_{660}) absorbs red light; P_{fr} (or P_{730}) absorbs far red light (see **fig A**). The absorption of light by one form of the pigment converts it reversibly into the other form.

This can explain the germination of seedlings noted above: a flash of red light produces biologically active P_{fr}, which triggers germination. But then a flash of far red light converts P_{fr} back to the inactive P_r before it has any effect.

As a seedling germinates, it makes P_r. As soon as it breaks through the surface of the soil and is exposed to red light, some of the new pigment is converted into P_{fr}. From that point onwards the two interconvertible forms exist in the plant.

The length of time it takes for one form of the pigment to be converted into the other depends on the light intensity. In low light intensity, it takes minutes; in high light intensity, it takes seconds. In the dark, P_{fr} is converted to P_r very slowly but no P_r is converted back. P_r is the more stable form of the pigment, but it is the P_{fr} that is biologically active.

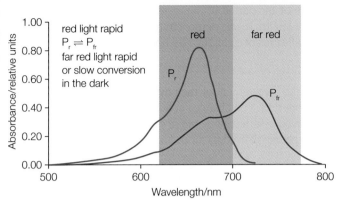

▲ **fig A** The conversion of phytochrome from one form to another has an important role in the coordination of plant growth and development. It is dependent on exposure to light. (Courtesy of *Science and Plants for Schools*.)

The balance between the two forms of phytochrome is affected by varying periods of light and dark. That then affects the plant metabolism, including flowering patterns. Sunlight contains more red light than far red light, so during daylight hours most of the phytochrome in a plant is in the far red form, P_{fr}. If the night period is long enough, all the phytochrome is converted back into the red form, P_r.

Phytochromes enable plants to respond to environmental cues such as change in day length. In some cases, phytochromes have a stimulating effect on growth in plants, in others they inhibit growth. Exactly how phytochromes influence the responses of the plant is still not fully understood, but as you will see in **Section 8B.7**, the evidence increasingly suggests that the presence of phytochromes stimulates the production of other growth regulators and plant hormones, resulting in the response to light.

DEVELOPING IDEAS ABOUT PHOTOPERIODISM

In temperate regions of the world such as the UK, the period of daylight can vary from about 9 to 15 hours throughout the year. The lengths of the days and nights give important environmental cues to living organisms, directing their growth, development and behaviour. The amount of time that an organism is exposed to light during a 24-hour period is known as the photoperiod. In plants, one of the most clearly affected activities is flowering, and scientists have developed models of how plants sense and respond to day-length cues.

Scientists found that day length appeared to be the environmental cue affecting flowering in many plants. Plants flowering when days are short and nights are long became known as **short-day plants (SDPs)**. SDPs include rice and cotton. Plants flowering in relatively long days and short nights are known as **long-day plants (LDPs)**, and these include oats and cabbages. It can be very difficult to decide whether a plant is a short- or long-day plant as the two groups merge. Some plants, such as cucumbers, tomatoes and pea plants, are unaffected by the length of the day and are known as **day-neutral plants (DNPs)**. These are usually plants that grow naturally in tropical regions where the day length is the same all year round. As a result, they are adapted to use different cues, such as the amount of available water, as the triggers for flowering.

Different flowering patterns allow plants to take advantage of different circumstances. In temperate regions, SDPs tend to flower in spring and autumn, when the light-shading canopy of leaves either has not developed or has fallen off. They also grow well near the equator, where the days are never longer than about 12 hours.

LDPs flower in the summer in temperate regions and are found further from the equator in areas in which in some seasons there are very long days.

Scientists eventually discovered that the length of the period of darkness is actually the environmental cue affecting flowering, not day length. It was demonstrated that if an SDP has the long night (period of darkness) interrupted by flashes of light, they do not flower.

HOW IS THE SIGNAL RECEIVED?

All the research on photoperiodism indicates the involvement of the phytochromes in the sensitivity of the flowering pattern of plants to the photoperiod. The changes in flowering patterns that can be caused by disturbing the dark periods can also be effected by red or far red light alone. Red light inhibits the flowering of SDPs, but if the red light is followed by far red light, the inhibition is removed. It is the balance of the P_r and P_{fr} that is key.

The current hypothesis is that, in SDPs, the biologically active molecule P_{fr} inhibits flowering, and a lack of P_{fr} allows flowering to occur. During long periods of darkness, the levels of P_{fr} fall, as it is almost all converted to P_r. This allows flowering to take place.

In LDPs, the situation is reversed, and it appears that high levels of P_{fr} stimulate flowering. The nights are short so relatively little P_{fr} is converted back to P_r. As a result, relatively high P_{fr} levels are maintained all the time, stimulating flowering.

DNPs evolved in tropical conditions in which the levels of P_r and P_{fr} are similar all year round, so even in temperate regions they do not respond to changes in day length by flowering. Other factors trigger their flowering.

Scientists know that phytochromes are only part of the story and control of flowering is very complex, but this provides a useful model to work with.

The detection of the photoperiod seems to occur in the leaves of the plant. In the 1930s, scientists first hypothesised the presence of a plant hormone known as **florigen**. They thought that plants produced florigen in response to the changing levels of phytochromes and the plant transport system carried it to the flower buds. The evidence included the following findings.

- If the whole plant is kept in the dark, apart from one leaf which is exposed to the appropriate periods of light and dark, flowering occurs as normal. A plant kept in total darkness does not flower (see **fig B**, experiment A).
- Using the same experimental set-up, if the photoperiodically exposed leaf is removed immediately after the stimulus, the plant does not flower. If the leaf is left in place for a few hours, it does flower.
- If two or more plants are grafted together and only one is exposed to appropriate light patterns, all the plants will flower (see **fig B**, experiment B).
- In some species, if a light-stimulated leaf from one plant is grafted onto another plant, the new plant will flower.

Experiment A

Cocklebur, a short-day plant, will not flower if kept under long days and short nights.

If even one leaf is masked for part of the day – thus shifting that leaf to short days and long nights – the plant will flower; note the burrs.

burrs (fruit)

masked leaf

Experiment B

1 Graft five cocklebur plants together and keep under long days and short nights, with most leaves removed.

2 Induce a leaf by long nights/short days.

3 If a leaf on a plant at one end of the chain is subjected to long nights, all of the plants will flower.

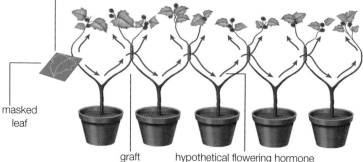

masked leaf

graft

hypothetical flowering hormone spreads through all the grafted plants

▲ **fig B** Evidence like this helps us understand how the photoperiodic response in the leaves is transferred to the flowering regions of the plant.

For years no one could isolate the theoretical hormone and so the florigen theory lost support. However, recently scientists have shown that, when a leaf is exposed to a given amount of light and dark, a particular form of mRNA is produced in the leaf, linked with a gene associated with flowering (the FT gene or Flowering Locus T). It is known as FTmRNA. It was thought that a large molecule like FTmRNA could not be florigen, as it would not be able to leave the cell. Now scientists have shown that FTmRNA can move from cell to cell to the transport tissues through the plasmodesmata. They have also shown that FTmRNA travels from the leaves in which it is formed to the apex of the shoot, where other genes associated with flowering are activated. So at the moment, it looks as though FTmRNA is the chemical known as florigen.

CHECKPOINT

1. (a) Define phytochromes.

    (b) Long-day plants and short-day plants need different length periods of light and darkness to trigger flowering. Summarise the role of phytochrome in the response of plants to differences in day length.

2. ▶ Explain how the evidence presented in **fig B** supports the idea of the plant hormone florigen. Take each bullet point in turn and explain its relevance.

    SKILLS   INTERPRETATION

3. Produce a flowchart to show how phytochrome and florigen interact to bring about flowering in a plant.

SUBJECT VOCABULARY

red light light with a wavelength of 620–700 nm, which is detected by plants using phytochromes
far red light light with a wavelength of 700–800 nm, which is detected by plants using phytochromes
phytochrome a plant pigment that reacts with different types of light, and as a result affects the responses of the plant
short-day plants (SDPs) plants that flower when days are short and nights are long
long-day plants (LDPs) plants that flower when days are long and nights are short
day-neutral plants (DNPs) plants whose flowering is not affected by the length of time they are exposed to light or dark
florigen plant hormone which appears to be involved in the photoperiodic response; it may be FTmRNA

The whole shape and form of a plant is dependent on its need for light. What part do phytochromes play in this process? Plants that are grown in the dark or are heavily shaded by other plants become **etiolated**. This means they grow rapidly, using up food reserves in an attempt to reach the light. As a result, the plants become tall and thin, with fragile, pale stems, long internodes (the stem between the leaf nodes), and small, pale, yellowish leaves because no chlorophyll is formed (see **fig A**). Etiolation seems to be a survival mechanism. All of the resources of the plant go into growing up towards the light needed for photosynthesis. Once the plant reaches the light, growth slows and the leaves turn green as chlorophyll forms.

This is similar to the changes that take place as a seed germinates and grows. Almost all seeds germinate under the ground, so the early stages of growth occur in the dark and are etiolated. The changes that occur after a plant becomes etiolated, and the reverse of etiolation when germinating seedlings break through the soil, appear to be controlled by phytochrome.

GERMINATION AND THE CHANGE IN PLANTS

Phytochrome is synthesised as P_r. A seedling that emerges from a seed underground only contains P_r because it has not been exposed to light. The early seedling that emerges from the seed has a cotyledon or a hooked apical shoot and shows typical characteristics of etiolation (see **fig A**). These include:

- rapid stem lengthening but very little thickening; the seedling grows as tall as possible and as fast as possible to reach the light
- relatively little root growth, just enough to act as an anchor and obtain water
- no leaf growth; the leaves are small and folded, so no energy is wasted producing leaf tissue that is useless underground
- no chlorophyll; the seedling is white or pale yellow, so no energy is wasted producing chlorophyll that is useless in the dark.

Once the tip of the new shoot breaks through the soil surface into the light, a series of changes takes place:

- the elongation of the stem slows down
- the stem straightens
- the cotyledons and/or first leaves open
- chlorophyll forms and the seedling begins to photosynthesise.

▲ **fig A** You can clearly see the difference between the normal seedlings on the left, which have emerged into the light, and the etiolated seedlings on the right, which have been grown in the dark. This demonstrates the importance of phytochromes on the development of young plants.

The changes that occur the moment the germinating seedling is exposed to light are controlled by phytochrome interconversion. In the seed, there is plenty of P_r but no P_{fr}. Without P_{fr}, the internodes grow but the leaves do not and no chlorophyll forms. Once the plant is exposed to light, P_r is rapidly converted to P_{fr} which accumulates quickly. P_{fr} inhibits the lengthening of the internodes so internode growth slows. It stimulates leaf development and the production of chlorophyll. The leaves open and the seedling becomes green and begins to photosynthesise. These changes start even before the seedling breaks through the surface of the soil because a little light penetrates through the surface of the soil and begins the transformation. As a result, the chloroplasts are maturing and the seedling is often green and ready to photosynthesise the moment it emerges through the soil.

PHYTOCHROMES AND TROPISMS

Until recently, scientists thought that the changes in plants controlled by phytochrome and the responses of plants to light controlled by auxins and other hormones were quite separate processes. However, research increasingly shows a link between phytochrome and both phototropisms (plant movements in response to one-sided light) and geotropisms (plant movements in response to an asymmetrical pull of gravity). For example, it appears that phototropisms cannot occur in very young shoots until phytochrome has been activated, and that geotropisms are also dependent on phytochrome actions.

PHYTOCHROME AS A TRANSCRIPTION FACTOR

Scientists are only just beginning to understand how phytochrome can change so many things, from triggering flowering to the production of chlorophyll and the growth of the stem. More and more evidence suggests that P_{fr} acts as a transcription factor, which is involved in switching genes on and off in plant cell nuclei.

Researchers have produced recombinant DNA (see **Section 8C.1**) linking the genes for the production of phytochrome to a gene for the production of **green fluorescent protein (GFP)**, originally from jellyfish. By inserting these hybrid genes into plant cells, the scientists produced plants with fluorescent phytochrome.

If scientists kept seedlings in the dark, the fluorescence linked to the inactive P_r was detected evenly through the cytoplasm of the cells. If they exposed the seedlings to red light, this labelled P_r was converted to labelled P_{fr} and the scientists observed that the fluorescence moved into the nucleus of the cells. Further, in the nucleus the fluorescent P_{fr} appeared as specks linked to the chromosomes.

Here is the current model for the way in which phytochrome works.

- When P_r is converted into P_{fr} in the presence of light, it moves into the nucleus through the pores in the nuclear membrane.
- In the nucleus, it binds to a nuclear protein known as the phytochrome-interacting factor 3 (PIF3).
- PIF3 is a known transcription factor.
- PIF3 only binds to P_{fr}. It does not bind to P_r.
- PIF3 only activates gene transcription and the formation of mRNA if it is bound to P_{fr}.

EXAM HINT

You should expect to be tested on what you learned about factors that affect gene expression in **Section 3C.3 (Book 1: IAS)**.

The hypothesis is that by binding to PIF3, P_{fr} activates different genes and thus controls different aspects of growth and development in plants. Scientists need to continue this research until we fully understand how phytochrome has its effect, but our models for the control and coordination of plants are becoming more sophisticated and more integrated all the time.

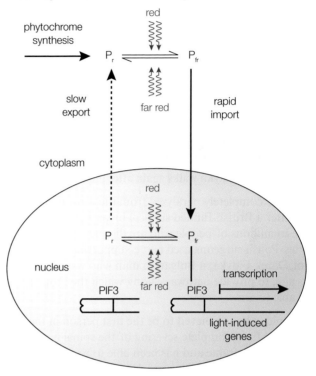

▲ **fig B** Model of P_{fr} as a transcription factor in plants (Courtesy of Science and Plants for Schools).

CHECKPOINT

1. Draw a diagram to illustrate our current model of how the interconversion of phytochrome affects a germinating seedling.

2. Describe the effects of red light (P_{fr} build-up) and far red light (P_r build-up) on different parts of a plant. You may find it useful to produce a table summary.

SUBJECT VOCABULARY

etiolated the form of plants which are grown in the dark, with long internodes, thin stems, small or unformed leaves and white or pale yellow in colour

green fluorescent protein (GFP) the product of a gene often used as a marker in the production of recombinant DNA

THE PARALYSED MAN WHO WALKED AGAIN

SKILLS ANALYSIS, INTERPRETATION, CREATIVITY, INTELLECTUAL INTEREST AND CURIOSITY, SELF-EVALUATION, COMMUNICATION

Neurones make and lose connections all the time in the brain as learning takes place and memories are formed and lost. But if the spinal cord is severed, the neurones simply do not grow and join together again and the affected individual will be paralysed below the break. This used to be considered irreversible …

NEWSPAPER EXTRACT

Paralysed man Darek Fidyka walks again after pioneering surgery

A man who was completely paralysed from the waist down can walk again after a British-funded surgical breakthrough which offers hope to millions of people who are disabled by spinal cord injuries. Polish surgeons used nerve-supporting cells from the nose of Darek Fidyka, a Bulgarian man who was injured four years ago, to provide pathways along which the broken tissue was able to grow.

The 38-year-old, who is believed to be the first person in the world to recover from complete severing of the spinal nerves, can now walk with a frame and has been able to resume an independent life, even to the extent of driving a car, while sensation has returned to his lower limbs.

Professor Geoffrey Raisman, whose team at University College London's institute of neurology discovered the technique, said: 'We believe that this procedure is the breakthrough which, as it is further developed, will result in a historic change in the currently hopeless outlook for people disabled by spinal cord injury.'

The surgery was performed by a Polish team led by one of the world's top spinal repair experts, Dr Pawel Tabakow, from Wroclaw Medical University, and involved transplanting olfactory ensheathing cells (OECs) from the nose to the spinal cord. OECs assist the repair of damaged nerves that transmit smell messages by opening up pathways for them to the olfactory bulbs in the forebrain. Relocated to the spinal cord, they appear to enable the ends of severed nerve fibres to grow and join together – something that was previously thought to be impossible. While some patients with partial spinal injury

have made remarkable recoveries, a complete break is generally assumed to be unrepairable.

Raisman said: 'The patient is now able to move around the hips and on the left side he's experienced considerable recovery of the leg muscles. He can get around with a walker and he's been able to resume much of his original life, including driving a car. He's not dancing, but he's absolutely delighted.'

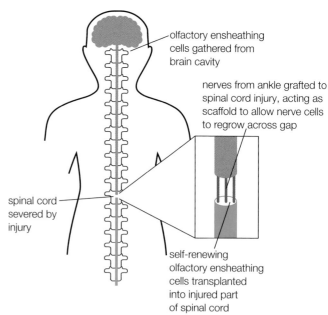

olfactory ensheathing cells gathered from brain cavity

nerves from ankle grafted to spinal cord injury, acting as scaffold to allow nerve cells to regrow across gap

spinal cord severed by injury

self-renewing olfactory ensheathing cells transplanted into injured part of spinal cord

▲ **fig A** This diagram shows the procedure on Mr Fidyka's spine. Darek Fidyka has been able to drive an adapted car and to walk again using a frame after receiving the pioneering treatment which repaired his severed spine.

Since this article was written, Darek has made further progress. He has regained control of his bladder and bowels, and can ride a tricycle.

SCIENCE COMMUNICATION

This article demonstrates the style of writing used in many online news stories.

1 This is a sensational story: even the doctors and scientists involved in the treatment seemed excited by what happened to Darek Fidyka. Explain how this online version of a popular newspaper sets out to:

   (a) convey the ground-breaking nature of the scientific development

   (b) give a clear explanation of the procedure.

2 Comment on whether you think the article succeeds in communicating both the excitement and the science. Justify your opinion.

3 (a) Explore online and find other descriptions of this same medical breakthrough. Describe **four** points that are common to all the accounts you look at, and **two** points that occur in some but not all of them.

   (b) Suggest reasons for the differences in the way the story is reported in different places.

INTERPRETATION NOTE

Being able to communicate scientific ideas to a wide audience is an important skill, particularly when the science has implications for decision making in society.

BIOLOGY IN DETAIL

Now let us examine the biology. You already know about the structure and functions of the cell membrane, diffusion, active transport, the structure and function of the mammalian nervous system and how nerve impulses are transmitted along neurones and between neurones across synapses. This knowledge will help you answer the questions below.

4 Describe the structure of the spinal cord and use this description to explain how, in a spinal injury:

   (a) the amount of paralysis depends on the site of the injury

   (b) function above the injury is usually unaffected

   (c) there may be a loss of motor function, a loss of sensory function, or both

   (d) autonomic functions such as bladder and bowel control may be lost as well as consciously controlled movements.

5 Some other news articles published at the same time as this story of Darek Fidyka's recovery described the cells used in the process as 'cells from the nose' or 'stem cells'. Both of these descriptions are inaccurate. Explain why.

6 Sometimes people with partial spinal cord injuries can recover to some extent. When the spinal cord is completely severed, there is usually no recovery. Suggest reasons for this based on what you know of nerve cells and using what you have learned from the case of Darek Fidyka.

ACTIVITY

- Spinal cord injuries can and do happen to anyone of any age, although they are relatively rare in children.
- Use the internet and any other resources available to investigate the main causes of spinal injuries in your country.
- Investigate further the current research into repairing spinal injuries.
- Use your findings to help you either produce a scientific poster or web page summarising the state of spinal injury research today or write a letter to the research funding organisations arguing for protected funding for spinal injury research.
- Use online resources as well as scientific magazines and books. Evaluate your sources before using them and reference all sources used.

8B EXAM PRACTICE

1 (a) (i) How many cells are estimated to be working together in the human brain? [1]

    A 100

    B 100 thousand

    C 100 million

    D 100 billion

(ii) What is the corpus callosum? [1]

    A a band of axons connecting the left and right sides of the cerebellum

    B a band of axons connecting the left and right sides of the cerebrum

    C a round body in the brain that controls the body movements

    D a region of the cerebrum that controls conscious thought

(iii) Complete the table below describing the functions of parts of the human brain. [4]

Area of brain	Function
cerebrum	
	control heart rate
	thermoregulation
cerebellum	

(b) The diagram below shows a human brain seen from the side.

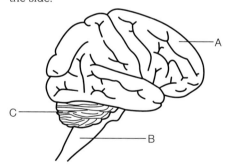

(i) Name the parts labelled **B** and **C**. [2]

(ii) Explain why the part labelled **A** is folded. [2]

(Total for Question 1 = 10 marks)

2 (a) How do plant hormones travel to their targets? [1]

    A in the phloem

    B in the xylem

    C in the sclerenchyma

    D through plasmodesmata

(b) The statements in the table below refer to some effects of **two** groups of plant growth substances, auxins and gibberellins. Copy the table. If the statement is correct place a tick (✓) in the appropriate box and if the statement is incorrect place a cross (✗) in the appropriate box. [5]

Effect	Auxins	Gibberellins
promote cell elongation		
promote root formation in cuttings and calluses		
promote fruit growth		
inhibit lateral bud development		
promote the breaking of dormancy in seeds		

(Total for Question 2 = 6 marks)

3 Plants can detect and respond to environmental stimuli. Cocklebur is a plant that flowers after it has been exposed to a sufficiently long period of darkness. The minimum length of time in darkness needed to stimulate flowering is called the critical period.

(a) What type of plant is cocklebur? [1]

    A a long-day plant

    B a day-neutral plant

    C a short-day plant

    D a dark-critical plant

(b) An investigation was carried out into the effect of light and dark periods on cocklebur flowering.

Four plants, **A**, **B**, **C** and **D**, were exposed to light and dark periods of different length. The presence or absence of flowers was recorded after several weeks. The diagram below shows the pattern of light and dark periods for these plants and the effect on flowering.

Plant	Time/hours						Flowers present
	0	4	8	12	16	20	
A							Yes
B							No
C							No
D							No

Key

☐ Light

▨ Dark

(i) What is the critical period for flowering of cocklebur plants? [1]

    A 20 hours

    B between 19 and 20 hours

    C 8 hours

    D between 7 and 8 hours

(ii) Using the information in the diagram and your own knowledge of photoreceptors, explain why plant **B** has not flowered.   [4]

(c) In a further investigation, plants **E** and **F** were exposed to six hours of darkness each day. Part of a leaf on plant **F** was covered so that the leaf experienced 10 hours of darkness each day. The diagram below summarises the results of this investigation.

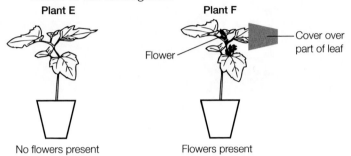

No flowers present   Flowers present

(i)   Using your knowledge of photoreceptors, explain why plant **F** flowered.   [3]
(ii) Explain the purpose of plant **E** in this investigation.   [2]

(d) Discuss benefits to plants of being able to respond to changes in day length.   [3]

(Total for Question 3 = 14 marks)

4 (a) (i) Which row correctly describes the peripheral nervous system?   [1]

	Peripheral nervous system	**Voluntary nervous system**	**Autonomic nervous system**
A	sensory neurones only	motor neurones only	sensory neurones only
B	sensory neurones only	motor neurones only	motor neurones and sensory neurones
C	motor neurones and sensory neurones	sensory neurones only	motor neurones only
D	sensory and motor neurones	motor neurones only	motor neurones only

(ii) The changes in the size of the pupil is a cranial reflex action. Copy and complete the table below to describe the stages in this reflex action when in bright light.   [4]

Stimulus	Receptor	Coordination	Effector	Response
		centres in midbrain and oculomotor nerve		

(iii) Explain the benefit of the pupil reflex.   [3]

(b) Discuss the role of the autonomic nervous system in controlling heart rate.   [5]

(Total for Question 4 = 13 marks)

5 A student investigated the effects of the plant hormones auxin and gibberellin on plant growth. He selected 10 healthy growing shoots. Five shoots were coated with gibberellin paste, and five were coated with auxin paste. He measured the length of the lateral shoots each day for 18 days. His results are shown in the graph below.

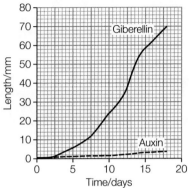

(a) Calculate the percentage increase in length at day 18 of shoots treated with gibberellin compared to those treated with auxin. Show your working.   [2]

(b) The student concluded that gibberellins have a greater effect on the growth of the lateral shoots than auxin. Suggest why the student's conclusion may not be valid.   [3]

(c) Suggest a suitable control for this experiment.   [1]

(d) Explain why the student used five shoots in each sample.   [3]

(Total for Question 5 = 9 marks)

TOPIC 8 COORDINATION, RESPONSE AND GENE TECHNOLOGY

8C GENE TECHNOLOGY

Genetic modification (GM) is a widely used technique. Many scientists are convinced that producing GM plants with added nutritional benefits or containing vaccines against serious diseases is the key to the future of the human race. Almost 80% of the soya beans grown around the world are now genetically modified and around 190 million hectares (over 20% of the available arable land) is used to grow GM crops. However, some people are inserting genes into plants and animals that have less obvious uses. Starlight Avatar is a type of decorative, glow-in-the-dark houseplant made by introducing DNA from luminescent bacteria into the chloroplast genomes of a houseplant. The developers hope that eventually houseplants might become a form of lighting in every home!

In this chapter, you will discover how genes from one organism are introduced into the genetic material of an organism of a different species. You will consider the types of change that can be introduced to crops, and their potential advantages for human and animal food security. You will look at the role played by microarrays and bioinformatics in our growing understanding of genomes and genetic modification. Finally, you will consider why the use of genetically modified commercial crops and animals has caused so much public debate.

MATHS SKILLS FOR THIS CHAPTER

- Translate information between graphical, numerical and algebraic forms (*e.g. use of GM crops globally*)

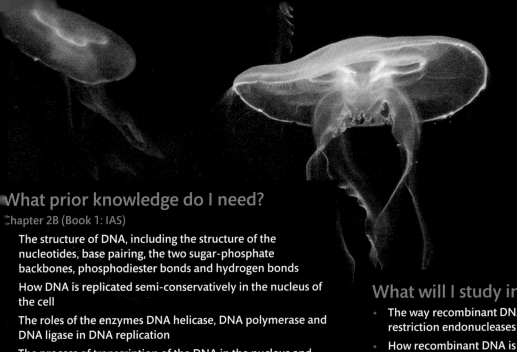

What prior knowledge do I need?

Chapter 2B (Book 1: IAS)

- The structure of DNA, including the structure of the nucleotides, base pairing, the two sugar-phosphate backbones, phosphodiester bonds and hydrogen bonds
- How DNA is replicated semi-conservatively in the nucleus of the cell
- The roles of the enzymes DNA helicase, DNA polymerase and DNA ligase in DNA replication
- The process of transcription of the DNA in the nucleus and translation at the ribosome
- The nature of the genetic code

Chapter 6A

- The aseptic techniques used in culturing organisms
- The principles and techniques involved in culturing microorganisms, including the use of different media
- Different methods for measuring the growth of a bacterial culture, including cell counts, dilution plating and mass and optical methods

What will I study in this chapter?

- The way recombinant DNA is produced, including the role of restriction endonucleases and DNA ligase
- How recombinant DNA is inserted into other cells using a variety of vectors
- How antibiotic resistance markers and replica plating can be used to identify recombinant cells
- The genetic modification of soya beans and how it has been used to improve production, including altering the balance of fatty acids to prevent the oxidation of soya products
- The use of 'knockout' organisms to investigate gene function
- Why the widespread use of genetic modification of major commercial crops and other transgenic processes have caused public debate of their advantages and disadvantages

What will I study later?

Beyond IAL

- The process of gene editing, for example using CRISPR Cas9
- The production of artificial DNA and artificial organisms
- The use of gene technology in alleviating major human genetic disease

LEARNING OBJECTIVES

■ Understand how recombinant DNA can be produced, including the roles of restriction endonucleases and DNA ligase.
■ Understand how recombinant DNA can be inserted into other cells.

Genetic engineering, **genetic modification** and **gene editing** are the names we give to changing the genetic material of an organism, usually by inserting genes from one organism into the genetic material of another organism. We already use gene technology in many ways but its full potential benefits remain to be explored.

PRODUCING RECOMBINANT DNA

DNA that is formed artificially by combining genetic material from different organisms is known as **recombinant DNA**. Bacteria are the most widely used genetically engineered organisms, and the basic processes used to produce transformed bacteria are common to all genetic modification.

Artificial copies of a desired gene can be made by taking an mRNA molecule transcribed from the gene and using it to produce the correct DNA sequence. This uses the enzyme **reverse transcriptase**. It reverses the transcription process to produce **complementary DNA (cDNA)** which can act as an artificial gene.

Alternatively, restriction endonucleases are used to cut DNA strands into small pieces that are easier to handle (see **Section 6C.5**). Each type of endonuclease will cut DNA only at specific (restricted) sites within a particular DNA sequence. Some restriction endonucleases can cut the DNA strands in a way that leaves a few base pairs longer on one strand than the other, forming a **sticky end**. Sticky ends make it easier to attach new pieces of DNA to them. Sticky ends attach to other compatible sticky ends so, for example, you can join together the sticky ends of DNA fragments cut by EcoR1, but not sticky ends cut by two different enzymes.

The next step is to integrate the new gene into a vector. Plasmids, the circular strands of DNA found in bacteria (see **Section 3A.4 (Book 1: IAS)**), are frequently used as vectors to carry the DNA into a host bacterial cell. DNA ligases are used as 'genetic glue' to join pieces of DNA together (see **fig A**). This is the original method used to insert the chosen DNA, cut from an organism or made artificially, into another piece of DNA that will carry it into the host cell. Once the plasmid is incorporated into the host nucleus, it forms part of the new recombinant DNA of the genetically engineered or transformed organism. In a bacterial host it usually remains as an autonomous plasmid capable of independent replication, as bacteria do not have a true nucleus. Successfully transformed cells can be identified, isolated and cultivated on an industrial scale so that the proteins they make can be harvested for human use. This technology is developing constantly and gene editing is becoming easier, cheaper, faster and more sophisticated.

EXAM HINT

The sticky ends will attach to each other because the pairs of bases are complementary. Remember that your IAS knowledge will be tested at IAL and you must be very familiar with the structure of DNA (see **Section 2B.3–5 (Book 1: IAS)**). In particular the strict base-pairing rules are important.

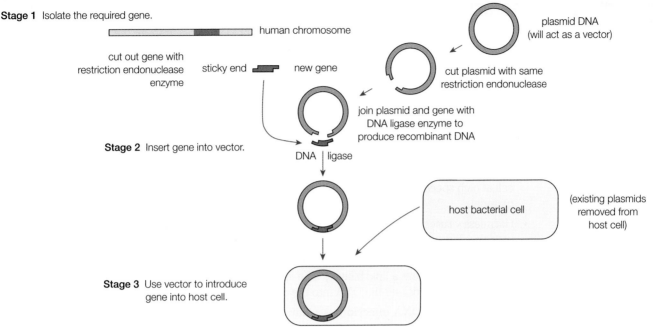

Stage 1 Isolate the required gene.

Stage 2 Insert gene into vector.

Stage 3 Use vector to introduce gene into host cell.

The modified bacterium will now produce a different protein as the new gene is expressed and causes synthesis of the protein.

▲ **fig A** The main stages of the traditional method of inserting a new gene into a bacterium. New technologies such as gene editing have made the process even more sophisticated.

IDENTIFYING TRANSFORMED ORGANISMS

One bacterium looks very like another, therefore scientists transfer special marker genes with the desired DNA so they can identify the microorganisms in which transformation has taken place. These marker genes are usually genes that make a bacterium dependent on a particular nutrient, or which cause the organism to fluoresce in UV light.

We can use a process known as **replica plating** to identify recombinant cells. This involves growing identical patterns of bacterial colonies on agar plates (Petri dishes) with different media. It allows us to identify colonies that cannot survive without a particular nutrient. These are the bacteria which have been genetically modified (GM). The main stages of the process are shown in **fig B**.

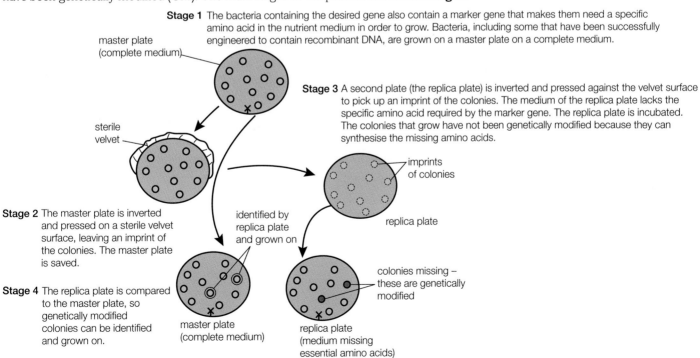

Stage 1 The bacteria containing the desired gene also contain a marker gene that makes them need a specific amino acid in the nutrient medium in order to grow. Bacteria, including some that have been successfully engineered to contain recombinant DNA, are grown on a master plate on a complete medium.

master plate (complete medium)

sterile velvet

Stage 3 A second plate (the replica plate) is inverted and pressed against the velvet surface to pick up an imprint of the colonies. The medium of the replica plate lacks the specific amino acid required by the marker gene. The replica plate is incubated. The colonies that grow have not been genetically modified because they can synthesise the missing amino acids.

imprints of colonies

replica plate

Stage 2 The master plate is inverted and pressed on a sterile velvet surface, leaving an imprint of the colonies. The master plate is saved.

identified by replica plate and grown on

Stage 4 The replica plate is compared to the master plate, so genetically modified colonies can be identified and grown on.

master plate (complete medium)

replica plate (medium missing essential amino acids)

colonies missing – these are genetically modified

▲ **fig B** Replica plating allows us to identify genetically modified bacteria.

LEARNING TIP

A colony of bacteria on an agar plate have all grown from one single bacterium. Therefore, they are all genetically identical. They are clones.

VECTORS

Vectors play a key role in the formation of recombinant DNA. They transfer the required gene, with any marker genes, into the new cells. A successful vector targets the right cells, ensures that the desired gene is incorporated into the host genetic material so it can be activated (transcribed and translated) and does not have any adverse side-effects.

Plasmids are particularly useful as vectors in the formation of GM bacteria and in the formation of GM plants. However, other vectors are needed to carry new DNA into animal cells, especially human cells, and into some types of plant cell. Scientists are trying a number of techniques with varying degrees of success. They include the following.

- **Gene guns** are used to shoot DNA carried on very small gold or tungsten pellets (balls) into the cell at high speed. Some cells survive this treatment and accept the DNA as part of the genetic material.

- Harmless viruses can be engineered to carry a desirable gene and then used to infect an animal's cells thus introducing the desirable DNA.

- **Liposome wrapping** is a technique in which the gene to be inserted is wrapped in liposomes (spheres formed from a lipid bilayer). The liposomes fuse with the target cell membrane and can pass through it to deliver the DNA into the cytoplasm.

- **Microinjection (DNA injection)** is a way of introducing DNA by injecting it into a cell through a very fine micropipette. This is manipulated using a micromanipulator, because even the steadiest hand would tremble enough to destroy the cell. The method is not very efficient because many cells have to be injected before one accepts the DNA successfully. However, it is the method that has resulted in most successful transgenic animals.

Once a vector is inside the cell, the next challenge is for the new DNA to reach the right place. This is proving very difficult, particularly when liposomes are used as the vectors. Estimates suggest that only about 1 in every 1000 genes that enter a cell in a liposome enter the nucleus to be transcribed. Researchers are trying to modify the vectors so that the new genes enter the cells and then the nuclei more effectively. So far, viruses are much better at taking DNA into the nucleus than other methods, because inserting DNA into the genetic material of other organisms is an important part of how viruses reproduce (see **Section 6A.2**). However, viruses that are used in genetic engineering can cause an immune response in some people in the same way that pathogenic viruses stimulate an immune response. Consequently, much of the research into genetic modification of human cells is focusing on non-viral vectors such as liposomes. This is because they cause fewer side-effects and potential immune responses, even though they are not as effective at transferring the DNA.

KNOCKOUT ORGANISMS: SILENCING GENES

Genetic modification does not always involve adding a new active gene to the genome of an organism. In some situations, scientists want to remove or silence a gene. For example, **knockout organisms** are widely used and are very important in genetic research. In a knockout organism, scientists silence (knock out) one or more genes, so they no longer function. They do this by inserting a new gene similar to the gene to be investigated, but which makes the original DNA sequence impossible to read. As a result, the original gene is silenced and cannot make a protein. Knockout genes are usually accompanied by marker genes to show that they have been incorporated.

Research can use knockout organisms to identify the function of a gene. Genome sequencing identifies many genes, but scientists often do not know the function of these genes in an organism. Knocking them out and observing the result can help to make their function clear.

We can also use knockout organisms to investigate disease and test potential treatments. We can knock out genes that are known to be non-functioning in human diseases to create animal models of the disease. These animal models are invaluable for progressing our understanding of human disease and ways of treating it.

CHECKPOINT

1. Draw a flow diagram showing the main stages of the formation of recombinant DNA.
2. Explain the importance of replica plating in producing a culture of GM bacteria.
3. Suggest advantages and disadvantages in the use of plasmids, viruses, gene guns and liposomes as vectors in the production of recombinant DNA organisms.

SUBJECT VOCABULARY

genetic engineering/genetic modification/gene editing the insertion of genes from one organism into the genetic material of another organism or changing the genetic material of an organism

recombinant DNA new DNA produced by genetic engineering technology that combines genes from the DNA of one organism with the DNA of another organism

reverse transcriptase an enzyme used to make artificial copies of a desired gene by taking an mRNA molecule transcribed from the gene and using it to produce the correct DNA sequence

complementary DNA (cDNA) DNA which can act as an artificial gene, made by reversing the transcription process from mRNA using reverse transcriptase

sticky end the name given to the area of base pairs left longer on one strand of DNA than the other by certain restriction endonucleases, making it easier to attach new pieces of DNA

replica plating a process used to identify recombinant cells that involves growing identical patterns of bacterial colonies on plates with different media

gene guns a technique to produce recombinant DNA by shooting the desired DNA into the cell at high speed on very small gold or tungsten pellets (balls)

liposome wrapping a technique for producing recombinant DNA that involves wrapping the gene to be inserted in liposomes, which combine with the target cell membrane and can pass through it to deliver the DNA into the cytoplasm

microinjection (DNA injection) a technique for producing recombinant DNA that involves injecting DNA into a cell through a very fine micropipette

knockout organism an organism with one or more genes silenced (knocked out) so they no longer work; they are often used to identify the function of a gene, to investigate disease and to test potential treatments

LEARNING OBJECTIVES

■ Know how drugs can be produced using genetically modified organisms (plants, animals and microorganisms).

▲ **fig A** Plating cultures of bacteria on agar containing (or lacking) particular nutrients allows us to identify the microorganisms containing the marked, genetically engineered plasmids because they are the only ones able to survive.

Pharmacogenomics is an area of drug development and medical treatment that is a new technology not yet in regular use. We are using our growing knowledge of the human genome in the development of genetically modified organisms to produce medicines and vaccines to treat human conditions, but much more work is still needed in genetic medicine.

MICROORGANISMS AND GENETIC MODIFICATION

Genetic modification most commonly uses microorganisms for a number of reasons. They are relatively easy and cheap to culture and, because they reproduce so rapidly, a transferred gene is copied very rapidly when the microorganisms are allowed to replicate in ideal conditions. It is very important to be able to tell which bacteria contain the desired, recombinant piece of DNA. Because one bacterium looks very like another, scientists transfer special marker genes with the desired DNA that make it easy to identify those microorganisms in which a successful transformation has occurred (see **Section 8C.1**). The bacteria which are identified by the markers (see **fig A**) can then be cultured on a large scale in industrial fermenters, and the proteins which they make are harvested.

There is an increasing number of chemicals made by genetically engineered organisms. They include antibiotics such as penicillin, hormones such as growth hormone and insulin, and enzymes.

MICROORGANISMS AND HUMAN INSULIN

People with Type 1 diabetes cannot make the insulin needed for their bodies to work properly. Everyone with Type 1 diabetes, and some people who have Type 2 diabetes, need regular injections of insulin to keep them healthy.

Historically, the source of insulin for people with diabetes was the pancreases of animals killed for meat. Although the insulin from these animals is similar to human insulin, it is not quite the same. This caused problems for some patients, because their immune systems reacted to the foreign antigens on the animal insulin. Also, the supply of insulin was not always reliable because it depended on how many animals were killed for meat.

In recent years, we have used biotechnology to develop a way of manufacturing human insulin using microorganisms (see **fig B**). The process was difficult because the insulin molecule is made up of two polypeptide chains. Scientists in the 1980s overcame this problem by introducing a synthesised gene for each insulin chain into different bacteria. They then cultured these bacteria in huge numbers. The mixture from the giant fermenters needs additional processing to separate the microorganisms and the desired end products from the rest of the mixture. The technique produces two pure protein chains which are oxidised to join them together. The resulting chemical is genetically identical to human insulin and appears to work in exactly the same way in the body. Its purity is guaranteed, which makes it easier to calculate doses accurately. For the majority of people with diabetes, it represents a great advantage.

Even more recently, scientists have developed a synthetic gene that mimics the normal human gene for insulin. This enables a single type of engineered bacteria to produce proinsulin. At the end of the process, enzymes convert proinsulin to insulin.

Using microorganisms in this way removes the problems of uncertain supply and provides a constant, convenient and pure source of a human hormone. This is a clear example of a positive effect in human medicine that results from genetic engineering and new biotechnology.

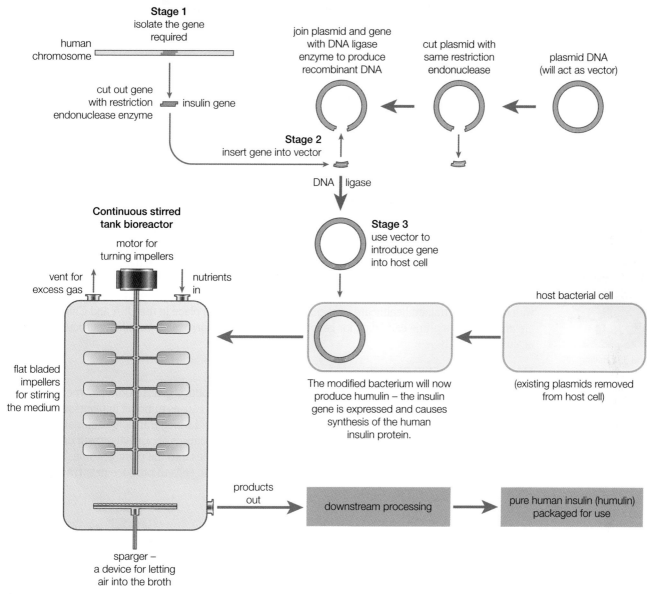

fig B This diagram summarises the type of process by which human insulin is harvested from genetically modified bacteria. The production of any drugs from genetically modified microorganisms is similar to this process.

THE VALUE OF GENETICALLY MODIFIED DRUGS

For most parents, the birth of a baby is followed by years of pleasure and amazement at the way their children grow. But some children grow more slowly. They may lack **growth hormone (GH)**, also known as **somatotrophin**. This hormone is secreted by the pituitary gland and stimulates growth.

For many years, doctors overcame the problem by giving children regular injections of pituitary extract, made from pituitary glands taken from human cadavers (dead bodies). The extract was expensive and always in short supply. Genetic engineering has solved the problem of supply; genetically modified organisms now produce the hormone in commercial amounts.

However, the solution came too late for some unfortunate children. Lack of care in the extraction of the pituitary glands led to contamination of the drug by brain tissue from individuals who had died as a result of the very rare Creutzfeldt–Jakob disease, a fatal brain disease. (A form of Creutzfeldt–Jakob disease was linked to bovine spongiform encephalopathy (BSE), or 'mad cow disease' in cattle.) Sadly, some people were treated with contaminated pituitary extract before the new hormone produced by genetically modified microorganisms was available. They died of Creutzfeldt–Jakob disease.

EXAM HINT

You may need to discuss the risks and benefits of using genetically modified organisms to produce molecules such as insulin or growth hormone.

▲ **fig C** Around 350 million people globally are already infected with hepatitis B. If genetically modified bananas could be grown carrying the hepatitis B vaccine, it would save many lives and be a great alternative to injections.

People sometimes express concern about the risks of using genetically modified organisms in medicines or food. It is wise to remember that risks can come from a wide variety of sources, and that genetically modified organisms can reduce some of these risks.

BANANA VACCINES?

Scientists have successfully used genetically modified microorganisms to make certain human proteins such as insulin and somatotrophin. However, prokaryotes do not possess the biochemistry to make some of the more complex human proteins. Now scientists are working on ways to introduce useful human genes into eukaryotic cells such as yeast, plants and even mammals.

Many scientists hope that transgenic plants will become an important weapon in the worldwide fight against disease. As you saw in **Chapter 6B**, vaccination is a very effective way of eliminating serious diseases. However, less economically developed countries cannot always afford vaccines. Vaccines usually require storage in a fridge, and in some countries few people have guaranteed access to a fridge. There can also be cultural and practical difficulties with seeing healthcare workers and allowing them to vaccinate children. If we can genetically modify plants or plant products such as bananas, potatoes or carrots to carry vaccines against human diseases such as infant diarrhoea or hepatitis B, we could solve many of these problems. Local communities could grow the relevant plants relatively cheaply and there is no need for cool storage. The children could be protected from deadly diseases with no need for trained healthcare workers. Trials are already in place in the USA and China. Bananas are emerging as a prime candidate for genetic modification to carry plant vaccines, particularly against hepatitis B.

MAKING TRANSGENIC PLANTS

The bacterium *Agrobacterium tumefaciens* is usually used to introduce genes from one type of plant into another or even from an animal into a plant. *A. tumefaciens* causes tumours in plants which are known as crown galls. It contains a plasmid called the Ti plasmid which transfers bacterial genetic information directly into the plant DNA. This is what normally causes abnormal growth of the plant cells, but modified plasmids can be used to carry beneficial genes into the plant genome. Then, by the process of plant cloning, we can use the modified transgenic cells to produce new **transgenic plants** (see **fig D**).

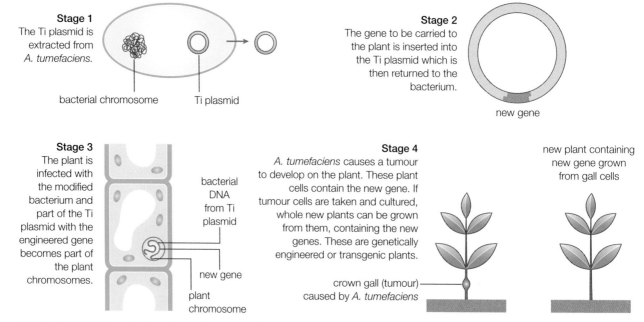

Stage 1
The Ti plasmid is extracted from *A. tumefaciens*.

bacterial chromosome   Ti plasmid

Stage 2
The gene to be carried to the plant is inserted into the Ti plasmid which is then returned to the bacterium.

Ti plasmid

new gene

Stage 3
The plant is infected with the modified bacterium and part of the Ti plasmid with the engineered gene becomes part of the plant chromosomes.

bacterial DNA from Ti plasmid

new gene

plant chromosome

Stage 4
A. tumefaciens causes a tumour to develop on the plant. These plant cells contain the new gene. If tumour cells are taken and cultured, whole new plants can be grown from them, containing the new genes. These are genetically engineered or transgenic plants.

crown gall (tumour) caused by *A. tumefaciens*

new plant containing new gene grown from gall cells

▲ **fig D** The basic method used to produce genetically modified plants, including those modified to produce human medicines.

GENETICALLY MODIFIED ANIMALS

It is very difficult to transfer new DNA into eukaryotic cells of animals but some attempts have already been successful. Some of the most exciting and recent research involving genetic engineering has been the work on inserting human genes into the tissues of sheep and cattle to produce transgenic animals. In 1991, the first transgenic sheep was born, a ewe called Tracey who produced the human protein alpha-1-antitrypsin (AAT) in her milk. This protein is missing in people who suffer from a genetic condition affecting their livers and lungs, causing emphysema to develop at a very early age.

The production of proteins using **transgenic animals** involves introducing a copy of a human gene which codes for the desired protein into the genetic material of an egg of a different animal species. As well as the gene for the specific protein there is also a promoter sequence which makes sure the gene will be expressed in the mammary gland of the lactating female. The fertilised egg, now a developing transgenic embryo, is then replaced inside a surrogate (substitute) mother, is born and grows to maturity. When the animal is mature and produces milk, that milk is harvested, purified and the human protein extracted. So far, scientists have succeeded in producing transgenic sheep, cows, pigs, rabbits and mice, all capable of producing human proteins in their milk. When we use large animals such as cows and sheep, large volumes of milk, and therefore human proteins, can be achieved, and the number of animals can be increased by breeding or cloning.

EXAMPLES OF DRUGS FROM TRANSGENIC ANIMALS

Transgenic animals have so far produced more than 20 different human proteins, and some of them are already in therapeutic use or being trialled.

- Factor VII and Factor IX are both important components of the human blood clotting cascade which can be missing in haemophilia and other blood clotting diseases. These factors are being harvested from transgenic milk.
- Alpha-1-antitrypsin is the protein produced by Tracey. The milk of such transgenic sheep contains up to 35 g of human protein in every litre, a very high yield. The milk is very expensive, it costs several thousand dollars per litre. Trials are in progress on treatment using the transgenic protein and the idea is that as the numbers of transgenic animals increase, the price will eventually decrease. However, developing these products commercially has presented major problems. Most of the transgenic sheep were destroyed when the company that produced them met financial difficulties.
- Activated protein C for treating deep vein thrombosis is also being tested.

As we begin to understand the mysteries of the human genome in increasing detail, we should be able to identify the genetic sequence for every chemical in our body. If genetically modified organisms can make many of our faulty molecules for us, whether they are bacteria, plants or animals, the impact on future health and medicine could be enormous.

EXAM HINT

Remember that genetic modification and production of human proteins by other organisms is only possible because the genetic code is universal.

CHECKPOINT

1. Draw a flow diagram to explain the process by which a drug is made using a genetically modified bacterium.

2. Compare the production of a drug from a genetically modified plant with that from a genetically modified microorganism.

3. ▶ Investigate the use of genetically modified animals to produce human medicines. Select one drug and evaluate the success of this process so far.

SKILLS   DECISION MAKING, INTEGRITY, COMMUNICATION

SUBJECT VOCABULARY

growth hormone (GH) the hormone secreted by the pituitary gland which stimulates growth
somatotrophin another name for growth hormone
transgenic plants plants which have been genetically modified to produce proteins from another organism
transgenic animals animals which have been genetically modified to produce proteins from another organism, often a human being

■ Know how microarrays can be used to identify active genes.
■ Understand what is meant by the term bioinformatics.

Scientists need to be able to identify individual genes for many reasons, including genetic modification, gene silencing, identifying gene mutations which cause specific diseases or analysing the patterns of genes and diseases within a population. Scientists have developed one very useful laboratory tool which allows them to detect thousands of active genes at the same time, the **microarray**.

USING MICROARRAYS

When genes are active, they are expressed. Messenger RNA (mRNA) is produced from active genes and used as a template for the production of amino acid chains on the surface of the ribosomes (see **Section 2B.6 (Book 1: IAS)**). If there is a mutation in a gene, a different form of mRNA will be produced. However, a gene contains a very large number of DNA bases, and there are many places where a mutation can occur. A microarray is a tool which scientists use widely to show if a DNA sample from an individual contains any mutations. These samples are also known as 'DNA chips'.

Basically, a DNA microarray is a slide on which there are thousands of spots. Each spot is in a specific position and contains a known DNA sequence. The mRNA samples are then collected. Scientists usually use a reference sample with a known gene sequence and an experimental sample. Typically, this will be from an individual with a particular disease which may have a genetic element, for example a form of cancer.

Reverse transcriptase enzymes convert the mRNA into cDNA (see **Section 8C.1**). Each sample is given a fluorescent label. Usually, the known sample is given a green fluorescent label and the experimental sample is given a red fluorescent label. The labelled DNA samples are mixed together and applied to the microarray slide, where they bind to the matching DNA probes (see **fig A**). This process is called **hybridisation**. After hybridisation, the microarray is scanned to measure the fluorescent light produced by the different spots. If both samples are expressing a gene equally, the light will appear yellow (a mixture of red and green light). If the experimental sample is expressing more than the control, the spot will appear red. If the sample is expressing less than the control, the spot will appear green. We can collect very large amounts of data using this technology. The analysis of the data provides detailed information about gene profiles, the causes of many diseases and the effectiveness of some treatments.

MICROARRAYS AND BREAST CANCER

Some forms of breast cancer respond very well to treatment with drugs which block oestrogen receptors or prevent oestrogen synthesis. Doctors can use microarrays to show the level of expression of the gene for oestrogen receptors in individual patients. If it is high, then the patient will respond well to these drugs and they will increase the chances of successful treatment. If the patient shows low levels of expression of the gene for oestrogen receptors, then the oestrogen-blocking drugs will not be effective and are not given. In this case, different approaches to treatment must be used.

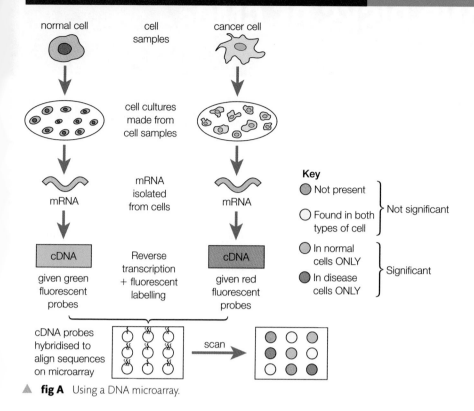

▲ **fig A** Using a DNA microarray.

BIOINFORMATICS

DNA technologies such as microarrays, DNA sequencing and DNA profiling generate huge amounts of data, far more than was ever produced in the past during scientific investigations. Individual scientists or even teams of scientists could not possibly analyse these data to search for patterns. This is why the new science of **bioinformatics** is so important. Bioinformatics is the development of the software and computing tools needed to organise and analyse raw biological data. This includes the development of algorithms, mathematical models and statistical tests which can help us interpret the enormous quantities of data that are generated. Using bioinformatics, we can process and use the information generated using microarrays and other forms of DNA analysis. Bioinformatics enable scientists and doctors to use the information from these DNA techniques to adapt the care of individual patients and learn more about the biology of many different organisms.

> **EXAM HINT**
>
> Remember that one gene may be hundreds of base pairs long and any organism may have many thousands of genes. To compile and store all the possible information gained from many species would be impossible without bioinformatics.

CHECKPOINT

1. Produce a flow diagram to show the main stages in carrying out a DNA microarray to identify a particular gene.

2. Describe bioinformatics and explain why this area of science is so important to scientists in the 21st century.

SUBJECT VOCABULARY

microarray a very useful laboratory tool which allows scientists to detect thousands of active genes at the same time

hybridisation the process by which labelled DNA samples bind to the matching DNA probes on a microarray slide

bioinformatics the development of the software and computing tools needed to organise and analyse large amounts of raw biological data (e.g. the results from microarray analysis of DNA)

8C   **4 BENEFITS AND RISKS OF GMOs**

SPECIFICATION REFERENCE

8.22

■ Understand the risks and benefits associated with the use of genetically modified organisms.

The pharmaceutical and food industries now use genetically modified (GM) organisms (also known as GMOs) to make a range of useful chemicals. However, as you have seen, prokaryotes do not possess the biochemistry to make some of the more complex human proteins. But scientists have introduced useful human genes into eukaryotic cells, including yeast (a single-celled fungus), plants and some mammals. Transgenic plants growing vaccines may become an important weapon in the worldwide fight against disease. Scientists also hope that the modification of plants will help us overcome some of the challenges of climate change and make it possible to grow enough food of the right nutritional value to feed the ever-growing world population. In this section, you will look at some more examples of the ways in which people are using genetically modified organisms, and consider some of the risks and benefits of this new technology.

GENETIC MODIFICATION OF CROPS

Plants, particularly cereal crops, are the main staple diet (the food normally eaten) of most human beings. The problem is that these crops are easily harmed by disease, pests, poor growing conditions and natural disasters such as floods and drought. Traditional plant breeding has resulted in many improvements but it is a slow process. Genetic modification can introduce many useful characteristics much faster. Many of the first commercially produced GM crops were modified to make life better for the producers and sellers of food, for example, by giving longer shelf life (the length of time a product can be kept) or herbicide resistance. The big second wave of GM plants delivered improvements for the consumer, such as improving the nutritional content of the food and making crops more resistant to adverse conditions. Examples include the following.

- Flood-resistant rice: complete immersion in water destroys rice crops, but global warming is causing more severe flooding. Scientists have developed strains of GM rice that can resist being totally covered in flood water for up to 3 weeks and still produce around 80% of the normal quantities of rice. These went from lab to field in 2 years and are already preventing people from dying from lack of food in some areas.

- Pesticide resistance: globally, 20% or more of all crops grown are lost to pests. If we could reduce this there would be a lot more food for everyone. Genetic modification of crop plants can produce plants that make their own pesticide within their leaves. This means farmers do not need to use chemical pesticides, which are expensive and can harm the environment. However, there are concerns that the pesticide genes will spread into wild plants, and insects and fungi will become resistant to the plant chemicals.

- Changing the nutrient values of plants: by genetically modifying crop plants, scientists have the potential to change the balance of different chemicals in the crop. If we can increase the vitamin or protein content of food (e.g. vitamin A in golden rice), or change the balance of fats, we can greatly increase the nutritional value of the food for the people who eat it. It may even be possible to grow plants that contain vaccines against deadly diseases (see **Section 8C.2**).

GENETIC MODIFICATION OF SOYA BEANS

Soya beans (known as soybeans in the US) are a major food crop globally. They are grown as a source of food rich in protein for people, and as a source of oils for commercial use in the food industry. They are used in animal feed, as a biofuel, and have many uses in the cosmetics industry.

It is estimated that 90–95% of all the soya beans grown in the United States, a major producer, are now genetically modified. Scientists have used techniques including *A. tumefaciens* and gene guns to add new genetic material to soya beans. The main genetic modifications of soya beans include the following.

- Herbicide resistance: soya bean production is badly affected by competition with weeds, so farmers spray herbicides to destroy the weeds. However, soya plants are also killed by many broad-leafed plant herbicides. Scientists genetically modified soya plants to make them resistant to common weed killers. These GM soya beans are used very widely across the areas of the world which grow soya beans and production has significantly increased as a result.

- Fatty acid balance: oil from soya beans is widely used in food production, but traditional soya oil contains relatively little oleic and stearic acids and relatively large amounts of linoleic acid. Linoleic acid oxidises easily, so soya oil tends to oxidise and is no longer good to eat. Also, linoleic acid is a polyunsaturated fatty acid, but oleic acid is a monounsaturated fatty acid, and studies have suggested that oleic acid is better for human health than linoleic acid. Scientists have produced GM soya beans that have a radically different balance of fatty acids, with a lot of oleic acid and less linoleic acid (see **fig A**). This has two benefits: the oil lasts longer and may be healthier.

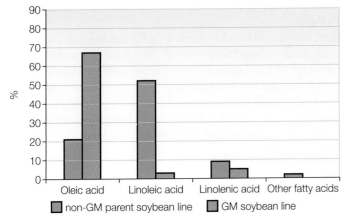

▲ **fig A**   The impact of genetic modification on the fatty acid balance in soya bean oil.

HUMAN GENE THERAPY

It is a very exciting to think that in future we may be able to block the faulty genes that cause some devastating diseases in people. However, the research necessary to find an effective treatment is long and not always successful. For example, scientists hope cystic fibrosis might be cured by inserting healthy versions of the *CFTR* gene into the cells of the lungs but progress so far has been slow.

Most gene therapy is performed on normal body cells (somatic cells) and so it is known as somatic cell gene therapy. Even if that person's treatment is successful, they will still transfer their faulty alleles to their children. A potential solution to this problem is to alter the germ cells so that the faulty genes are no longer transferred. This could be done in the very early embryo immediately after *in vitro* fertilisation. The new individual would be free of the disease and their children would not risk inheriting the disease.

This may sound like a good idea, and some people argue that if the technology is available it should be used to benefit as many people as possible. However, many other people are very concerned about possible risks and some people are offended by the idea itself. It is still not certain what effect this action could have on an early embryo and the impact on the individual might not become clear until much later.

It seems very positive to aim to remove the risk of genetic disease, but it could be difficult to know at what point to stop. Some people could be prepared to pay a great deal of money for unethical treatment. For example, it raises serious questions about manipulating genes, not simply to remedy disease but to enhance longevity, change skin colour or increase adult height or intelligence. At the moment, most germ-line research is banned in most countries.

RISKS AND BENEFITS

There has been a great deal of public debate all round the world about GM technology. In some parts of the world (e.g. the Americas and Australia) people have few problems with the use of GM plants and foods. In other areas, including the UK and many parts of Europe, there is still uncertainty.

This area of science is very complex and many people do not understand it fully. Much of their knowledge comes from the media, and most journalists, even science correspondents, are not scientists. It is not easy to obtain the evidence, nor is it easy to understand. Science is full of uncertainties, but people like things to be clear and simple. As a result many people, including some of the politicians who make laws about research, are arguing without all the evidence and often with an incomplete understanding of the science. Interestingly, the use of GM bacteria to produce human medicines such as insulin raises relatively little debate. In this case, the benefits to many people are very clear, and there appear to be very few risks associated with it. However, genetic modification of food plants, animals and humans divides opinions. Issues often raised about GM technology include the following.

- In the early days of genetic modification, antibiotic resistance was often used as a marker gene. People were concerned about the risk that these genes for antibiotic resistance might get into wild plant populations and add to the problems of antibiotic resistance. Scientists listened to these concerns and now widely use different marker genes (for fluorescence or a particular nutrient requirement) instead of antibiotic resistance.

- Concern about infertile seeds. Scientists usually include marker genes in GM plants to make it easy to identify them. These marker genes may include alleles that ensure the plant is infertile, so that the plants cannot reproduce. This removes the risk of them spreading into the environment. However, it does also mean that farmers must buy fresh supplies of transgenic plant seeds each year for planting. This is more of a concern in the case of food crops than for plants which are being modified to use as medicines, but some of the same issues arise. The big worry is that this will seriously disadvantage people in poorer countries, for whom the benefits from modified crops are most needed.

- Worries over the risk of eating alien DNA in GM food plants. People forget that we eat many different types of DNA every day and have enzymes specifically to digest it. We may even eat the organism that donated the new gene anyway. For many people, the risk/benefit balance shifts dramatically when GM food contains vital medicines or important vitamins. In this case too, there is growing evidence that eating GM food carries no risks and many benefits so, in time, this may no longer be a cause for concern (see **fig B**).

▲ **fig B** As millions of people eat GM crops every year without any adverse side-effects, it is becoming clear that the benefits of GM crops, with high yields and/or increased nutritional value, are greater than the risks.

- Environmental concerns about the risk of gene transfer from GM plants and animals to wild species. As seen above, people worried about the risk of antibiotic marker genes and other alien DNA spreading into wild populations and causing problems. As we collect more evidence that there is relatively little contamination of wild plants by genetically modified crops, concerns diminish.

- Strong objections to the use of animals and other organisms for scientific research. Some people feel that genetic modification threatens the rights of the modified organism. Some people believe that no benefits to people are more important than the risk to animals of being used in scientific research.

- A real concern that gene technology will be the property of a few companies in developed countries. Some people fear that the new advances may be largely biased towards the needs of the richer countries. However, the work on plant-based vaccines and flood-and-drought-resistant rice is being made available at very low cost in the places it is needed, which is helping to counter that fear and ensure the benefits of GM technology are felt by the people who need it most.

Today the use of transgenic plants in medicines such as vaccines is relatively new and unknown, so concerns about risks in the general population are at a low level. Perhaps if media coverage becomes negative, worries will grow. However, it seems likely that, in the same way as using GM microorganisms to produce human medicines, people will accept the benefits of this technology, seeing those benefits as being greater than any possible risks.

As the scientific evidence builds, scientists are increasingly convinced of the safety and value of GM crops. However, as with other GM food, some people still have concerns about possible risk both for the humans that use them and for the animals that are used to produce them. Despite this, for the great majority of the world, food security is often an issue and infectious disease and malnutrition still kill huge numbers every year. GM crops and medicines may well offer the difference between survival and death.

This chapter shows you some of the areas of concern about gene technology which are being debated. Despite these concerns, many people, including most scientists, feel that the enormous benefits resulting from GM technology developments in food production, medicine development and potential therapies are much greater than any risks or ethical obstacles which may arise. When people express concerns about the risks of using GM organisms in medicines or food, it is wise to remember that risks can come from a wide variety of sources, including starvation, eating poor quality food and infectious diseases. In the future, the hope is that GM organisms will help us reduce some of these risks, making them part of the solution, not the problem.

EXAM HINT

Many potential risks have been associated with genetic modification. Don't forget that genetic modification is still a new science. One of the biggest concerns is about the unknown: we simply do not know what the effects of genetic modification may be in several generations' time.

EXAM HINT

When you are asked to discuss the risks and benefits of GM, make sure you present a balanced argument. Give equal treatment to some benefits and some risks and present evidence to support your arguments wherever possible.

CHECKPOINT

1. Discuss the use of GM bacteria in the production of drugs.
2. Produce a report or a poster to summarise the debate on the use of GM in global crop production.
3. ▶ 'Genetic modification of the fatty acid balance in soya beans offers advantages to both producers and consumers.' Explain this statement.
4. ▶ Many scientists are convinced that GM crops will be the solution to the problem of feeding the growing world population. Suggest **three** reasons for their optimism.

SKILLS   CREATIVITY

CAN GM CROPS SAVE THE WORLD?

SKILLS CRITICAL THINKING, ANALYSIS, INTERPRETATION, CREATIVITY, INNOVATION, INTELLECTUAL INTEREST AND CURIOSITY, INITIATIVE, COMMUNICATION, TEAMWORK, COOPERATION

Here is a quote from a document produced by the UAE Minister for Water and Environment in conjunction with the United Nations Food and Agriculture Organization (the FAO). Since this document was produced, a number of drought-resistant GM crops, including maize, have been successfully launched and used in a number of African countries and other areas of the world, but progress remains slower than scientists had hoped.

ADAPTED FROM UN FOOD AND AGRICULTURE ORGANIZATION

Conclusions

Plant biotechnology offers an unprecedented opportunity to address some of the world's most serious issues, including hunger, poverty and disease. This is because biotechnology can circumvent the species barriers that prevent useful traits being introduced into plants by conventional breeding. By transferring genes from bacteria, fungi, animals and sexually incompatible plants into our food crops and medicinal plants, it is possible to improve their agronomic traits and provide them with additional metabolic abilities.

Genetic engineering and biotechnology provide good opportunities for the investment and improvement of food security and food production in the Arab countries; however, they have their own challenges that need to be considered by the Center.

The following questions are relevant to be asked not only as related to Arab countries, but also for the developing countries in general.

- What opportunities exist for biotechnology to contribute toward improving agricultural productivity, expanding markets, and stimulating employment and income generation in Arab countries, and what are the constraints that limit capturing these opportunities and in using biotechnological approaches?

- What challenges do these countries face in realizing these opportunities and in mitigating the risks associated with the use of biotechnology?

Considering the important challenges that are encountered by the agricultural sector and its sustainability in the Arab countries, it seems that the new biotech crop applications offer enormous potential benefits in the second decade of commercialization, 2006–2015, in terms of meeting increased food, feed and fibre demands in the Arab World and contributing to more prosperity for both producers and consumers. Of particular importance are the genes for drought tolerance that are under development in both the private and public sector. The genes for drought tolerance are genes that very few farmers in the world can afford to be without and this is particularly true for the rainfed dryland areas that typify much of

the land in the Arab countries for which ICARDA (International Center for Research in the Dry Areas) has a regional mandate, and where biotech research is undertaken on drought tolerance. The first commercial variety with drought tolerance is expected to be drought-tolerant maize in the US in 2011. The drought genes have already been introduced into several crops and early field tests are underway; for example, drought-tolerant wheat is being field-tested in Australia.

In this regard, it is evident that the decision to invest in agricultural biotechnology is timely and appropriate and is of great strategic importance at a time when the new technologies can contribute to:

- an increased sustainable supply of the most affordable and nutritious supply of food, feed and fibre, which is critical for facilitating prosperity for both producers and consumers in the Arab States

- sustainable crop production in the dry-land areas to alleviate poverty of the rural poor who are farmers and the rural landless who are dependent on agriculture for their livelihoods

- speeding the crop breeding that will mitigate the new challenges associated with climate change, when droughts will become more severe and prevalent, temperature changes will be more variable, and when agriculture which produces up to 30% of greenhouse gases must be part of the solution rather than part of the problem.

Source: *http://www.fao.org/docrep/012/al310e/al310e06.pdf* (for the UAE section of the report highlighted here)

SCIENCE COMMUNICATION

The extract on the previous page is from a document produced by ministers from many countries working with the United Nations Food and Agriculture Organization (FAO).

1 (a) Who do you think is the intended audience for this document?

 (b) Summarise the main message of the extract on the previous page in a single paragraph.

 (c) What information would you expect to find in the rest of this document?

BIOLOGY IN DETAIL

Now let us examine the biology. You already know how genetically modified organisms are produced and how they can be given helpful characteristics from other organisms. This knowledge will help you answer the questions below.

2 (a) Give **three** potential benefits of genetically modified organisms highlighted in the article.

 (b) State which genes are highlighted as being of particular importance and explain why.

3 Discuss why the genes for drought resistance are so important, both in the Arab world and globally.

4 (a) Suggest potential benefits of the introduction of genetically modified organisms into the agriculture of a community.

 (b) Discuss some of the concerns expressed which might delay the introduction of GMOs in a country and give biological reassurances.

ACTIVITY

Salt-tolerant plants

• Water is vital for plants from date palms to cassava to grow and produce good crop yields. Many countries affected by drought conditions for part or all of the year do have access to sea water but the salt kills most plants.

• Investigate biosaline agriculture (the development of salt-resistant plants) using genetic modification and other technologies.

• Use a smart phone to make a news report on the potential of this science to make food supplies more sustainable locally and/or globally.

• Use online resources as well as scientific magazines and books. Evaluate your sources before using them and reference all sources used.

1 (a) There are a number of ways in which recombinant bacteria can be identified. Which of the following techniques might cause ethical objection? [1]

1 replica plating to test nutrient requirements

2 antibiotic resistance marker genes

3 ultraviolet fluorescence

A only technique 1

B techniques 2 and 3

C only technique 3

D only technique 2

(b) Several restriction endonucleases cut after a guanine base. Using this information, look at these sequences and decide which group of sequences would result in the formation of sticky ends. [1]

1 GGATCC

2 AGCT

3 GAATTC

A sequences 1, 2 and 3

B only sequences 1 and 3

C only sequences 2 and 3

D only sequences 1 and 2

(c) The flow diagram below shows how a genetically modified organism may be produced by inserting a human gene into a bacterium.

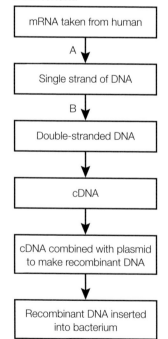

(i) Name the enzymes used at points A and B on the flow diagram. [2]

(ii) Describe how the cDNA and the plasmid could be combined together. [3]

(Total for Question 1 = 7 marks)

2 Production of cheese involves coagulating (clotting) milk protein (casein) using the enzyme chymosin. Traditionally, chymosin was collected from the stomachs of calves. Now it can be commercially produced using yeast cells. The diagram below shows some of the main stages in the production of chymosin using yeast cells.

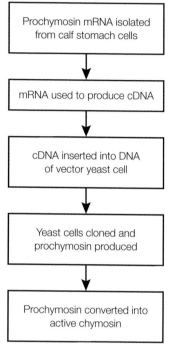

(a) Cheese produced using chymosin cultured from yeast cells is acceptable to most vegetarians. State **one** reason why some consumers may still have concerns about eating cheese prepared in this way. [1]

(b) (i) With reference to the diagram above, outline how enzymes are used in the production of chymosin using yeast cells. [4]

(ii) During the process described in the diagram, not all the yeast cells would be successfully modified. Describe **one** method that could be used to identify the modified cells. [2]

(Total for Question 2 = 7 marks)

3 Cotton plants are used to produce the cotton from which cloth is made. They are grown in certain parts of the world, such as India and the USA. Cotton farming is an important way in which people earn their living. Normally, cotton plants need to be sprayed with chemical insecticides to kill insect pests. Recently, genetically modified (GM) cotton plants have been developed which produce a natural insecticide of their own. This insecticide kills the insect pests but is harmless to humans.

A field of cotton plants showing the 'bolls' of cotton fibres which are collected to make cloth. The bolls are the fruit of the plant and contain seeds.

(a) Describe **three** advantages to cotton farmers and to the environment of growing genetically modified cotton. [3]

(b) Explain how the genetic modification of plants is similar to but distinct from conventional plant breeding. [4]

(c) For many people, genetic modification of plants remains controversial. In 2007, the European Union decided not to lift the ban on GM crops in Europe although they are widely grown in the USA and India.

State whether you are for or against the growing of GM crops. Explain why you hold this view. Use scientific knowledge and make reference to social or ethical issues to support your explanation. [4]

(Total for Question 3 = 11 marks)

4 (a) (i) Complete the table below showing how enzymes are involved in the process of genetic modification. [3]

Name of enzyme	Process
ligase	join sections of DNA together
DNA polymerase	
	cut DNA at specific sequence
reverse transcriptase	

(ii) Which of the following is the process by which DNA ligase can join **two** sections of DNA? [1]
A repairing the peptide bond
B repairing the phosphodipeptide bond
C repairing the phosphodiester bond
D repairing the phosphate bond

(b) (i) What is a plasmid? [1]
A a short length of DNA
B a vector used to insert DNA into a yeast cell
C a circular piece of RNA
D a circular piece of DNA

(ii) What is a vector? [1]
A a short length of DNA
B a mechanism to insert DNA into cells
C a mechanism to insert RNA into cells
D a circular piece of DNA

(c) Suggest a suitable vector for each of the following four genetic engineering procedures.
(i) Transforming bacteria to produce human insulin. [1]
(ii) Engineering rice plants to make vitamin A. [1]
(iii) Treating cystic fibrosis in the lungs of a human patient. [1]
(iv) Producing copies of a large human gene inside bacteria. [1]

(d) The list below represents steps in the process to engineer bacteria capable of producing human insulin. Give the correct order for steps R to W. [5]

R use reverse transcriptase to make cDNA and add sticky ends

S incubate plasmid with restriction enzyme

T extract mRNA from human pancreatic cells

U heat shock bacteria in calcium chloride solution and add plasmid

V incubate cDNA with cut plasmid

W add ligase enzyme

(Total for Question 4 = 15 marks)

MATHS SKILLS

In order to be able to develop your skills, knowledge and understanding in biology, you will need to have developed your mathematical skills in a number of key areas. This section gives more explanation and examples of some key mathematical concepts you need to understand. Further examples relevant to your International A Level Biology studies are given throughout the book and in Book 1.

USING LOGARITHMS

CALCULATING LOGARITHMS

Many formulae in science and mathematics involve powers. Consider the equation:

$$10^x = 62$$

We know that the value of x lies between 1 and 2, but how can we find a precise answer? The term *logarithm* means index or power, and logarithms allow us to solve such equations. We can take the 'logarithm base 10' of each side using the **log** button of a calculator.

WORKED EXAMPLE 1

$10^x = 62$

$\log_{10}(10^x) = \log_{10}(62)$

$x = 1.792392...$

We can calculate the logarithm using any number as the base by using the **log** button.

WORKED EXAMPLE 2

$2^x = 7$

$\log_2(2^x) = \log_2(7)$

$x = 2.807355...$

Many equations relating to the natural world involve powers of e. We call these exponentials. The logarithm base e is referred to as the natural logarithm and denoted as **ln**.

USING LOGARITHMIC PLOTS

An earthquake measuring 8.0 on the Richter scale is much more than twice as powerful as an earthquake measuring 4.0 on the Richter scale. This is because the units involved in measuring earthquakes use the concept of logarithm scales in charts and graphs. This helps us to accommodate enormous increases (or decreases) in one variable as another variable changes.

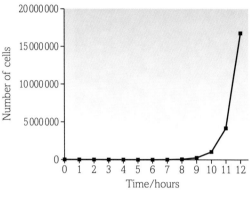

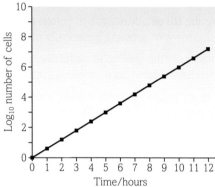

▲ **fig A** Logarithmic scales are useful when representing a very large range of values, such as in the case of bacterial growth.

SELECTING AND USING A STATISTICAL TEST

In Book 1, you learned how to use measures of average (mean, median and mode) and measures of dispersion (range and standard deviation). We will now look at some statistical tests that can be used to analyse data.

DECIDING ON A NULL HYPOTHESIS

When you use a statistical test, you need to be clear what you are testing for. We do this by setting a null hypothesis, which is the thing you are trying to prove or disprove. You should clearly state your null hypothesis before you begin a test.

After running your chosen statistical test, you will be left with a number known as the observed value. To know whether or not you should accept your null hypothesis, you need to compare the observed value with a table of critical values. To find the correct value to compare it with, you will usually need to calculate the 'degree of freedom' for your results.

In essence, the critical values tell you whether or not you can accept your null hypothesis. If your observed result fits with the 5% significance level given in the table, you can be 95% certain that your null hypothesis is true – and you should therefore accept it. If not, reject it!

CHOOSING A TEST

There are many different statistical tests that can be used. The tests that you may meet in your IAL Biology studies are shown in **table A**.

Using the Spearman's rank correlation coefficient to test for correlation

Step 1: State the null hypothesis

The null hypothesis for this test is: 'r_s is equal to zero, meaning that there is no correlation between the two variables.'

Step 2: Calculate the observed value

The formula for the Spearman's rank correlation coefficient is:

$$r_s = 1 - \frac{6\sum d^2}{n(n^2 - 1)}$$

where d is the difference in rank between each pair of variables, n is the number of pairs and the symbol $\sum$ means 'the sum of'. To use the Spearman's rank correlation coefficient, find the difference in rank between each pair, square these differences and add them all together. Now substitute this value into the formula in place of $\sum d^2$.

Step 3: Decide whether or not to accept the null hypothesis

To decide whether you can accept the null hypothesis, you need to compare you observed value of r_s with a critical values table.

Find the critical value that corresponds to the 5% significance level for your number of pairs. If your value of r_s (ignoring whether it is positive or negative) is less than the critical value, accept the null hypothesis. If not, reject it.

Using the chi squared test

Step 1: State the null hypothesis

The null hypothesis for this test is: 'The observed results are consistent with the expected distribution, meaning that differences between observed and expected results are due to chance.'

Step 2: Calculate the observed value

The formula for the chi squared test is:

$$\chi^2 = \sum \frac{(O - E)^2}{E}$$

where O is your observed result, E is the expected result and the symbol $\sum$ means 'the sum of'. For each result, you need to find the difference between the observed and expected values, square this difference then divide by the expected value. You then add together the results of all these calculations to get the observed value for χ^2.

Step 3: Decide whether or not to accept the null hypothesis

To decide whether you can accept the null hypothesis, you need to compare your observed value of χ^2 with a critical values table. To do this you need to know the degree of freedom, which for the chi squared test is the number of categories minus 1:

$$df = n - 1$$

Find the critical value that corresponds to the 5% significance level for your calculated degree of freedom. If your value of χ^2 is less than the critical value, accept the null hypothesis. If not, reject it.

Using the Student's *t*-test

It might seem obvious that you could easily compare the means of the two categories by simply subtracting one from the other. However, can you be sure that the means calculated from the two sample data sets are representative of the whole population? The Student's *t*-test takes into account the degree of overlap between the two sets of data and allows you to judge whether any difference between the means is statistically significant or just due to chance.

Step 1: State the null hypothesis

The null hypothesis for this test is: 'The means of the interval variable for the two categories are equal.'

Step 2: Calculate your observed value

In order to calculate t, you first need to find the mean $\bar{x}$ and the variance s^2 for both of the categories.

$$\bar{x} = \frac{\sum x}{n}$$

$$s^2 = \frac{\sum x^2 - \frac{(\sum x)^2}{n}}{n - 1}$$

You can now use the following formula to calculate the value of t:

$$t = \frac{\bar{x}_1 - \bar{x}_2}{\sqrt{\frac{s_1^2}{n_1} + \frac{s_2^2}{n_2}}}$$

where:

$\bar{x}_1$ = mean of the first set of data

$\bar{x}_2$ = mean of the second set of data

s_1^2 = variance of the first set of data

s_2^2 = variance of the second set of data

n_1 = number of items in first set of data

n_2 = number of items in second set of data.

NAME	PURPOSE	EXAMPLE	NULL HYPOTHESIS
SPEARMAN'S RANK CORRELATION COEFFICIENT	To test whether or not two variables display correlation.	Test whether or not there is a correlation between finishing positions in a race and age of runners.	There is no correlation between the two variables. (Spearman's rank correlation coefficient equals zero.)
CHI SQUARED TEST	To test how likely it is that any differences between observed and expected results are due to chance.	Test whether or not a ratio of phenotypes from mating supports a particular inheritance model.	The observed results are consistent with the expected distribution. (The differences between observed and expected results are due to chance.)
STUDENT'S *t*-TEST	To test whether or not the difference in the mean value of a variable for the two categories is significant.	Test whether or not a drug treatment has been effective compared with a placebo.	The means of the interval variable for the two categories are equal.

table A Statistical tests for IAL Biology

Step 3: Decide whether or not to accept the null hypothesis

To decide whether you can accept the null hypothesis, you need to compare your observed value of t with a critical values table. To do this you need to know the degree of freedom, which for the Student's t-test is the total number of data values minus 2: $df = n_1 + n_2 - 2$

Find the critical value that corresponds to the 5% significance level for your calculated degree of freedom. If your value of t is less than the critical value, accept the null hypothesis. If not, reject it.

APPLYING YOUR SKILLS

You will often find that you need to use more than one maths technique to answer a question. In this section, we will look at three example questions and consider which maths skills are required and how to apply them.

WORKED EXAMPLE 3

Microorganisms can be grown quickly using a fermenter. If the conditions of the fermenter are set correctly (for example, there are enough nutrients and sufficient space to grow) the division of bacteria can become exponential. One such microorganism, a species of yeast called S. cerevisiae, can multiply every 100 minutes.

(a) *Assuming there is no cell death, and two yeast cells are introduced into the fermenter, how many cells will there be after 300 minutes?*

(b) *Write an equation to show how the population of yeast over a period of time can be calculated, using d to represent the number of divisions.*

(c) *Using your equation from (b), calculate the population of yeast generated after 32 cell divisions in a fermenter.*

(d) *How many cell divisions would have taken place if the population of yeast was only allowed to reach 65 536 cells?*

(a) In 300 minutes, the cells will divide three times. After the first division, the population will have doubled to four from the two original yeast cells. The second division will double the population again, to eight cells. The third division will double the population yet again, resulting in 16 cells.

(b) If we started with just one cell, the number of cells after time d would be given by 2^d. However, because we are starting with two cells, we need to double this, so $N_d = 2 \times 2^d$ where N_d is the number of cells after d divisions. Using the rules of indices, we could also write this as $N_d = 2^{(d+1)}$.

(c) To find the number of cells after 32 divisions, we can just substitute $d = 32$ into our equation:

$$N_{32} = 2^{(32+1)} = 2^{33} = 8\,589\,934\,592 \text{ cells}$$

(d) To find the number of divisions necessary to produce 65 536 cells, we can substitute $N_d = 65\,536$ into our equation:

$$2^{(d+1)} = 65\,536$$

Now we take logarithm base 2 of both sides to simplify the equation:

$$\log_2(2^{(d+1)}) = \log_2(65\,536)$$

$$d + 1 = 16$$

$$d = 15$$

So there will be 65 536 cells after 15 divisions.

WORKED EXAMPLE 4

A student investigated the effect of changing the substrate (glucose and fructose) on the rate of respiration in yeast. Identical samples of yeast were used in a respirometer.

(a) *Explain why a Student's t-test would be the most appropriate statistical test to see whether there is a difference in the rate of respiration caused by different substrates.*

(b) *Is there a significant difference between the rate of respiration using glucose and fructose?*

SUBSTRATE	RATE OF RESPIRATION / AU							MEAN
	TRIAL 1	TRIAL 2	TRIAL 3	TRIAL 4	TRIAL 5	TRIAL 6	TRIAL 7	
glucose	25	27	34	18	21	26	28	25.6
fructose	17	35	42	19	35	22	44	30.6

(a) The Student's t-test can be used to determine whether there is a statistically significant difference between the means of two categories (substrates fructose and glucose).

(b) To answer this question, we need to perform a Student's t-test.

Step 1: State the null hypothesis
Our null hypothesis is: 'There is no difference in the rate of respiration using glucose or fructose.'

Step 2: Calculate the observed value
We have already been given the means of the two data sets:

$\bar{x}_{glucose} = 25.6$

$\bar{x}_{fructose} = 30.6$

In order to calculate the variance of the data sets, we need to calculate the sum of squared values, $\sum x^2$, and the sum of the values squared, $(\sum x)^2$.

$\sum x^2$ for glucose $= (25)^2 + (27)^2 + (34)^2 + (18)^2 + (21)^2 + (26)^2 + (28)^2$
$= 4735$

$\sum x^2$ for fructose $= (17)^2 + (35)^2 + (42)^2 + (19)^2 + (35)^2 + (22)^2 + (44)^2$
$= 7284$

$(\sum x)^2$ for glucose $= (25 + 27 + 34 + 18 + 21 + 26 + 28)^2 = 32\,041$

$(\sum x)^2$ for fructose $= (17 + 35 + 42 + 19 + 35 + 22 + 44)^2 = 45\,796$

We can now calculate the variance (s^2) of each data set:

$$s^2_{glucose} = \frac{\left[4735 - \left(\frac{32\,041}{7}\right)\right]}{6} = 26.29$$

$$s^2_{fructose} = \frac{\left[7284 - \left(\frac{45\,796}{7}\right)\right]}{6} = 123.62$$

And so we can now calculate t:

$$t = \frac{\bar{x}_{glucose} - \bar{x}_{fructose}}{\sqrt{\frac{s^2_{glucose}}{n_{glucose}} + \frac{s^2_{fructose}}{n_{fructose}}}}$$

$$t = \frac{(25.6 - 30.6)}{\sqrt{\left[\left(\frac{26.29}{7}\right) + \left(\frac{123.62}{7}\right)\right]}}$$

$$t = \frac{5}{\sqrt{[(3.76) + (17.66)]}}$$

$$t = \frac{5}{4.63}$$

$$t = 1.08$$

Step 3: Decide whether or not to accept the null hypothesis
The number of degrees of freedom is $n_1 + n_2 - 2 = 12$

The critical value for a 5% significance level with 12 degrees of freedom is 2.18.

The observed value of 1.08 is less than the critical value. Therefore we accept the null hypothesis that there is no significant difference in the rate of respiration using glucose or fructose.

WORKED EXAMPLE 5

The managers of a nature reserve wanted to increase the diversity of grasses and other plants in their meadow. The usual management strategy was to mow the meadow. It was suggested that allowing sheep to graze may help to increase diversity. The meadow was divided into two plots. One plot was mowed while sheep were allowed to graze the other plot. After one season, the diversity was assessed by measuring the frequency (% cover) of certain plants.

The results are shown in the table below:

SPECIES	MEAN % COVER	
	MOWING	GRAZING
A	3	8
B	2	7
C	4	8
D	2	1
E	19	6

(a) *The diversity of the mown plot was calculated as 2.39.*

   Calculate the biodiversity of the grazed field. Use the formula:
   $$D = \frac{N(N-1)}{\sum n(n-1)}$$

(b) *Which field was more diverse?*

(c) *Use a suitable statistical test to assess whether or not there is a significant change in the diversity as a result of allowing sheep to graze the meadow.*

(a) $D = \dfrac{30 \times (30-1)}{(8 \times 7) + (7 \times 6) + (8 \times 7) + (1 \times 0) + (6 \times 5)}$

   $D = 4.73$

(b) The grazed field is more diverse.

(c) *Step 1: State the null hypothesis*
   The null hypothesis is: 'The observed frequencies of plants are the same as the expected frequencies.'

   Step 2: Calculate the observed value
   The formula for the chi squared test is:
   $$\chi^2 = \frac{\sum(O - E)^2}{E}$$

   where O is the result for the grazed field, E is the result for the mowed field and symbol Σ means 'the sum of'.

   $$\chi^2 = \frac{(8-3)^2}{3} + \frac{(7-2)^2}{2} + \frac{(8-4)^2}{4} + \frac{(1-2)^2}{2} + \frac{(6-19)^2}{19} = 21.81$$

   Step 3: Decide whether or not to accept the null hypothesis

   The number of degrees of freedom is $n - 1 = 5 - 1 = 4$

   The critical value for a 5% significance level with 4 degrees of freedom is 9.49.

   The observed value is greater than the critical value. Therefore, we must reject the null hypothesis. The difference between the observed and expected frequencies is not down to chance.

PREPARING FOR YOUR EXAMS

IAS AND IAL OVERVIEW

The Pearson Edexcel International Advanced Subsidiary (IAS) in Biology and the Pearson Edexcel International Advanced Level (IAL) in Biology are modular qualifications. The IAS can be claimed on completion of the International Advanced Subsidiary (IAS) units. The IAL can be claimed on completion of all the units (IAS and IA2 units).

- International AS students will sit three exam papers. The IAS qualification can either be standalone or contribute 50% of the marks for the International Advanced Level.
- International AL students will sit six exam papers, the three IAS papers and three IAL papers.

The tables below give details of the exam papers for each qualification.

IAS Papers	Unit 1 Molecules, Diet, Transport and Health	Unit 2 Cells, Development, Biodiversity and Conservation	Unit 3 Practical Skills in Biology I
Topics covered	Topics 1–2	Topics 3–4	Topics 1–4
% of the IAS qualification	40%	40%	20%
% of the IAL qualification	20%	20%	10%
Length of exam	1 hour 30 minutes	1 hour 30 minutes	1 hour 20 minutes
Marks available	80 marks	80 marks	50 marks
Question types	multiple choice short open open response calculation extended writing	multiple choice short open open response calculation extended writing	short open open response calculation
Mathematics	A minimum of 10% of the marks across all three papers will be awarded for mathematics at Level 2 or above		

IAL Papers	Unit 4 Energy, Environment, Microbiology and Immunity*	Unit 5 Respiration, Internal Environment, Coordination and Gene Technology*	Unit 6 Practical Skills in Biology II
Topics covered	Topics 5–6	Topics 7–8	Topics 5–8
% of the IAL qualification	20%	20%	10%
Length of exam	1 hour 45 minutes	1 hour 45 minutes	1 hour 20 minutes
Marks available	90 marks	90 marks	50 marks
Question types	multiple choice short open open response calculation extended writing	multiple choice short open open response calculation extended writing	short open open response calculation
Mathematics	A minimum of 10% of the marks across all three papers will be awarded for mathematics at Level 2 or above		

* These papers will include synoptic questions that may draw on two or more different topics.

EXAM STRATEGY

ARRIVE EQUIPPED

Make sure you have all of the correct equipment needed for your exam. As a minimum you should take:

- pen (a black ballpoint pen is best)
- pencil (HB)
- rule (ideally 30 cm)
- eraser / rubber (make sure it's clean and doesn't smudge the pencil marks or rip the paper)
- calculator (scientific).

ENSURE YOUR ANSWERS ARE LEGIBLE

Your handwriting does not have to be perfect but the examiner must be able to read it. When you're in a hurry it's easy to write key words that are difficult to decipher.

PLAN YOUR TIME

Note how many marks are available on the paper and how many minutes you have to complete it. This will give you an idea of how long to spend on each question. Be sure to leave some time at the end of the exam for checking answers. A rough guide of a minute a mark is a good start, but short answers and multiple-choice questions may be quicker. Longer answers might require more time.

UNDERSTAND THE QUESTION

Always read the question carefully and spend a few moments working out what you are being asked to do. The command word used will give you an indication of what is required in your answer. It can be useful to highlight key words in the question.

Be scientific and accurate, even when writing longer answers. Use the technical terms you've been taught.

Always show your working for any calculations. Marks may be available for individual steps, not just for the final answer. Also, even if you make a calculation error, you may be awarded marks for applying the correct technique.

MAKE THE MOST OF GRAPHS AND DIAGRAMS

Diagrams and sketch graphs can often earn marks more easily and quickly than written explanations, but only if they are carefully drawn and fully annotated.

- If you are asked to read a graph, pay attention to the labels and numbers on both the x- and y-axes. Remember that each axis is a number line.
- If asked to draw or sketch a graph, always ensure you use a sensible scale and label both axes with quantities and units. If plotting a graph, use a pencil and draw small crosses or dots for the points.
- Diagrams must always be neat, clear and fully labelled or annotated.

CHECK YOUR ANSWERS

For open-response and extended-writing questions, check the number of marks that are available. If three marks are available, have you made three distinct points?

For calculations, read through each stage of your working. Substituting your final answer into the original question can be a simple way of checking that the final answer is correct. Another simple strategy is to consider whether the answer seems sensible. Pay particular attention to using the correct units.

SAMPLE EXAM ANSWERS

QUESTION TYPE: MULTIPLE CHOICE

The cerebellum is an area of the human brain. What is the main function of the cerebellum?

A *Maintaining homeostasis* ☐

B *Control of breathing and heart rate* ☐

C *Decision making and memory* ☐

D *Control of balance and coordination* ☐   (1)

This question relies on recall of knowledge. While this question requires a choice from a list of statements, other questions may be based on a choice of rows from a table or a list of letters linked to labels on a diagram.

Question analysis

- Multiple-choice questions may require simple recall, as in this case, but sometimes a calculation or some other form of analysis will be required.

- In multiple-choice questions you are given the correct answer along with three incorrect answers (called distractors). You need to select the correct answer and put a cross in the box of the letter next to it.

- The three distractors supplied will feature the answers that you are likely to arrive at if you make typical or common errors. For this reason, multiple-choice questions aren't as easy as you might at first think. If possible, try to answer the question before you look at any of the answers.

- If you change your mind, put a line through the box with the incorrect answer (⊠) and then mark the box for your new answer with a cross(⊠).

- If you have any time left at the end of the paper, multiple-choice questions should be put high on your list of priority for checking answers.

Student answer

C Decision making and memory ⊠

As with many multiple-choice questions, the answer choices contain alternatives that can be easily confused. This may be because the words themselves are similar or because the alternatives refer to key terms linked to one topic. To avoid being misdirected to an incorrect answer, one approach is to try to think of the answer initially without reading the choices given and then look for your answer in the list. Some students skim read notes and learn to recognise technical terms only by the initial letters, which can lead to errors. If you sometimes have difficulty with spelling or recognising key words, try reading them out loud one syllable at a time when preparing for exams.

COMMENTARY

This is an incorrect answer because:

- The student has confused the two brain areas of the cerebellum and the cerebrum. These are similar-sounding words. While areas of the cerebrum are responsible for decision making and memory, the main function of the cerebellum is control of balance and coordination (answer D).

QUESTION TYPE: SHORT OPEN

Describe how the bacterium Mycobacterium tuberculosis *is transmitted from person to person.* (2)

The command word 'describe' requires you to give an account and provide the reasoning behind it. As there are up to 2 marks available, the answer should have two clear points relating directly to the question. The key word here is 'transmitted' – how the disease is passed from one host to another. Comments on how the pathogen then invades tissues would not gain credit.

Question analysis

- Short open questions usually require simple short answers, often one word. Generally they will require simple recall of the biology you have been taught.

- Short open questions require succinct and clear answers. They may be worth 1 or 2 marks. For 1-mark questions it is not always necessary to write in full sentences. For a 2-mark question there may be credit for two distinct points or alternatively for one main idea with further detail or elaboration.

Student answer

The bacterium *Mycobacterium tuberculosis* causes tuberculosis in human lungs. It is spread by water droplets in the air.

The student has wasted time and space noting what disease *Mycobacterium tuberculosis* causes. This does not answer the question and will not gain a point. On short open questions it is easy to fill the available space with comment that is superfluous and then fail to add sufficient relevant detail.

COMMENTARY

This is an average answer because:

- Water droplets in air are correctly identified as the mode of transmission and would be credited with 1 mark.

- Further detail, such as droplets caused by sneezing and coughing, would be needed for 2 marks.

QUESTION TYPE: OPEN RESPONSE

A laboratory experiment was carried out to investigate the effect of different wavelengths of light on the rate of photosynthesis in the pondweed Elodea. *Pieces of* Elodea *of equal mass were placed in experimental chambers and exposed to light at wavelengths in the red, blue, yellow and green regions of the spectrum. The oxygen produced in the first hour was collected for each of the colours of light, and this was used as an indication of the rate of photosynthesis. The experiment was repeated several times and the mean and standard deviation for each light treatment were calculated.*

Using the results from the graph in fig A, describe the results of this investigation. (3)

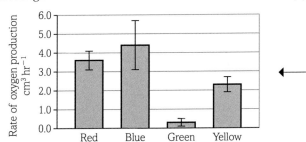

The command word here is 'describe', so you will need to give an account of the results shown. You do not need to explain the scientific reasons behind any patterns. It is important to note the command word carefully in graph-based questions, as you may be asked to analyse, explain or compare and contrast results, all of which require a very different approach.

fig A The mean rate of oxygen production by the pondweed *Elodea* under light of different wavelengths. The error bars show the standard deviation of the means.

Question analysis

- Questions based on graphs are frequently encountered. Make sure that you check the axes' scales carefully before describing results or reading from graphs, as scales are often more complex at A level.

- In questions that ask for a description, start by stating the general pattern or overall trends. Then develop your statements, giving more detail on the extent of any differences and more detail of the patterns shown. In some questions, credit is also given for manipulating the data; this is not simply reading figures from the graph but actually using figures to make a comparison, usually by way of a simple calculation.

Student answer

The blue light produced the highest rate of oxygen production. However, the error bars for blue light show a large degree of overlap with those of red light, so there may be no real difference between the results from these two treatments. The standard deviation for blue light was greater than for the other treatments, which indicates that the results from this treatment may be less reliable. The lowest rate of oxygen production was shown with green light, which had very little photosynthesis. Yellow light was associated with an intermediate rate of oxygen production, having approximately half the rate shown by blue light.

The student has given a detailed answer and has considered both the mean values and the standard deviation in describing the results of the experiment and has correctly avoided explaining any differences seen.

COMMENTARY

This is a good answer because:

- There are at least three clear points made about the patterns shown in the results.
- The answer is logically structured.
- The general pattern and extent of any difference between treatments is clear, and there has been some attempt to manipulate the data.

QUESTION TYPE: EXTENDED WRITING

* Explain how nerve impulses are propagated along an axon once an action potential has been initiated. (6)

The command word here is 'explain', so there must be some reasoning and detail of how an impulse is conducted along the axon. A simple description of an action potential would not suffice.

Question analysis

- In questions marked with an asterisk (*), marks will be awarded for your ability to structure your answers logically showing how the points that you make are related or follow on from each other where appropriate.

- Four marks are available for making valid points in response to the question. In mark schemes, these are referred to as 'indicative marking points'. To gain all four marks, your answer needs to include six indicative working points. The remaining two marks are awarded for structuring your answer well, with clear lines of reasoning and linkages between points.

- In extended writing questions, as in any question, look at the number of marks and try to think of at least that number of distinct points to make. In longer questions it may help to plan out your answer in brief bullet points on a blank space on the paper. This will help you to order your points more logically.

- It is important that you use correct scientific terminology to demonstrate your biological knowledge. Vague answers will only achieve the lowest mark band.

Student answer

When an action potential is initiated, this causes localised currents along the axon membrane. These currents cause depolarisation of the adjacent section of membrane. This causes gated sodium channels to open. The ions move in and there is positive feedback on the sodium channels. A wave of depolarisation and repolarisation repeats itself along the axon and so travels along it. If the axon is myelinated then conduction will be faster.

The answer demonstrates adequate knowledge and some relevant biological facts have been selected and used. There is some linkage of facts and a reasonably logical structure, and all material is relevant to the question. As such, this answer would achieve a moderate mark band.

COMMENTARY

This is an average answer because:

- While there are at least six points made in this answer, some are not adequately explained and it is not clear if the student really understands all of the underlying biological processes. The student does not explain the movement of ions that results in depolarisation and repolarisation. Sodium ions should always be referred to (not simply sodium). The nature and mechanism of saltatory conduction in particular is not explained.

- The points made are all relevant, but the student does not consider the concept of the refractory period and how this results in propagation in one direction only.

QUESTION TYPE: CALCULATION

The following results were recorded from an investigation into the effect of temperature on the rate of respiration in mung beans.

Temperature/°C	Volume of oxygen absorbed/mm³						Standard deviation	Time taken/s	Rate/mm³s⁻¹
	trial 1	trial 2	trial 3	trial 4	trial 5	mean			
10	3.0	4.2	3.5	4.1	3.9		0.49	120	
20	6.2	7.1	6.8	7.3	8.3	7.14	0.69	120	0.060
30	13.0	12.3	15.0	13.2	14.1	13.52	2.09	120	0.113

(a) *Complete the table by calculating the mean and the rate for the bean at 10 °C.* (2)

(b) *The Student's t-test can be used to determine whether sets of data are similar or different.*

The value of t *is given by the formula:*

$$t = \frac{\bar{x}_1 - \bar{x}_2}{\sqrt{\dfrac{s_1^2}{n_1} + \dfrac{s_2^2}{n_2}}}$$

The student wanted to show that the results for 20 °C and 30 °C were statistically different. Use the Student's t-test to calculate the value of t. *Show your working.* (2)

> The first part of this question should be straightforward as you need to calculate a mean and a rate. Always check your answer to see that it seems right compared to the other values given in the table. If you are unsure, you can check your approach by working through the values on a completed row. In the second part of the question, you will need to use the formula provided. Don't forget that standard deviation needs to be squared to give s^2.

Question analysis

- The command word is calculate. This means that you need to obtain a numerical answer to the question, showing relevant working. If the answer has a unit, this must be included.
- Always have a go at calculation questions. You may get some small part correct that will gain credit.
- In multipart questions such as this, you may be awarded marks for applying the correct method, even if earlier stages (for example, completing the table) are incorrect.
- Always do a 'common sense' check on answers by looking to see if your values seem in line with the data as a whole.
- Make sure that you understand any symbols used, such as the 'square root' symbol ($\sqrt{}$) here.
- Take an approved calculator into every exam and make sure that you know how to use it.

Student answer

10	3.0	4.2	3.5	4.1	3.9	3.74	0.49	120	0.031

$t = 6.48$

> The student has answered this uncomplicated question correctly. Make sure that you also understand how standard deviation (s) and variance (s^2) relate to one another.

COMMENTARY

This is an average answer because:

- The student has filled in the table and calculated the value of t correctly, so this answer would gain all 4 marks. However, the student would have been wise to show the main steps of their working out. This is because, had they made an error in their calculation, they still might have gained some credit for demonstrating the correct mathematical approach.

> You should understand how the value of t is used, including how to calculate the degrees of freedom and how to interpret the significance.

QUESTION TYPE: PRACTICAL PLANNING

A student wanted to carry out an investigation on the effect of aluminium ions on the growth of lawn grass. He carried out a pilot study in which he grew seeds from a commercial grass species mix in the laboratory. Liquid growth medium was used, to which aluminium sulfate was added at a variety of concentrations that were representative of clean to moderately polluted ground water. Three germinating seeds were grown in each concentration. The student allowed the seedlings to grow for a few weeks then measured the length of the longest shoot. His pilot study results are shown in the table below.

fig B Diagram of a grass seedling showing roots and shoots.

Aluminium ion concentration/ micromoles dm^{-3}	Shoot length/mm
50	98
50	81
50	90
200	80
200	89
200	89
400	77
400	86
400	83

When measuring the seedlings, the student noticed that the amount of root growth seemed to be stunted at higher aluminium ion levels, while shoots seemed relatively unaffected.

He decided that his final project would investigate the relative amount of roots and shoot growth at different aluminium ion concentrations.

His hypothesis was: 'The higher the aluminium concentration, the smaller the roots will be relative to the shoots.'

Devise a plan that would allow this hypothesis to be tested. Your plan should extend and modify methods from the pilot study in order to allow valid and reliable results to be gathered. (6)

The command word here is 'devise'. This means that you must plan or invent a procedure from existing principles or ideas. There are lots of clues to suitable methods given in the text, but no marks will be awarded for simply repeating these – you will need to do some fine-tuning or amending to gain credit. A data table is given but you are not asked to do anything with the figures; you can therefore assume that the table must provide further prompts and ideas for your plan. Again, you should critically assess what has been done already and modify where necessary. The diagram (fig B) also gives some hints: it shows that there are several roots and shoots to each seedling and that roots have a complex shape.

Question analysis

- A practical style question can address detailed planning.
- Questions in Unit 6 can test skills that you have learned in all previous units.

- This question is based on a recommended additional practical in Topic 4. You may or may not have carried out this practical. However, the main principles of planning will have been covered in other practical work.
- When a question asks you to plan an experiment, you should start by thinking about each type of variable. For instance: How could the independent variable (in this case aluminium ion concentration) be changed – how many different levels and what range? Will you need a control (a null or comparison level of the independent variable)? How will the dependent variable be measured and will any calculations or conversion of data be required? What variables should be controlled and how? Once you have considered these points, make sure you list any equipment that you will need (not beakers or spatulas, only special items such as microscopes, water baths or reagents). Finally, you will need to think about sample size and repeats.

Student answer

At least five different concentrations of aluminium ions should be used. For example: 100, 200, 300, 400 and 500 micromoles dm^{-3}. At least five grass seedlings should be grown in each concentration. Factors that should be controlled include time, light levels and the concentration of mineral ions in the growth medium. The amount of growth solution used should also be the same in each case. The length of the longest shoot and longest root should be measured with a ruler to the nearest millimetre after 3 weeks of growth. A mean should be calculated for each concentration, and graphs of shoot length and root length against aluminium ion concentration drawn and compared.

The answer would gain credit for a suitable number/range of concentrations, for considering repeats and for stating several variables that should be controlled. A stronger answer would include more detail of how variables might be controlled, especially light (for example, by using a light box with a 12/12 hour dark cycle, or by stating the details of randomised design layout). The 'amount' of growth solution is also not specific enough – volume and concentration are more suitable terms. Other suitable controlled variables might include temperature, grass species or pH. The diagram shows that root length might be difficult to measure, even for just the longest root. In addition, simple length does not take into account the overall shoot and root size of the seedling; total shoot mass and total root mass for each seedling would be a more valid measurement. The student has not given a method for determining the relative size of roots to shoots; root mass could be expressed as a percentage (or as a decimal ratio) of shoot mass, for example.

COMMENTARY

This is an average answer because:

- The student has thought about many of the key variables but has not given sufficient detail for some.
- A key requirement of the question has been missed: a measure of the relative size of roots and shoots was required, rather than simply the separate size of each.
- There is no mention of a control: seeds should also be grown without added aluminium ions, for comparison.

COMMAND WORDS

The following table lists the command words used across the IAS/IAL Science qualifications in the external assessments. You should make sure you understand what is required when these words are used in questions in the exam.

Add/Label
Requires the addition of labelling to stimulus material given in the question, for example, labelling a diagram or adding units to a table.

Assess
Give careful consideration to all the factors or events that apply and identify which are the most important or relevant. Make a judgement on the importance of something, and come to a conclusion where needed.

Calculate
Obtain a numerical answer, showing relevant working. If the answer has a unit, this must be included.

Comment on
Requires the synthesis of a number of factors from data/information to form a judgement. More than two factors need to be synthesised.

Compare and contrast
Looking for the similarities **and** differences between two (or more) things. Should not require the drawing of a conclusion. Answer must relate to both (or all) things mentioned in the question.

The answer must include at least one similarity and one difference.

Complete/ Record
Requires the completion of a table/diagram/ equation.

Criticise
Inspect a set of data, an experimental plan or a scientific statement and consider the elements. Look at the merits and/or faults of the information presented and back judgements made.

Deduce
Draw/reach conclusion(s) from the information provided.

Derive
Combine two or more equations or principles to develop a new equation.

Describe
Give an account of something. Statements in the response need to be developed, as they are often linked, but do not need to include a justification or reason.

Determine
The answer must have an element that is quantitative from the stimulus provided, or must show how the answer can be reached quantitatively.

Devise
Plan or invent a procedure from existing principles/ideas.

Discuss
Identify the issue/situation/problem/argument that is being assessed within the question.

Explore all aspects of an issue/situation/problem.

Investigate the issue/situation/problem etc. by reasoning or argument.

Draw
Produce a diagram either using a ruler or drawing freehand.

Estimate
Give an approximate value for a physical quantity or measurement or uncertainty.

Evaluate
Review information then bring it together to form a conclusion, drawing on evidence including strengths, weaknesses, alternative actions, relevant data or information. Come to a supported judgement of a subject's qualities and its relation to its context.

Explain
An explanation requires a justification/ exemplification of a point. The answer must contain some element of reasoning/ justification; this can include mathematical explanations.

Give/State/ Name
All of these command words are really synonyms. They generally all require recall of one or more pieces of information.

Give a reason/ reasons
When a statement has been made and the requirement is only to give the reasons why.

Identify
Usually requires some key information to be selected from a given stimulus/resource.

Justify
Give evidence to support (either the statement given in the question or an earlier answer).

Plot
Produce a graph by marking points accurately on a grid from data that is provided and then drawing a line of best fit through these points. A suitable scale and appropriately labelled axes must be included if these are not provided in the question.

Predict
Give an expected result or outcome.

Show that
Prove that a numerical figure is as stated in the question. The answer must be to at least 1 more significant figure than the numerical figure in the question.

Sketch
Produce a freehand drawing. For a graph, this would need a line and labelled axes with important features indicated; the axes are not scaled.

State what is meant by
When the meaning of a term is expected but there are different ways in which these can be described.

Suggest
Use your knowledge and understanding in an unfamiliar context. May include material or ideas that have not been learned directly from the specification.

Write
When the questions ask for an equation.

GLOSSARY

abiotic factors the non-living elements of the habitat of an organism

absolute refractory period the first millisecond or so after the action potential during which it is impossible to re-stimulate the fibre, the sodium ion channels are completely blocked and the resting potential has not been restored

absorption spectrum a graph showing the amount of light absorbed by a pigment against the wavelength of the light

abundance the relative representation of a species in a particular ecosystem

acetyl coenzyme A (acetyl CoA) the 2C compound produced in the link reaction which feeds directly into the Krebs cycle, combining with a 4C organic acid to form a 6-carbon compound

acetylcholine (ACh) neurotransmitter in the parasympathetic nervous system, the synapses of motor neurones, and cholinergic synapses in the brain

acetylcholinesterase an enzyme found within the post-synaptic membrane of cholinergic nerves that breaks down acetylcholine in the synapses after it has triggered a post-synaptic potential

ACFOR scale a simple scale used to describe the abundance of a species in a given area

acquired immunodeficiency syndrome (AIDS) the disease which results from the destruction of the T helper cells as a result of infection with HIV

actin one of the proteins which form the contracting mechanism of muscle cells

action potential when the potential difference across the membrane is briefly reversed to about +40 mV on the inside with respect to the outside for about 1 millisecond

action spectrum a graph demonstrating the rate of photosynthesis against the wavelength of light

actomyosin the chemical produced when cross-bridges form between actin and myosin during muscle contraction

adenosine diphosphate (ADP) a nucleotide formed when a phosphate group is removed from ATP, providing energy to drive reactions in the cell

adenosine triphosphate (ATP) a nucleotide that acts as the universal energy supply in cells. It is made up of the base adenine, the pentose sugar ribose and three phosphate groups

adrenaline hormone produced by the adrenal glands which stimulates the fight or flight response

adrenergic nerves nerves using noradrenaline as their synaptic neurotransmitter

aerobic respiration the form of cellular respiration that occurs in the mitochondria in the presence of oxygen

agglutination the grouping together of cells caused when antibodies bind to the antigens on pathogens

agranulocytes leucocytes with round nuclei but without granules in their cytoplasm; they are involved in the specific immune response to infection

allopatric speciation an evolutionary process that occurs when populations become physically or geographically isolated

amplified the process by which DNA is replicated repeatedly (using the polymerase chain reaction) to produce a much bigger sample

anaerobic respiration the form of cellular respiration that occurs in the cytoplasm when there is no oxygen present

antagonistic pairs muscles which work in opposition to each other, pulling in opposite directions

anthropogenic produced by people

anti-retroviral drugs drugs which are effective against retroviruses such as HIV

antibiotic a drug that either destroys microorganisms or prevents them from growing and reproducing

antibiotic resistant a microorganism that is not affected by an antibiotic (even one that may have been effective in the past)

antibodies glycoproteins that are each produced in response to a specific antigen

antidiuretic hormone (ADH) hormone produced in the hypothalamus and stored in the posterior pituitary that increases the permeability of the distal tubule and the collecting duct of the kidney to water, reducing the amount of urine produced but increasing the concentration of the urine

antigen-presenting cell (APC) a cell displaying an antigen/MHC protein complex

antigens glycoproteins, proteins or carbohydrates on the surface of cells, toxins produced by bacterial and fungal pathogens, and some whole viruses and bacteria that are recognised by white blood cells during the specific immune responses to infection; they stimulate the production of an antibody

arteriovenous shunt a system which closes to allow blood to flow through the major capillary networks near the surface of the skin, or opens to allow blood along a 'shortcut' between the arterioles and venules, so it does not flow through the capillaries near the surface of the skin

artificial active immunity when the body produces its own antibodies to an antigen acquired through vaccination

artificial passive immunity when antibodies are extracted from one individual and injected into another (e.g. one form of the tetanus vaccine)

ATPase the enzyme which catalyses the formation and breakdown of ATP, depending on the conditions

atrioventricular node (AVN) a group of cells stimulated by the wave of excitation from the SAN and the atria; it imposes a delay before transmitting the impulse to the bundle of His

attenuated pathogens viable pathogens that have been modified so that they do not cause disease but still cause an immune response that results in the production of antibodies and immunity

autonomic (involuntary) nervous system the involuntary nervous system; autonomic motor neurones control bodily functions that the conscious area of the brain does not normally control

autotrophic organisms that make complex organic compounds from simple compounds in the environment

auxins plant hormones that act as powerful growth stimulants (e.g. indoleacetic acid, IAA) and are involved in apical dominance, stem and root growth, and tropic responses to unilateral light

axon the long nerve fibre of a motor neurone, which carries the nerve impulse

B cells lymphocytes that are made and mature in the bone marrow; once they are mature, they are found in the lymph glands and free in the body

B effector cells lymphocytes that are made and mature in the bone marrow which divide to form the plasma cell clones

B memory cells lymphocytes that are made and mature in the bone marrow and that provide the immunological memory to a specific antigen; they allow the body to respond very rapidly to the same pathogen carrying the same antigen a second time

bacterial flora the combination of different species of microorganisms found in or on a specific region of the body

bactericidal kills bacteria

bacteriostatic inhibits the growth of bacteria

baroreceptors mechanoreceptors in the aorta and carotid arteries that are sensitive to pressure changes

basophils leucocytes with a two-lobed nucleus; they produce histamines involved in inflammation and allergic reactions

belt transect when two tapes are stretched out and the ground between them surveyed or a tape stretched out and quadrats are taken at regular intervals

binary fission asexual reproduction in bacteria in which the bacteria split in half

biofuels fuels produced directly or indirectly from biomass

bioinformatics the development of the software and computing tools needed to organise and analyse large amounts of raw biological data (e.g. the results from microarray analysis of DNA)

biomes the major ecosystems of the world

biosphere all of the areas of the surface of the Earth where living organisms survive

biotic factors the living elements of a habitat that affect the ability of a group of organisms to survive there

bleaching the photochemical breakdown of visual pigments (e.g. rhodopsin to opsin and retinal)

blood–brain barrier a barrier formed by the endothelial cells that line the capillaries of the brain which are very tightly joined together, making it difficult for bacteria to cross into the brain but also making it difficult for therapeutic drugs to enter the brain

bone the strong, calcium-rich tissue which is the main component of the vertebrate skeleton

brain the area of the CNS in which information can be processed and from which instructions can be issued as required to give fully coordinated responses to a range of situations

bundle of His a group of conducting fibres in the septum of the heart

calibration checking the measurement values given by one system of measurement against another of known accuracy

Calvin cycle a series of enzyme-controlled reactions that take place in the stroma of chloroplasts and result in the reduction of carbon dioxide from the air to bring about the synthesis of carbohydrate

capsid the protein coat of a virus

capsomeres the repeating protein units that make up the capsid of a virus

carbon cycle a series of reactions by which carbon is constantly recycled between living things and the environment

carbon sink a reservoir where carbon dioxide is removed from the atmosphere and locked into organic or inorganic compounds

cardiac muscle muscle which makes up the heart

cardiac output a measure of the volume of blood pumped by the heart per minute, calculated by multiplying cardiac volume by heart rate

cardiac volume the volume of blood pumped at each heartbeat

cardiovascular control centre centre in the medulla oblongata of the brain that receives information from a number of different receptors and controls changes to the heart rate and the cardiac volume through parasympathetic and sympathetic nerves

carotenoids photosynthetic pigments consisting of orange carotene and yellow xanthophyll

cartilage hard but flexible skeletal tissue that often acts as a shock absorber and prevents wear in joints

causal relationship one event happens as a direct result of another, with a clear mechanism by which one factor causes a given change

cellular respiration the process by which food is broken down to yield ATP (a source of energy for metabolic reactions)

central nervous system (CNS) a specialised concentration of nerve cells which process incoming information and which send out impulses through motor neurones, which carry impulses to the effector organs

cerebellum the area of the brain that coordinates smooth movements; it uses information from the muscles and the ears to control balance and maintain posture

cerebral hemispheres the two parts of the cerebrum, joined by the corpus callosum

cerebrum the area of the brain responsible for conscious thought, personality, control of movement and much more

chemiosmosis the process that links the electrons that are passed along the electron transport chain to the production of ATP, by the movement of hydrogen ions through the membrane along electrochemical, concentration and pH gradients

chemiosmotic theory the model developed by Peter Mitchell to explain the production of ATP in mitochondria, chloroplasts and elsewhere in living cells

chemoreceptors sensory nerve cells (or organs) that respond to chemical stimuli

chi squared test a statistical test that enables you to determine whether there is a statistically significant association between the distribution of two species

chlorophyll _a_ a blue-green photosynthetic pigment, found in all green plants

chlorophyll _b_ a yellow-green photosynthetic pigment

chloroplast envelope the outer and inner membranes of a chloroplast including the intermembrane space

cholinergic nerves nerves using acetylcholine as their synaptic neurotransmitter

chondrocytes the cells that form cartilage

climate the average weather in a relatively large area (such as a country) over a long period of time

climate change a large-scale change in global or regional weather patterns that happens over a period of many years

climatic climax community the only climax community possible in a given climate

climax community a self-sustaining community with relatively constant biodiversity and species range. It is the most productive group of organisms that a given environment can support long term

clonal selection the selection of the cells that carry the right antibody for a specific antigen

clone a group of genetically identical cells which are all produced from one cell

Clostridium difficile a type of bacterium that often exists in the intestines and causes no problems unless it becomes dominant as a result of antibiotic treatment that has removed or damaged the normal gut flora

cobra venom (α-cobratoxin) a substance made by several species of cobra that binds reversibly to acetylcholine receptors in post-synaptic membranes in motor neurones, preventing the production of a post-synaptic action potential

collecting duct takes urine from the distal tubule to be collected in the pelvis of the kidney; it is the region of the kidney in which most of the water balancing needed for osmoregulation takes place

colonisation the process by which new species spread to new areas

communicable diseases caused by pathogens which can be passed from one organism to another

community all the populations of all the different species of organisms living in a habitat at any one time

complementary DNA (cDNA) DNA which can act as an artificial gene, made by reversing the transcription process from mRNA using reverse transcriptase

computed tomography (CT) scans scans using thousands of tiny beams of X-rays which are passed through an area of the body such as the head to produce an image of the brain

cones photoreceptors found in the fovea of the retina which contain the visual pigment iodopsin; they respond to bright light, give great clarity of vision and colour vision

convergence when several sensory receptors all synapse with one sensory neurone so the neurotransmitters add together to trigger an action potential in the sensory neurone

correlation a strong tendency for two sets of data to vary together

countercurrent multiplier a system that produces a concentration gradient in a living organism; energy from cellular respiration is required

cyclic AMP (cAMP) a compound formed from ATP that is produced when peptide hormones such as ADH and adrenaline bind to membrane receptors, and acts as a second messenger in cells

cyclic photophosphorylation a process that drives the production of ATP; light-excited electrons from PSI are taken up by an electron acceptor and transferred directly along an electron transport chain to produce ATP, with the electron returning to PSI

cytochrome oxidase an enzyme in the electron transport chain which receives the electrons from the cytochromes and is reduced as the cytochromes are oxidised, with the production of a molecule of ATP

cytochromes members of the electron transport chain; they are protein pigments with an iron group (similar to haemoglobin) which are reduced by electrons from reduced FAD, which is reoxidised, with the production of a molecule of ATP

cytokines molecules which signal between cells; they have several roles in the immune system, including stimulating other phagocytes to move to the infection site

day-neutral plants (DNPs) plants whose flowering is not affected by the length of time they are exposed to light or dark

deamination the removal of the amino group from excess amino acids in the ornithine cycle in the liver; the amino group is converted into ammonia and then to urea, which can be excreted by the kidneys

death phase/decline phase when reproduction has almost ceased and the death rate of cells is increasing so that the population number falls

decarboxylases enzymes that remove carbon dioxide from a molecule

decomposers the final trophic level in any set of feeding relationships; these are the microorganisms such as bacteria and fungi that break down the remains of animals and plants and return the mineral nutrients to the soil

dehydrogenases enzymes that remove hydrogen from a molecule (during oxidation reactions)

denature to cause permanent changes in a protein by too high a temperature

dendrites the thin, finger-like extensions from the cell body of a neurone that connect with neighbouring neurones

dendrochronology the dating of past events using tree ring growth

dendron the long nerve fibre of a sensory neurone, which carries the nerve impulse

density-dependent factors factors affecting the number of organisms occupying a niche which are dependent on the number of organisms in a specific area

density-independent factors factors affecting the number of organisms occupying a niche which are the same regardless of population size

depolarisation the condition of the neurone when the potential difference across the membrane is briefly reversed during an action potential, with the cell becoming positive on the inside with respect to the outside for about 1 millisecond

dilution plating a method used to obtain a culture plate with a countable number of bacterial colonies

distal tubule the section of the nephron after the loop of Henle that leads into the collecting duct where some of the balancing of the water needs of the body occurs

distribution where a species of organism is found in the environment and how it is organised

DNA profiling the identification of repeating patterns in the non-coding regions of DNA

DNA sequencing/gene sequencing the analysis of the individual base sequence along a DNA strand or an individual gene

DNA viruses viruses that have DNA as the genetic material

dopamine the neurotransmitter produced by nerve cells in the substantia nigra, which is closely involved in the control and coordination of movement

dopamine agonists chemicals that bind to dopamine receptors in brain synapses and mimic the effect of dopamine

ecosystem an environment including all the living organisms interacting within it, the cycling of nutrients and the physical and chemical environment in which the organisms are living

ecstasy (MDMA: 3,4-methylenedioxy-N-methylamphetamine) an illegal drug that acts as a stimulant and has a psychotropic effect by blocking the serotonin reuptake transport system

edaphic factors factors that relate to the structure of the soil

effector cells specialised cells that bring about a response if stimulated by a neurone

effectors systems (usually muscles or glands) that work to reverse, increase or decrease changes in a biological system

electrocardiogram (ECG) technology used to investigate the rhythms of the heart by producing a record of the electrical activity of the heart

electron transport chain a series of electron-carrying compounds along which electrons are transferred in a series of oxidation/reduction reactions, driving the production of ATP

endocrine glands glands without ducts (ductless glands) that produce hormones and release them directly into the bloodstream

endotherms animals that warm their bodies through metabolic processes at least in part and usually have a body temperature higher than the ambient temperature

endotoxins lipopolysaccharides that are an integral part of the outer layer of the cell wall of Gram-negative bacteria and act as toxins to other cells

envelope a coat around the outside of a virus derived from lipids in the host cell

eosinophils leucocytes important in the non-specific immune response against parasites, in allergic reactions and inflammation, and in developing immunity to disease

ethanol an organic chemical with the formula C_2H_5OH produced as a result of anaerobic respiration (fermentation) in fungi and some plant cells

etiolated the form of plants which are grown in the dark, with long internodes, thin stems, small or unformed leaves and white or pale yellow in colour

excitatory post-synaptic potential (EPSP) the potential difference across the post-synaptic membrane caused by an influx of sodium ions into the nerve fibre as the result of the arrival of a molecule of neurotransmitter on the receptors of the post-synaptic membrane; this makes the inside more positive than the normal resting potential, increasing the chance of a new action potential

exocrine glands glands that produce chemicals (e.g. enzymes) and release them along small tubes or ducts

exocytosis the energy-requiring process by which a vesicle fuses with the cell surface membrane so the contents are released to the outside of the cell

exons the coding regions of DNA (the genes)

exotoxins soluble proteins produced and released into the body by bacteria as they metabolise and reproduce in the cells of their host; these proteins act as toxins in different ways

expiratory centre region of the ventilation centre involved in voluntary exhalation (breathing out)

expiratory reserve volume (ERV) the volume of air that you can force out above the normal expired tidal volume

extensors muscles that extend (stretch or open) a joint

extrapolate apply already known trends to unknown situations to predict what will happen

FAD (flavin adenine dinucleotide) a hydrogen carrier and coenzyme; in cellular respiration, FAD receives hydrogen to form reduced FAD driving the production of ATP

far red light light with a wavelength of 700–800 nm, which is detected by plants using phytochromes

fast twitch muscle fibres muscle fibres which contract very rapidly and strongly and fatigue quickly; they have relatively low levels of myoglobin and low numbers of mitochondria

fever a raised body temperature, often in response to infection

flexors muscles that flex (close or bend) a joint

florigen plant hormone which appears to be involved in the photoperiodic response; it may be FTmRNA

food chain a simple way of modelling the feeding relationships between a series of organisms in an ecosystem

forensic entomology the study of insect life relating to crime

forensic science the application of scientific techniques to the investigation of a crime

fovea area of the retina packed with cones which provides colour vision and great visual acuity

functional magnetic resonance imaging (fMRI) scans scans which monitor the uptake of oxygen in different brain areas, making it possible to watch the different areas of the brain in action while people conduct different tasks

ganglion (plural: ganglia) a collection of nerve cell bodies outside of the central nervous system

gene guns a technique to produce recombinant DNA by shooting the desired DNA into the cell at high speed on very small gold or tungsten pellets (balls)

generation time the time span between bacterial divisions

generator potential a graded response to a stimulus across the synapse of a sensory receptor

genetic engineering/genetic modification/gene editing the insertion of genes from one organism into the genetic material of another organism or changing the genetic material of an organism

gibberellins plant hormones that act as growth regulators, particularly in the internodes of stems, by stimulating elongation of the growing cells; they also promote the growth of fruit and are involved in breaking dormancy in seeds and in germination

global warming a measurable increase in the temperature of the Earth's atmosphere or temperature at the surface of the Earth

gluconeogenesis the synthesis of glucose from non-carbohydrates

glyceraldehyde 3-phosphate (GALP) a 3-carbon sugar produced in the Calvin cycle using reduced NADP and ATP from the light-dependent stage; GALP is the key product of photosynthesis and is used to replace the RuBP needed in the first step of the cycle, in glycolysis and the Krebs cycle, and in the synthesis of amino acids, lipids, etc. for the plant cells

glycerate 3-phosphate (GP) a 3-carbon compound thought to be the result of breakdown of a theoretical highly unstable 6-carbon compound formed as a result of the reaction between RuBP and carbon dioxide in the Calvin cycle

glycolysis the first stage of cellular respiration, which occurs in the cytoplasm and is common to both aerobic and anaerobic respiration

grana layers of thylakoid membranes within a chloroplast

granulocytes leucocytes with granules that absorb stain in the cytoplasm of the cells; this makes them visible under the microscope; they have lobed nuclei and are involved in the non-specific responses to infection

green fluorescent protein (GFP) the product of a gene often used as a marker in the production of recombinant DNA

greenhouse effect the process by which gases in the Earth's atmosphere absorb and re-radiate the radiation from the Sun, which has been reflected from the Earth's surface, maintaining a temperature at the surface of the Earth that is warm enough for life to exist

greenhouse gases gases found in the atmosphere, including carbon dioxide, methane and water vapour, which are involved in the greenhouse effect

grey matter the cell bodies of neurones in the CNS

gross primary productivity (GPP) in plants, the rate at which light from the Sun catalyses the production of new plant material, measured as $g\,m^{-2}\,year^{-1}$, $g\,C\,m^{-2}\,year^{-1}$ or $kJ\,m^{-2}\,year^{-1}$

growth hormone (GH) the hormone secreted by the pituitary gland which stimulates growth

habitat the place where an organism lives

habituation diminishing of an innate response to a frequently repeated stimulus

haemocytometer a thick microscope slide with a rectangular indentation and engraved (marked) grid of lines that is used to count cells

herd immunity when a high proportion of a population is immune to a pathogen, usually through vaccination, thus lowering the risk of infection to all, including those not vaccinated

heterotrophic organisms that obtain complex organic molecules by feeding on other living organisms or their dead remains

histamines chemicals released by the tissues in response to an allergic reaction

HIV positive someone who has antibodies to HIV in their blood, indicating that they have been infected with the virus and so are at risk of passing it on to other people

homeostasis the maintenance of a state of dynamic equilibrium in the body, despite changes in the external or internal conditions

hormones organic chemicals which are produced in endocrine glands, are released into the blood and travel through the transport system to parts of the body where they cause changes, which may be extensive or very targeted; hormones are usually either proteins, parts of proteins such as polypeptides, or steroids

hospital-acquired infections (HAIs) infections that are acquired by patients while they are in hospitals or care facilities; these infections may be the result of poor hygiene or the result of antibiotic treatment; the pathogens may be antibiotic resistant

human immunodeficiency virus (HIV) a retrovirus that causes AIDS

hybridisation the process by which labelled DNA samples bind to the matching DNA probes on a microarray slide

hydrogen acceptor a molecule which receives hydrogen and becomes reduced in cell biochemistry

hypothalamus a small area of brain directly above the pituitary gland that controls the activities of the pituitary gland and coordinates the autonomic (unconscious) nervous system

immune response the specific response of the body to invasion by pathogens

immunisation the process of protecting people from infection by giving them passive or active artificial immunity

immunoglobulins antibodies

individual counts a measure of the number of individual organisms in an area

inflammation a common, non-specific response to infection involving the release of histamines from mast cells and basophils; this causes the blood vessels to dilate producing local heat, redness and swelling

inhibitory post-synaptic potential (IPSP) the potential difference across the post-synaptic membrane caused by an influx of negative ions as the result of the arrival of a molecule of neurotransmitter on the receptors of the post-synaptic membrane; this makes the inside more negative than the normal resting potential, decreasing the chance of a new action potential

inoculation the process by which microorganisms are transferred into a culture medium under sterile conditions

inspiratory capacity (IC) the volume that can be inspired from the end of a normal expiration; IC = VT + IRV

inspiratory centre main area of the ventilation centre involved in the control of breathing in; breathing out is usually passive

inspiratory reserve volume (IRV) the volume of air that you can take in above the normal inspired tidal volume

interferons chemicals produced by cells in very small amounts when invaded by viruses; interferons act to prevent the viruses invading other cells

interglacials the relatively warm periods between ice ages

interspecific competition competition between different species within a community for the same resources

intraspecific competition competition between members of the same species for a limited resource

intrinsic rhythmicity the intrinsic (internal) rhythm of contraction and relaxation in the cardiac muscle of the heart

introns the large, non-coding regions of DNA that are removed before messenger RNA is translated into proteins

iodopsin the visual pigment in the cones

knockout organism an organism with one or more genes silenced (knocked out) so they no longer work; they are often used to identify the function of a gene, to investigate disease and to test potential treatments

Krebs cycle a series of biochemical steps that leads to the complete oxidation of glucose, resulting in the production of carbon dioxide, water and relatively large amounts of ATP

lactate ions of a 3-carbon compound (lactic acid) which is the end-product of anaerobic respiration in mammals

lag phase when bacteria are adapting to a new environment and are not reproducing at their maximum rate

lamellae extensions of the thylakoid membranes which connect two or more grana and act as a supporting skeleton in the chloroplast; they maintain a working distance between the grana so that these receive the maximum light and function as efficiently as possible

latent the state of the non-virulent virus within the host cell

leaching the loss of minerals from soil as water passes through rapidly

leucocytes white blood cells which are larger than erythrocytes and can squeeze through tiny blood vessels as they can change their shape; there are around 4000–11 000 leucocytes per mm^3 of blood and there are several different types which carry out different functions in the body

levodopa (L-dopa) a precursor of dopamine which can cross the blood–brain barrier and has been in use since the 1960s to treat Parkinson's disease

lidocaine a drug used as a local anaesthetic that works by blocking the voltage-gated sodium channels in post-synaptic membranes in sensory neurones, preventing the production of an action potential

ligaments elastic tissue which forms joint capsules

light-dependent reactions the reactions that take place in the light on the thylakoid membranes of the chloroplasts; the reactions produce ATP and break down water molecules in a photochemical reaction, providing hydrogen ions to reduce carbon dioxide and produce carbohydrates

light-independent reactions the reactions that use the reduced NADP and ATP produced by the light-dependent stage of photosynthesis in a pathway known as the Calvin cycle; this occurs in the stroma of the chloroplast and results in the reduction of carbon dioxide from the air to cause the synthesis of carbohydrates

limiting factor the factor needed for a reaction to progress that is closest to its minimum value

line transect a way of collecting data more systematically; a tape is stretched between two points and every individual plant (or animal) that touches the tape is recorded

link reaction the reaction needed to move the products of glycolysis into the Krebs cycle

lipopolysaccharides chemicals made up of a combination of lipids and polysaccharides (complex carbohydrates)

liposome wrapping a technique for producing recombinant DNA that involves wrapping the gene to be inserted in liposomes, which combine with the target cell membrane and can pass through it to deliver the DNA into the cytoplasm

liquid culture growing microorganisms in a nutrient broth in a flask or test tube rather than on an agar plate

log phase/exponential phase when the rate of bacterial reproduction is close to or at its theoretical maximum, repeatedly doubling in a given time period

long-day plants (LDPs) plants that flower when days are long and nights are short

lymphocytes small leucocytes with very large nuclei that are vitally important in the specific immune response of the body; they make up the main cellular components of the immune system; they are made in the white bone marrow of the long bones

lysogeny the period when a virus is part of the reproducing host cell, but does not affect the host adversely

lysozymes enzymes found in tears and other body secretions that are capable of destroying microbial cell walls

macrophages cells that engulf pathogens by phagocytosis as part of the specific immune system

magnetic resonance imaging (MRI) scans scans produced using magnetic fields and radio waves to image the soft tissues; they produce images showing much finer detail than CT scans

major histocompatibility complex (MHC) proteins proteins that display antigens on the cell surface membrane

mast cells cells found in the connective tissue below the skin and around blood vessels; they release histamines when the tissue is damaged

medulla oblongata (medulla) the most primitive part of the brain that controls reflex centres controlling functions such as the breathing rate, heart rate, blood pressure, coughing, sneezing, swallowing, saliva production and peristalsis

methicillin-resistant *Staphylococcus aureus* (MRSA) a strain of *S. aureus* that is resistant to several antibiotics, including methicillin

micro-satellite a section of DNA with a 2–6 base sequence repeated 5 to 100 times

microarray a very useful laboratory tool which allows scientists to detect thousands of active genes at the same time

microclimate a small area with a distinct climate that is different from the surrounding areas

microhabitat a small area of a habitat

microinjection (DNA injection) a technique for producing recombinant DNA that involves injecting DNA into a cell through a very fine micropipette

mini-satellite a section of DNA with a 10–100 base sequence repeated 50 to several hundred times

monoamine oxidase B (MAOB) inhibitors drugs which inhibit the enzyme monoamine oxidase B (MAOB), which breaks down dopamine in brain synapses; thus, MAOB inhibitors reduce the destruction of the dopamine made by the cells

monocytes the largest of the leucocytes, they can pass from the blood into the tissues to form macrophages

motor neurones neurones that carry impulses from the central nervous system to the effector organs

myelin sheath a fatty insulating layer around some vertebrate neurones produced by the Schwann cell

myofibril a very long contracting fibre in skeletal muscle cells

myogenic contracts without an external stimulus

myoglobin a pigment with a high affinity for oxygen found in the muscle

myosin one of the proteins which form the contracting mechanism of muscle cells

NAD (nicotinamide adenine dinucleotide) a coenzyme that acts as a hydrogen acceptor

natural active immunity when the body produces its own antibodies to an antigen encountered naturally

natural passive immunity when antibodies made by the mother are passed to the baby via the placenta or mother's milk

negative feedback systems systems for maintaining a condition, such as the concentration of a substance, within a narrow range; receptors detect a change in conditions and, as a result, effectors are stimulated to restore the equilibrium

nephrons microscopic tubules that make up most of the structure of the kidney

nerve impulses the electrical signals transmitted through the neurones of the nervous system

nerves bundles (groups) of nerve fibres which may be all axons from motor neurones, all dendrons from sensory neurones or a mixture of both in a mixed nerve

net primary productivity (NPP) the material produced by photosynthesis and stored as new plant body tissues; that is, NPP = GPP − R (where R = losses due to respiration)

neurones cells specialised for the rapid transmission of impulses throughout an organism

neurosecretory cells nerve cells that produce secretions from the ends of their axons; these secretions either stimulate or inhibit the release of hormones from the anterior pituitary, or are stored in the posterior pituitary and then later released as hormones

neurotransmitter a chemical which transmits an impulse across a synapse

neutralisation the action of antibodies in neutralising the effects of bacterial toxins on cells by binding to them

neutrophils the most common type of leucocyte; they engulf and digest pathogens by phagocytosis

niche the role of an organism within the habitat in which it lives

nicotine a drug found in cigarettes that mimics the effect of acetylcholine and binds to specific acetylcholine receptors in post-synaptic membranes known as nicotinic receptors

nitrogen cycle the recycling of nitrogen between living things and the environment by the actions of microorganisms

nodes of Ranvier gaps between the Schwann cells that enable saltatory conduction

non-cyclic photophosphorylation a process involving both PSI and PSII in which water molecules are broken into smaller units using light energy to provide reducing power to make carbohydrates and at the same time produce more ATP

noradrenaline neurotransmitter in the sympathetic nervous system and adrenergic synapses of the brain

null hypothesis the hypothesis that any differences between data sets are the result of chance

nutrient agar a jelly extracted from seaweed and used as a solid nutrient for culturing microorganisms, commonly used in Petri dishes

nutrient broth a liquid nutrient for culturing microorganisms, commonly used in flasks, test tubes or bottles

nutrient medium a substance used for the culture of microorganisms, which can be in liquid form (nutrient broth) or in solid form (usually nutrient agar)

opportunists/pioneer species species which are the first to colonise new or disturbed ecosystems

opsonins chemicals which bind to pathogens and label them so they are more easily recognised by phagocytes

opsonisation a process that makes a pathogen more easily recognised, engulfed and digested by phagocytes

optimum temperature the temperature at which an enzyme works most efficiently

ornithine cycle a series of enzyme-controlled reactions that convert ammonia from excess amino acids into urea in the liver

osmoreceptors sensory receptors in the hypothalamus that detect changes in the concentration of the blood plasma

osmoregulation the maintenance of the osmotic potential of the tissues of a living organism within a narrow range, by controlling water and salt concentrations

oxidation the removal of electrons from a substance (e.g. by the addition of oxygen or removal of hydrogen)

oxidative phosphorylation the oxygen-dependent process in the electron transport chain where ADP is phosphorylated

parasympathetic nervous system involves autonomic motor neurones which produce acetylcholine as their neurotransmitter and often have a relatively slow, inhibitory effect on an organ system; these neurones have very long myelinated preganglionic fibres that leave the CNS and synapse in a ganglion very close to the effector organ; postganglionic fibres are very short and unmyelinated

pathogens microorganisms that cause diseases

penicillin the first antibiotic discovered; it is bactericidal and affects the formation of bacterial cell walls

percentage cover the area covered by the above-ground parts of a particular species

peripheral nervous system the parts of the nervous system that spread through the body and are not involved in the central nervous system

phaeophytin a grey pigment which is produced by the breakdown of the other photosynthetic pigments

phagocyte cell which engulfs and digests other cells or pathogens

phagocytosis the process by which a cell engulfs another cell and encloses it in a vesicle to digest it

phagosome the vesicle in which a pathogen is enclosed in a phagocyte

photochemical reaction a reaction initiated by light

photolysis the breaking down of a molecule into smaller units using light

photorespiration the alternative reaction catalysed by RUBISCO in a low carbon dioxide environment which uses oxygen and releases carbon dioxide, making photosynthesis less efficient

photosynthesis the process by which living organisms, particularly plants and algae, capture the energy of the Sun using chlorophyll and use it to convert carbon dioxide and water into simple sugars

photosystem I (PSI) a combination of chlorophyll pigments which absorbs light of wavelength 700 nm and is involved in cyclic and non-cyclic photophosphorylation

photosystem II (PSII) a combination of chlorophyll pigments which absorbs light of wavelength 680 nm and is involved only in non-cyclic photophosphorylation

phytochrome a plant pigment that reacts with different types of light, and as a result affects the responses of the plant

pituitary gland a small gland in the brain that has an anterior lobe and a posterior lobe and produces and releases secretions that affect the activity of most of the other endocrine glands in the body

plagioclimax a climax community that is at least in part the result of human intervention

plasma cell clones clones of identical plasma cells that all produce the same antibody

plasma cells cells that produce antibodies to particular antigens at a rate of around 2000 antibodies per second

plasmid small, circular piece of DNA that codes for a specific characteristic of the bacterial phenotype

polarised the condition of a neurone when the movement of positively charged potassium ions out of the cell down the concentration gradient is opposed by the actively produced electrochemical gradient, leaving the inside of the cell slightly negative relative to the outside

polymerase chain reaction (PCR) the process used to amplify a sample of DNA (to make more copies of it very rapidly)

population a group of organisms of the same species, living and breeding together in a habitat

positive emission tomography (PET) scans scans produced by detecting the radiation given off by a radiotracer injected into a patient; computer analysis shows areas in which the radiotracer builds up, so detailed three-dimensional images of the inside of the body, including the brain, are formed

positive feedback systems systems in which effectors work to increase an effect that has triggered a response

predator an organism which hunts and eats other organisms (prey)

presynaptic membrane the membrane on the side of the synapse which receives the first impulse and from which neurotransmitters are released

prey an organism which is hunted and eaten by other organisms (predators)

primary consumers organisms that eat producers, either plants or algae

primary infection the initial stage of tuberculosis, when *M. tuberculosis* has been inhaled into the lungs, invaded the cells of the lungs and multiplied slowly, often causing no obvious symptoms of disease

producers organisms that make food by photosynthesis or chemosynthesis

provirus the DNA that is inserted into the host cell during the lysogenic pathway of reproduction in viruses

proximal tubule the first region of the nephron after the Bowman's capsule; it is here that over 80% of the glomerular filtrate is absorbed back into the blood

Purkyne tissue conducting fibres that penetrate down through the septum of the heart, spreading between and around the ventricles

pyramid of biomass a model of feeding relationships that represents the biomass of the organisms at each trophic level in a food chain

pyramid of energy a model of feeding relationships that represents the total energy store of the organisms at each trophic level in a food chain

pyramid of numbers a model of feeding relationships that represents the numbers of organisms at each trophic level in a food chain

pyruvate ions the end-product of glycolysis

quadrat a sample area used in practical ecology, often measured using a square frame divided into sections that you lay on the ground

radiotracer any radioactive isotope introduced into the body to study metabolic processes

receptor cells specialised neurones that respond to changes in the environment

recognition site specific base sequences where restriction endonucleases cut the DNA molecule

recombinant DNA new DNA produced by genetic engineering technology that combines genes from the DNA of one organism with the DNA of another organism

red light light with a wavelength of 620–700 nm, which is detected by plants using phytochromes

redox indicators chemicals which are different colours when they are oxidised and reduced; they can act as artificial hydrogen carriers to investigate respiration in yeast

reduced NAD NAD which has received a hydrogen atom in a metabolic pathway

reduction the addition of electrons to a substance (e.g. by the addition of hydrogen or removal of oxygen)

reduction/oxidation reactions (redox reactions) reactions in which one reactant loses electrons (is oxidised) and another gains electrons (is reduced)

reflex responses rapid responses that occur without conscious thought

reforestation the replanting of trees in an area where trees have been lost

refractory period the time it takes for ionic movements to repolarise an area of the membrane and restore the resting potential after an action potential

relative refractory period a period of several milliseconds after an action potential and the absolute refractory period during which an axon may be re-stimulated, but only by a much stronger stimulus than before

release-inhibiting factors substances that inhibit the release of hormones from the anterior pituitary

releasing factors substances that stimulate the release of hormones from the anterior pituitary

replica plating a process used to identify recombinant cells that involves growing identical patterns of bacterial colonies on plates with different media

residual volume (RV) the volume of air left in the lungs after the strongest possible expiration

respiratory quotient (RQ) the relationship between the amount of carbon dioxide produced and the amount of oxygen used when different respiratory substrates are used in cellular respiration

respiratory substrate the substance used as a fuel and oxidised during cellular respiration

respirometer a piece of apparatus used for measuring the rate of respiration in whole organisms or cultures of cells

resting potential the potential difference across the membrane of around -70 mV when the neurone is not transmitting an impulse

restriction endonucleases enzymes used to cut up strands of DNA at particular points in the intron sequences

retrovirus a type of RNA virus that controls the production of DNA corresponding to the viral RNA and inserts it into the host cell DNA

reverse transcriptase (1) an enzyme synthesised in the life cycle of a retrovirus which makes DNA molecules corresponding to the viral RNA genome (2) an enzyme used to make artificial copies of a desired gene by taking an mRNA molecule transcribed from the gene and using it to produce the correct DNA sequence

R_f value the ratio of the distance travelled by the pigment to the distance travelled by the solvent alone when pigments are separated by chromatography

rhodopsin (visual purple) the visual pigment in the rods

ribulose bisphosphate (RuBP) a 5-carbon compound that combines with carbon dioxide from the air in the Calvin cycle to fix the carbon dioxide and form a 6-carbon compound

ribulose bisphosphate carboxylase/oxygenase (RUBISCO) a rate-controlling enzyme that catalyses the reaction between carbon dioxide/oxygen and ribulose bisphosphate

rigor mortis temporary muscle contraction causing the body to become rigid after death

RNA viruses viruses that have RNA as the genetic material

rods photoreceptors found in the retina which contain the visual pigment rhodopsin; they respond to low light intensities, give black and white vision and are very sensitive to movement

saltatory conduction the process by which action potentials are transmitted from one node of Ranvier to the next in a myelinated nerve

sarcomere basic unit of muscle structure

sarcoplasm the cytoplasm of muscle cells

sarcoplasmic reticulum the equivalent of the endoplasmic reticulum, found in muscle cells

Schwann cell a specialised type of cell associated with myelinated neurones which forms the myelin sheath

sebum an oily substance produced by the skin which contains chemicals that inhibit the growth of microorganisms

secondary consumers animals that feed on primary consumers

secondary production the process of making new animal biomass from plant material that has been eaten

selective medium a growth medium for microorganisms containing a very specific mixture of nutrients, so only a particular type of microorganism will grow on it

selective reabsorption the process by which substances needed by the body are reabsorbed from the kidney tubules back into the blood

selective toxicity a substance that is toxic against some types of cells or organisms but not others

sense organs groups of receptors working together to detect changes in the environment

sensors/receptors specialised cells that are sensitive to specific changes in the environment

sensory neurones neurones that carry impulses from receptors about the internal or external environment into the central nervous system

serotonin the neurotransmitter in a group of cells in the brain stem which have axons that spread throughout the brain into the cortex, the cerebellum and the spinal cord, and have an extensive influence over much of the brain

short-day plants (SDPs) plants that flower when days are short and nights are long

short tandem repeats micro-satellite regions that are widely used in DNA identification of suspects; statistically, the chance of two people matching on 11 or more sites is so small that it is considered to be reliable evidence in court

significant not due to chance

sinoatrial node (SAN) a specialised group of cells in the right atrium with the fastest natural intrinsic rhythm; it generates a regular electrical signal and acts as the heart's own natural pacemaker to keep the heart beating regularly

skeletal muscle/striated muscle/voluntary muscle muscle with a striped appearance under the microscope; it moves the bones of the skeleton under voluntary control

sliding filament theory the theory that the actin and myosin filaments overlap during muscle contraction

slow twitch muscle fibres muscle fibres which contract and fatigue slowly; they contain many mitochondria, a lot of myoglobin and a rich blood supply

smooth muscle/involuntary muscle muscle which is under the control of the involuntary nervous system

sodium gates specific sodium ion channels in the nerve fibre membrane that open up, allowing sodium ions to diffuse rapidly down their concentration and electrochemical gradients

somatotrophin another name for growth hormone

Spearman's rank correlation coefficient a statistical tool used to test whether two variables are significantly correlated

spinal cord the area of the CNS that carries the nerve fibres into and out of the brain and also coordinates many unconscious reflex actions

SSRIs (selective serotonin reuptake inhibitors) antidepressant drugs that inhibit the reuptake proteins in the presynaptic membrane so more serotonin remains in the synaptic cleft and more impulses travel along the post-synaptic axon, reducing the symptoms of depression by producing a more positive mood and improving the ability to sleep

stalked particles structures on the inner mitochondrial membrane where ATP production occurs

stationary phase when the total growth rate is zero as the number of new cells formed by binary fission is equal to the numbers of cells dying due to factors including competition for nutrients, lack of essential nutrients, an accumulation of toxic waste products and possibly lack of, or competition for, oxygen

sterile a term used to describe something that is free from living microorganisms and their spores

sticky end the name given to the area of base pairs left longer on one strand of DNA than the other by certain restriction endonucleases, making it easier to attach new pieces of DNA

stroma the matrix which surrounds the grana and contains all the enzymes needed to complete the process of photosynthesis and produce glucose

Student's t-test a statistical test that allows you to judge whether any difference between the means of the two sets of data is statistically significant

substantia nigra the area of the midbrain involved in the control and coordination of movement; it is affected by Parkinson's disease

succession the process by which the communities of organisms colonising an area change over time

sustainability the production of a decent standard of living for everyone now, without compromising the needs of future generations or the ecosystems around us

sustainable resources resources which can be grown and used in a sustainable way

sympathetic nervous system involves autonomic motor neurones which produce noradrenaline as their neurotransmitter and often have a rapid response, activating an organ system; these neurones have very short myelinated preganglionic fibres that leave the CNS and synapse in a ganglion very close to the CNS; postganglionic fibres are long and unmyelinated

sympatric speciation an evolutionary process that occurs when organisms become reproductively isolated by mechanical, behavioural or seasonal changes

synapse the junction between two neurones that nerve impulses cross via neurotransmitters

synaptic cleft the gap between the pre- and post-synaptic membranes in a synapse

synaptic knobs the bulges (bumps) at the end of the presynaptic neurones in which neurotransmitters are made

synaptic vesicles membrane-bound sacs (tiny bags) in the presynaptic knob which each contain about 3000 molecules of neurotransmitter; they move to fuse with the presynaptic membrane after an impulse arrives in the presynaptic knob

synovial fluid fluid which lubricates the most mobile joints

T cells lymphocytes made in the bone marrow that mature and become active in the thymus gland

T helper cells lymphocytes that mature in the thymus gland and are involved in the process that produces antibodies against the antigens on a particular pathogen

T killer cells lymphocytes that mature in the thymus gland and produce chemicals that destroy pathogens

T memory cells very long-lived cells which constitute part of the immunological memory

temperature coefficient (Q_{10}) a coefficient showing the effect of temperature on the rate of a reaction

tendons inelastic tissue which joins muscles to bones

tertiary consumers animals that feed on secondary consumers (i.e. they eat other carnivores); they are usually the top predators in an area

tetanus the state of a muscle which is fully contracted and remains so for some time; the disease tetanus got its name because the muscles of the body go into tetanus and cannot relax, so you cannot breathe and this causes death

tetracycline a bacteriostatic antibiotic that inhibits protein synthesis

thermoregulation a homeostatic mechanism that enables organisms to control their internal body temperature within set limits

thermoregulatory centre region of the hypothalamus in the brain that acts as the thermostat of the body keeping it at a set temperature; it contains temperature receptors which cause nerve impulses to be sent to different parts of the body if the temperature of the blood flowing through the hypothalamus increases or decreases

threshold the point when sufficient sodium ion channels open for the rush of sodium ions into the axon to be greater than the outflow of potassium ions, resulting in an action potential

threshold level the level of stimulus that triggers a response

thylakoids membrane discs found in the grana of a chloroplast

tidal volume (VT) the volume of air that enters and leaves the lungs at each natural resting breath

tobacco mosaic virus (TMV) a virus that infects the leaves of tobacco plants and other closely related species, and causes a mosaic patterning which can reduce the yield

total lung capacity (TLC) the sum of the vital capacity and the residual volume

total viable cell count a measure of the number of cells that are alive in a specific volume of a culture

transgenic animals animals which have been genetically modified to produce proteins from another organism, often a human being

transgenic plants plants which have been genetically modified to produce proteins from another organism

tricyclic antidepressants (TCAs) antidepressant drugs that work by increasing the levels of serotonin and noradrenalin in the brain

trophic level a term which describes the position of an organism in a food chain or web and its feeding relationship with other organisms

tropisms plant growth responses to environmental cues; they are also known as tropic responses

tropomyosin a long chain protein molecule that wraps around the double actin chains and covers the myosin binding sites in the relaxed state

troponin protein molecules attached to tropomyosin; they change shape when attached to calcium ions, pulling on the tropomyosin and revealing the myosin binding sites on the actin strands

tubercule a localised mass of tissue containing dead bacteria and macrophages formed as a result of a healthy immune response to an infection by *M. tuberculosis*

tubular secretion the process by which inorganic ions are secreted into or out of the kidney tubules as needed to maintain the osmotic balance of the blood

turbid a term used to describe something that is opaque because of suspended matter

turbidimetry a method of measuring the concentration of a substance by measuring the amount of light passing through it

ultrafiltration the process by which fluid is forced out of the capillaries in the glomerulus of the kidney into the kidney tubule through the epithelial walls of the capillary and the capsule

vaccination the introduction of harmless forms of organisms or antigens by injection or mouth to produce artificial immunity

vasoconstriction the process by which the blood vessels become narrower by contraction of their muscle walls, which reduces blood flow

vasodilation the process by which the blood vessels become wider by relaxation of their muscle walls, which increases blood flow

vector living organisms or environmental factors which transmit pathogens from one host to another

ventilation centre centre in the medulla oblongata of the brain that receives information from a number of different receptors and controls changes to the rate at which you breathe

ventilation rate a measure of the volume of air breathed per minute, calculated by multiplying tidal volume by breathing frequency

virus attachment particles (VAPs) specific proteins (antigens) that target proteins in the host cell surface membrane

visual acuity the ability to see very clearly in sharp focus

vital capacity (VC) the total of the tidal volume and the inspiratory and expiratory reserves

voluntary nervous system involves motor neurones that are under voluntary or conscious control involving the cerebrum

weather the conditions in the atmosphere at a particular time (for example, if it is sunny or windy or rainy when you go outside)

white fibrous tissue inelastic connective tissue made up of bundles of collagen fibres

white matter the nerve fibres of neurones in the CNS

INDEX

References in **bold** are to definitions.